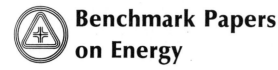

Benchmark Papers
on Energy

Series Editors:
R. Bruce Lindsay, Brown University
Mones E. Hawley, Professional Services International

PUBLISHED VOLUMES AND VOLUMES IN PREPARATION

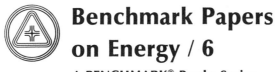

Benchmark Papers
on Energy / 6
A BENCHMARK® Books Series

THE CONTROL OF ENERGY

Edited by

R. BRUCE LINDSAY
Brown University

Dowden, Hutchinson & Ross, Inc.
Stroudsburg, Pennsylvania

Copyright © 1977 by **Dowden, Hutchinson & Ross, Inc.**
Benchmark Papers on Energy, Volume 6
Library of Congress Catalog Card Number: 77–24075
ISBN: 0–87933–292–1

79   78   77   1   2   3   4   5
Manufactured in the United States of America.

LIBRARY OF CONGRESS CATALOGING IN PUBLICATION DATA

Main entry under title:
The Control of energy.
   (Benchmark papers on energy; 6)
   Includes indexes.
   1. Power (Mechanics)—Addresses, essays, lectures. 2. Electric power—Ad-
dresses, essays, lectures. 3. Force and energy—Addresses, essays, lectures.
4. Feedback control systems—Addresses, essays, lectures. I. Lindsay, Robert
Bruce, 1900-
TJ163.9.C67   621   77–24075
ISBN 0–87933–292–1

Exclusive Distributor: **Halsted Press**
A Division of John Wiley & Sons, Inc.
ISBN: 0–470–99286–7

PREFACE

The Benchmark Papers on Energy constitute a series of volumes that makes available to the reader in carefully organized form important and seminal articles on the concept of energy, including its historical development, its applications in all fields of science and technology, and its role in civilization in general. This concept is generally admitted to be the most far-reaching idea that the human mind has developed to date, and its fundamental significance for human life and society is everywhere evident.

One group of volumes in the series contains papers bearing primarily on the evolution of the energy concept and its current applications in the various branches of science. Another group of volumes concentrates on the technological and industrial applications of the concept and its socioeconomic implications.

Each volume has been organized and edited by an authority in the area to which it pertains and offers the editor's careful selection of the appropriate seminal papers, that is, those articles which have significantly influenced further development of that phase of the whole subject. In this way every aspect of the concept of energy is placed in proper perspective, and each volume represents an introduction and guide to further work.

Each volume includes an editorial introduction by the volume editor, summarizing the significance of the field being covered. Every article or group of articles is accompanied by editorial commentary, with explanatory notes where necessary. Both an author index and a subject index are provided for ready reference. Articles in languages other than English are either translated or summarized in English. It is the hope of the publisher and editor that these volumes will serve as a working library of the most important scientific, technological, and social literature connected with the idea of energy.

The present volume *The Control of Energy* has been prepared by one of the series editors. Its aim is to draw attention to the importance of energy control in all its principal manifestations. The transfer and transformation of energy cannot be expected to carry out all the tasks demanded of them without methods of control, either manual or automatic. This has indeed been a significant problem from the time of the earliest uses of energy in daily life and industry. The same type of problem is found in the living organism, as is well illustrated by the concept of homeostasis. This volume contains a collection of papers illustrating the development of energy control methods from the early types of machines to the most sophisticated modern feedback devices. Attention is paid to all forms of energy save the nuclear variety. In the face of the relatively enormous amount of available literature, the choice of papers for inclusion has necessarily been arbitrary. It should be emphasized that the attempt has been made to adhere to the policy employed in earlier Benchmark volumes and lay stress on early seminal papers which in a certain sense paved the way for future developments. Hence the book has a definite historical flavor. It has been necessary to omit many important recent papers that will undoubtedly influence the development of the subject in the future. A few articles of outstanding review and tutorial character have been included. No apology is necessary for the inclusion of some material on cybernetics, the science of control in its most general form, even if the coverage has to be limited to the historical and philosophical aspects. The reader is cautioned not to be misled by the fact that he will not find the word *energy* stressed in most of the articles reproduced. Yet the concept of energy and its control is definitely involved in each one. It is hoped that the collection and the associated commentaries will help put the whole subject in its proper perspective.

I am deeply indebted to Patricia Galkowski, Caroline Helie, and their colleagues in the Sciences Library of Brown University for their gracious and indefatigable help in the location of source material. I am very grateful to Denise Perreault of the Department of Physics of Brown University for the typing of the translated articles as well as of the introduction and the commentaries. My longtime friend, Dean Harold Hazen of the Massachusetts Institute of Technology, made valuable suggestions with respect to the introduction. It is a pleasure to express my appreciation of the fine collaboration rendered me by the editorial staff of the publishers.

R. BRUCE LINDSAY

CONTENTS

Contents

PART IV: FEEDBACK AMPLIFIERS

PART V: CYBERNETICS

CONTENTS BY AUTHOR

INTRODUCTION:

THE CONTROL OF ENERGY

Energy is the premier concept of science and probably the most important single idea the human race has developed in its attempt to describe and understand experience. It is no exaggeration to say that all the phenomena we observe in nature, which involve things happening in time, can be described and understood in terms of the transfer or flow of energy from one place to another and/or the transformation of energy from one form to another. The fundamental principles governing such transfers and transformations, as for example the change of heat energy into mechanical energy as in an engine, are the laws of thermodynamics. The first law states that in any such transfer or transformation, the total quantity of energy must remain constant; this is the law of the conservation of energy. The second law says, in effect, that in every transformation of energy there is involved a loss in the ability to repeat this transformation. This is the law of increasing entropy, rather loosely stated. In applying these principles to scientific and technological problems it is necessary to consider the initiation and maintenance of the energy changes in question. These processes may be said to constitute the *control* of energy, and such control is obviously of enormous importance in all practical applications of the energy concept.

A simple illustration will serve to make this clear. A single match is sufficient to light a gas burner, thus initiating the chain or self-sustaining reaction we call the burning of gas. This reaction will continue by itself as long as enough gas and air are supplied to make possible the continued rapid oxidation of the gaseous components, which we refer to as combustion. Here a very small amount of energy (the burning of the match) initiates and controls the transformation of a large amount of energy. Such examples are present everywhere in our experience. Another simple but obviously important illustration is provided by the role of the spark plug and its spark in initiating the combustion of the fuel mixture in the cylinders of the car engine. The production of a large amount of mechanical energy through the expansion of the vapor is controlled here by the expenditure of a small amount of energy in the spark discharge. Still another example is found in the firing of a gun. The small amount of mechanical energy expended in the impact of the trigger hammer on the cartridge releases a large amount of chemical energy in the explosion of the powder that in turn is transformed into the mechanical energy communicated to the bullet.

In most practical cases of energy transfer and transformation it is important to be able to control the *rate* at which the processes take place. In a car the throttle valve or accelerator performs this function by arranging for the manual control of the amount of fuel mixture allowed to enter the cylinders. A relatively small amount of mechanical energy used in opening the valve can lead to the transformation of a large amount of chemical energy by combustion in the cylinders and thereby produce a change in the speed of the car. Moreover, by appropriate manipulation of the accelerator, the production of mechanical energy by the engine and hence the speed of the car can be controlled within narrow limits. This is possible both when the power load on the engine is increased as the car goes up hill or when the load is merely air and ground frictional resistance in level road driving. Without the possibility of such control, satisfactory automatic propulsion of the motor vehicle would be impossible. The automatic centrifugal governor devised in the 18th century by James Watt for the control of energy flow in a steam engine, to be discussed in detail later in this book, is another illustration of this type of energy control.

Another aspect of energy control is exhibited by any simple machine such as a lever or pulley system. Such a device transfers mechanical energy from one place to another. Here the conservation of energy demands that the amount of energy delivered per unit time (i.e., the power) at the output end of the machine shall equal that introduced at the input end, minus any dissipative losses in the machine itself. The control here consists in the first place in the actual transfer of the mechanical energy and more particularly in what may be termed its distribution. Thus one may introduce energy at one end of a pulley system by pulling *down* while the transferred energy at the output end is manifested in the pulling *up* or *raising* of a weight. Another simple example is provided by the distribution of heat energy in a hot air or hot water domestic heating system. Instead of having to produce the warmth at each spot where and when it is needed, we transform the chemical energy of the fossil fuel into heat in a central furnace and then provide its distribution by convection. The transfer of electrical energy by a transmission line is another example of the same sort of thing.

It is convenient therefore to look at energy control from three principal points of view. The first is that exemplified by the simple distribution of energy in space. The second is the control involved in the use of a relatively small amount of energy to initiate the transformation and/or transfer of a much larger amount, either reversibly or irreversibly. The third is the use of small amounts of energy to change reversibly the rate of energy transformation or flow.

It is important to emphasize that in all aspects of energy control, even if the form of the energy is unchanged as in a simple transfer process, what may be called the components of the energy may change. Thus in a machine designed to raise a heavy weight, a low-force input leads to a high-force output (at the expense of the speed of operation). In the transfer of energy by an alternating current transformer, low voltage electrical energy may be changed into high voltage energy and vice versa.

The simplest type of energy control is, of course, manual, as in the case of the throttle of a steam locomotive engine operated by the engine driver or the accelerator of a car operated by the car driver. In any electrical circuit network, like the ordinary household electrical distribution system, a simple switch is a form of manual energy control since through its activation, electrical energy may be made available to perform its various domestic tasks. A more sophisticated form of switch is the relay, which in its electromagnetic form employed in the early electrical telegraph, used a small current to energize an electromagnet that in turn opened or closed a switch controlling a much larger current. Such relays have in more recent times been replaced by electronic devices like thermionic vacuum tubes and transistors, as will be discussed in greater detail later in this book.

Manual control of energy transfer and transformation can be a very cumbersome and inconvenient method and certainly was so in the early days of the steam engine. This stimulated the search for automatic control, of which Watt's governor was an example. Such a device forms a part of the system it is designed to control and employs a small part of the energy transformed and utilized in the system to feed back to the energy source information that the load on the system demands a change in the rate of energy transfer

3

or transformation. Thus the automatic steam engine governor is designed so as to accommodate the flow of energy from the source, e.g., the steam in the boiler, to the external load on the engine (the mechanical power consuming pump or other machinery) in such a way as to maintain a constant speed of the moving parts of the engine. The main engine shaft is attached by a gear to a vertically rotating shaft at the top end of which two metal balls are freely attached by struts in such a way that as the angular speed of the shaft increases the two balls tend to move away from the shaft through the centrifugal effect connected with their increased angular speed. At their lower ends the balls are loosely connected with a sleeve that can move freely up and down the rotating shaft. This sleeve is connected by a linkage to the throttle valve that controls the flow of steam into the cylinder of the engine. If the load on the engine increases, the angular velocity of the governor shaft decreases. This brings the balls closer to the shaft, moving the sleeve down. Through the linkage, this increases the opening in the throttle valve, allowing more steam to flow from the boiler, thus increasing the power flow and consequently compensating for the drop in velocity. Conversely, if the load decreases and the engine starts to race, the angular speed of the governor shaft increases, the balls fly out farther from the shaft, the sleeve moves up and, through the linkage, decreases the opening in the throttle valve with a consequent decrease in the flow of steam and the transfer of energy to the engine. The speed of the engine is thus automatically controlled in spite of variations in the load. The effectiveness of the control is of course contingent on making the time of the response of the governor to variations in the load short compared with the time for the fluctuations in the speed. This automatic control device is called a *feedback* mechanism and provides a good example of what is generally called feedback control.

Another illustration of feedback control is the household thermostat, whose function it is to control the flow of heat energy into a space so as to keep the temperature constant. It feeds back to the central heater information about the temperature of the space in which it is located. If the temperature falls, the corresponding movement in the temperature sensitive element in the thermostat closes a switch that causes a current to flow so as to activate a motor which in turn operates the fuel injection system (or, in an old coal burner, opens the dampers) in such a way as to provide more fuel combustion and hence increase the flow of heat energy into the space. Conversely when the temperature rises above the desired level, the thermostat responds by opening the control switch, thus lowering the rate of combustion in the furnace and the flow of thermal energy into the space. Here again the time of response of the automatic feedback control is important. Ideally one would wish this time to be so short that the temperature fluctuation in the heated space would be negligible. However, with any actual mechanism there is bound to be some temporal variation.

Automatic feedback control is not limited to physical devices that transfer and transform energy. Living organisms, specifically human beings, also manifest this type of control. It is very important that they should do so because of the varying environmental conditions to which they are exposed. Thus in the case of the human organism, chemical energy in the food consumed is transformed into thermal, mechanical, and electrical energy in the body. In fact a large part of the mechanical energy produced in the body through muscular activity, flow of blood, and so on (i.e., all except that which the body transfers to other objects in its vicinity in the form of human labor) is ultimately transformed into heat in accordance with the mechanical equivalent of heat (1 calorie for each 4.2 joules of mechanical energy). In the face of this production some regulatory system evidently must exist in order to maintain the body temperature at a reasonably stable level, (namely an average of about 98.6°F or 37°C). It must be remembered that the body is an open system exposed to the environment and subject to environmental changes, such as humidity, pressure, and temperature. Internal changes both in the healthy and diseased body also tend to affect the body equilibrium. To counteract both external and internal influences that can affect equilibrium, the body possesses a set of regulatory mechanisms whose action the late Harvard physiologist Walter B. Cannon called *homeostasis*. To take one example, let us consider temperature equilibrium. When the body is exposed to an external environment at a much lower temperature than the body itself, it tends to lose heat, leading to a reduction in temperature. But then information as to the need for a compensating increase in the bodily heat production is fed back to the brain (probably the hypothalamus in the diencephalon) by the nervous system, which dictates compensatory action. This can take the form of an increase in the rate of metabolism by an increase in the breathing rate to promote more rapid oxidation, shivering, an increase in the level of blood sugar (connected with increased adrenal secretion), or all of these. Conversely, when the body is exposed to an environment at a higher temperature than itself, a similar compensatory mechanism induces restoration of equilibrium by increasing heat loss through such activities as panting, perspiration, and so on. In both cases, energy control is definitely involved. Of course, homeostasis cannot be expected to operate successfully over too wide a range in environmental conditions. Even when protected with the insulating control of heavy clothing a person can freeze to death in a sufficiently cold environment, and even without clothes can suffer heat stroke when the outside temperature is sufficiently high.

The automatic control of energy flow and transformation has become so important in modern life and industry that a special term—*servomechanism*—has been invented to describe the devices by which this control is carried out. To understand the significance of the word, consider again the behavior of the simple domestic thermostat. It is the function of this instrument to instruct the heating system to provide more or less heat depending on the fall or rise in the temperature of the room. But the amount of energy available through the instrument itself is not sufficient to activate the heating mechanism directly, i.e., to open or close dampers, or to start or stop a fuel-burning heating unit. All the simple thermostat can do is to close or open a simple switch demanding a very small amount of energy for its operation, e.g., the energy necessary for the slight bending of a bimetallic strip. This switch in turn turns off or on a mechanism, usually in the form of an electric motor, powered by an external source that enables the heating system to respond to the instruction given to it. This intermediate mechanism is the servomechanism, so-called from the Latin *servus*, a slave. It is clear that a servomechanism can take many forms. It may be a simple machine such as the foot brake in a motor car that transfers the low force energy of the foot on the brake pedal to the high torque energy produced by the brake shoes. Since this is a manually operated servomechanism there is no element of feedback in the mechanism itself. When the mechanism is instructed by the car driver to apply the brakes, it does so (assuming it is working properly). However, it is correct to say that there *is* feedback in the mind and muscle system of the driver. Thus, if he finds that he has brought the speed of the car down below the limit really desired, he merely lifts his foot from the brake pedal and applies it to the accelerator (we are assuming here automatic transmission). We may consider the whole system of car and driver as exhibiting automatic control with feedback, in which the mental and physical behavior of the driver enters as well as the physical action of the car.

To take another example, in automatic ship steering the compass is directly linked to a servomechanism that it instructs to keep

6

the ship on a predetermined course. This servomechanism controls the engine that drives the rudder. In a certain sense the whole system from compass to rudder may be considered a servomechanism, with its automatic control obviously involving feedback. If the course departs somewhat from that which is desired, information about this is fed back to the servomechanism that then takes appropriate action to correct the error.

The use of a device like a servomechanism to control the transfer and/or transformation of a relatively large amount of energy inevitably suggests the concept of an amplifier. The ordinary machine can of course act as a force or torque amplifier, and we have already mentioned its control function in the spatial distribution and direction of an energy supply. A power amplifier is a very important example of energy control. One of the simplest types from the standpoint of observation and ease of understanding its mode of operation is the standard acoustical resonator. This is a hollow tube or cavity with an orifice open to the surrounding air. Such a device has a natural frequency at which air vibrating back and forth in the orifice has the maximum amplitude and produces sound of maximum intensity. If a source of sound of appropriate frequency is brought close to the orifice its intensity will markedly increase: the device is said to resonate with respect to the source. Actually what happens is that the resonator extracts energy from the sound source faster than would be the case if the resonator were not there. The sound power or the intensity is therefore increased as long as the source can emit energy at that increased rate. We may say that the resonator *loads* the source. An even simpler case is that of a point source of sound in front of and near a large plane reflecting surface. The plane acts to load the sound source and hence cause it to emit sound energy at a faster rate than if the wall were not there. This produces an amplifying effect. In both cases the sound source has its energy exhausted faster than if it were isolated, unless this energy is continuously replenished by some device such as an electronic oscillator.

The enormously increased use of electrical energy in our modern industrial civilization is greatly facilitated by its control through electronic devices, of which the thermionic vacuum tube and the transistor are the most important examples. They permit the controlled extraction of oscillatory electrical energy from a source (e.g., a battery) in an electrical circuit. Such devices are electrical amplifiers. They are commonly classified as voltage, current, and power amplifiers. When equipped with appropriate power supplies they can transform high (or low) power, alternating current into high power, direct current. Their widespread use in radio, television, and other electronic devices is well known. They can also be equipped so as to involve feedback and hence ensure control of power transmission.

The automatic control of energy has in recent times been made more sophisticated by the association of high speed computers with servomechanisms. A computer can be programmed so as to provide continuous automatic control for attached devices and has introduced greater speed and flexibility into the control process.

All automatic control devices involve the transmission of signals, commonly called *information*, whose purpose is the communication of messages directing the performance of certain behavior. Thus in the centrifugal governor on a steam engine we may think of the change in load informing the governor, which in turn informs the throttle valve to take appropriate action. All homeostatic phenomena in the body involve the transmission of information to other parts of the body including the brain, which in turn informs the various organs what action to take. In a very real sense we may look upon the transmission of information as the fundamental basis of all control, no matter what it is that is being controlled. It was with this in mind that Norbert Wiener in the 1940s decided to apply the term *cybernetics* to the whole field of communication and control. In so doing he was essentially reviving the term which the French physicist Andre Marie Ampere first introduced in 1834 in his *Essai sur la Philosophie des Sciences*. Ampere took the term (in French, *cybernetique*) from the Greek meaning steersman, and applied it exclusively to the science of government. There is very little evidence that he felt inclined to generalize the idea in the sense Wiener gave it. Since all transmission of information involves the flow of energy, even if in minute amounts, we may rightfully use the word cybernetics to describe the field of energy control. Hence in this book we shall pay some attention to cybernetics. However most of the literature now available on this subject neglects the energy aspect, either because the energy quantities are so small or because it is more convenient to place emphasis on the transmission of signals as entities constituting the content of information. It is true that it has been found convenient to discuss the communication of information in terms of thermodynamics and hence energy and in particular entropy enter the subject of cybernetics. In this book we shall be chiefly concerned with the control of substantial amounts of energy and not with what may be called information theory proper.

BIBLIOGRAPHY

The bibliography of books on the subject of energy is lengthy. The following is a brief list of books and articles dealing in a general way with the subject matter of the present volume and suitable for the general reader.

Brown, G. S., and D. P. Campbell. *Principles of Servomechanisms*. New York: John Wiley & Sons, 1948.

Chalmers, Bruce. *Energy*. New York: Academic Press, 1963.

Guilbaud, G. T. *What Is Cybernetics?* London: Heinemann, 1959.

Hardie, A. A. *The Elements of Feedback and Control*. London: Oxford University Press, 1964.

Harrison, George Russell. *The Conquest of Energy*. New York: William Morrow and Company, 1968.

Illyn, V. A. Some Aspects of Cybernetics. *Cybernetica* **2**:203–214 (1959).

Lehninger, Albert L. *Bioenergetics*. New York: W. A. Benjamin, 1965.

Mayr, Otto. *The Origins of Feedback Control*. Cambridge, Mass.: MIT Press, 1969.

Putnam, Palmer G. *Energy in the Future*. New York: D. Van Nostrand, 1953.

Rau, Hans. *Solar Energy*. New York: Macmillan Co., 1958.

Summer, W. *Photosensitors—A Treatise on Photoelectric Devices and Their Application to Industry*. New York: Macmillan Company, 1958.

Sutton, George W., ed. *Direct Energy Conversion*. New York: McGraw-Hill, 1966.

Thirring, Hans. *Power Production*. London: George G. Harrap and Co., 1956.

West, J. C. *A Textbook of Servomechanisms*. London: English Universities Press, 1953.

Part I

DIRECT CONTROL OF ENERGY

Editor's Comments
on Papers 1 Through 13

The direct control of energy by man is as old as man himself, at any rate going back to the invention of the production of fire and to the development of machines. From the historical point of view we can best introduce the subject by concentrating on the machine as a device for the transfer of energy of a given form from one place to another with no loss in the energy save through frictional dissipation. It is indeed true that though the concept of energy was of course implicit in the behavior of the machines of antiquity they were described by their inventors and users (inadequately from our modern point of view) from the standpoint of force. Appreciation of the significance of energy as an all-important scientific concept developed slowly, as has been emphasized in *Energy: Historical Development of the Concept*, edited by R. B. Lindsay (Volume 1 of Benchmark Papers on Energy, Dowden, Hutchinson & Ross, Inc., Stroudsburg, Pennsylvania, 1975). It was only in the latter part of the 18th century and the early part of the 19th century that the energy concept attained sufficient clarity to be applied to machines. This is well discussed in the writings of Gaspard Gustave de Coriolis (1792–1843), the well-known French authority on theoretical and applied mechanics. In his memoir "Du Calcul de l'Effet des Machines" (1829) he set forth more clearly than had been done previously the fundamental character of a machine as a means of transferring and thus controlling the distribution of energy. We reproduce here in English translation as Paper 1 sections 25 and 26 of the memoir as an illustration of his point of view. It will be noted that Coriolis uses the word "work" (French, *travail*), where the modern notation would be "energy." This, however, leads to no loss of relevance or clarity.

From a philosophical point of view great interest attaches to a work by the distinguished French architect Jacques Lafitte (1884–1966) entitled *Reflexions sur la Science des Machines* (1932). This reflects a good grasp of the energy control significance of machines considered from the widest possible point of view. As an architect Lafitte brought to his subject a rather more general atti-

13

tude than that of most scientists and engineers. His book has been considered by the eminent French cybernetician G. T. Guilbaud (in his work *What is Cybernetics*, English translation, Heinemann, London, 1959) as being significant for cybernetics or the science of control in general. We here reproduce as Paper 2 an English translation of Part I of Lafitte's book. A commentarial summary of the rest of the contents is provided at the end of the translated extract.

The role of the machine as an energy transfer and distribution device has been of outstanding significance in the construction industry, in transportation, in manufacturing of all kinds, in domestic economy and, in fact, in practically every activity in which human beings engage. A book devoted to machines must necessarily be an encyclopedia. Here we must be contented with an aribtrary choice. As a good illustration of the use of the machine in transportation we reproduce as Paper 3 an article on the transmission system in a motor car. Among the host of appropriate articles on this topic we have chosen Oliver K. Kelley's paper "Operating Principles and Applications of the Fluid Coupling and Torque Converter to Automatic Transmissions" (1956). The development of automatic transmission for the motor car was a notable event in the automotive industry and a great boon to the driving public. Emphasis on automatic transmission as an energy distribution device and hence as an example of the control of energy will be appreciated by the careful reader of this paper. Mr. Kelley (b. 1904) was for many years a member of the General Motors Corporation Engineering Staff.

As has been mentioned in the Introduction, the action of an acoustic resonator is a good example of the control of sound energy, since such a resonator enables sound of a given frequency to be extracted from a source at a greater rate than would be the case if it were not in the immediate vicinity of the source. It thus acts as a simple power amplifier. To illustrate this, Paper 4 reproduces the section on the simple Helmholtz resonator from the book *Acoustics* (1930) by G. W. Stewart (1876–1956) and R. B. Lindsay (b. 1900). Professor Stewart was for many years professor of physics at the University of Iowa and did notable research in acoustics and x-ray scattering.

Mention of the acoustic resonator immediately suggests the filtration of sound as an example of the control of sound energy. An acoustic filter is a device that controls the flow of sound energy through a transmission line in such a way that only harmonic sounds in a definite frequency band are transmitted whereas others are refused passage. As a review of the subject of acoustic filtration we reproduce in Paper 5 the article "The Filtration of Sound I" (1938) by R. B. Lindsay.

From sound we pass to the control of light energy, which is also an important element in modern science and technology. As an illustration of the filtration of light we reproduce as Paper 6 the article "Fresnel Formulae Applied to the Phenomena of Nonreflecting Films" (1940) by Katherine B. Blodgett (b. 1898). This was one of the pioneer papers on light radiation control that led to important improvements in photographic lenses.

It is well known that ordinary light sources, even when commonly considered monochromatic, do not actually emit light of a single frequency or light that is completely coherent in phase. Through the action called optical pumping, in which the population of the higher energy states of the atoms in a given source can be greatly increased and emission stimulated by the transition from such states to the ground state, it has proved possible to secure very intense and coherent light of a practically single frequency. The device that accomplishes this is termed a *laser*, an acronym for *l*ight *a*mplification by the *s*timulated *e*mission of *r*adiation. It is no exaggeration to say that the invention of the laser has revolutionized the control of light energy. It would be impossible to list here all of its practical applications in the measurement of physical quantities, in communication by means of light beams and in holography. A good reference in this connection is *Lasers and Their Applications* by M. J. Beesley, published by Taylor and Francis, London and Halsted Press, N.Y., 1976. We reproduce as Paper 7 what is believed to be the first article on the development of a solid laser "Stimulated Optical Radiation in Ruby" (1960) by T. A. Maiman (b. 1927).

The combination of light and electricity has led to the development of an increasingly valuable energy control device known as the phototube, based on the photoelectric effect, in which light of appropriate frequency falling on a metal plate in a vacuum has the power to eject electrons. The current produced in this way can act as a relay for the release of large amounts of energy in another form, e.g., mechanical. Phototube applications are legion. As an interesting example of application to the control of fluid flow we reproduce as Paper 8 the article "Phototube Control of Fluid Flow" (1944) by R. C. McNickle, formerly of the Brooke Engineering Company of Philadelphia, Pennsylvania.

The development of the alternating current generator in the half century following Faraday's fundamental discovery of electromagnetic induction made it necessary to control the output of electrical energy from such a device and its distribution over great distances from the source. One of the earliest pieces of equipment devised for this purpose was the alternating current transformer by means of which high voltage, low current electrical energy can be transformed into low voltage, high current electrical energy and vice versa. We find it interesting here to go back to the historical review article by J. Dixon Gibbs, "The Distribution of Electrical Energy by Secondary Generators" (1885). This we reproduce as Paper 9. To this day the transformer has remained a vital element in the distribution and utilization of electrical energy. Gibbs was a well-known British electrical engineer who died in 1912. With the French engineer Lucien Gaulard (1850–1888), he was very closely connected with the development of the transformer.

The control of the large scale emission and detection of electrical oscillations was first made possible by the invention of the thermionic valve or tube, employing the emission of electrons from a glowing filament in a vacuum. The influence of this energy control instrument on radio communication and later in electronic devices of all kinds has been enormous and needs no recapitulation here. Though the thermionic emission effect was discovered by Thomas A. Edison (1847–1931) in 1883 there is no record of any practical application of it to the control of electrical energy in oscillatory form until the invention of the Fleming valve in 1905. We produce here as Paper 10 John Ambrose Fleming's article "The Conversion of Electric Oscillations into Continuous Currents by Means of a Vacuum Valve" (1905). This set the stage for the vast development of what we have learned to call electronics in the 20th century. John Ambrose Fleming (1849–1945) was a British physicist whose work had a great influence on radio communication.

Lee DeForest (1873–1961) greatly enhanced the applicability of the Fleming valve by introducing a third electrode in the tube in addition to the filament and the plate. This electrode is called the grid. He thus invented the three-electrode vacuum tube, whose role in the amplification, production, and detection of electrical oscillations was predominant in the early electronics industry. We reproduce here as Paper 11 DeForest's interesting 1920 historical survey of the development of the vacuum tube: "The Andion—Its Action and Some Recent Applications." Perhaps DeForest in his enthusiasm claimed too much for his invention, and competent authorities believe he did not wholly understand the control action of the grid. The fact remains that his invention stimulated an immense development in electronics in the early part of the 20th century. DeForest, trained as a physicist, proved to be one of the last of the great individual electrical inventors.

An important means of controlling the frequency of electrical oscillations was the piezoelectric resonator of Walter G. Cady (1874–1974). It is an interesting example of the influence of mechanical vibrations on electricity and has had an important role in radio communication. We reprint here as Paper 12 the major part of Cady's 1922 article "The Piezoelectric Resonator." Cady was a well-known American physicist and for many years a professor at Wesleyan University.

The 20th century has witnessed a decided acceleration in the practical application of basic research. In our study of the control of energy this is very well illustrated by the development of the transistor. Here theoretical and experimental investigation in solid state physics has beautifully paid off in the exploitation of the properties of certain semi-conductors for use in the amplification, control, and generation of electrical signals. The transistor was invented by J. Bardeen (b. 1908), W. H. Brattain (b. 1902), and W. Shockley (b. 1910) in 1948. In its many subsequent elaborations it has, to a considerable extent, replaced the thermionic vacuum tube as an electronic device in electrical energy control in radio, television, and computer construction and action. We reproduce here as Paper 13 the summary article by Bardeen and Brattain, "Physical Principles Involved in Transistor Action" (1948). This paper concludes the list of our selection of seminal papers emphasizing the direct control of energy. It is interesting to observe, indeed, that the concept of feedback mentioned in the Introduction already plays a role in transistor action. This concept will receive more detailed attention in the papers in Part Two and those following.

1

ON THE CALCULATION OF THE BEHAVIOR
OF MACHINES

G. G. Coriolis

*This excerpt was translated expressly for this Benchmark
volume by R. Bruce Lindsay, Brown University, from
pp. 26–30 of* Du calcul de l'effet des machines, *Paris:
Carilian-Goeury Libraire, 1829.*

[*Editor's Note:* In this memoir Coriolis applies the general principles of
mechanics to the behavior of machines. After a preliminary analytical
discussion in which he effectively derives what we now call the work-
kinetic energy theorem (in which work is defined essentially as in modern
mechanics), he goes on to provide some remarks about machines in
general, as follows].

Section 25. It follows from what has been said in the foregoing sections that the quantity we have called work (French, *travail*) is one that
we cannot increase by the employment of a machine. It is only the
force or the path described that a machine can increase or decrease. A
machine can resolve force and path in various ways; it can modify their
points of application and their direction; in a word, it can modify
everything involving force and path without ever being able to increase
the work done. The portion of the work that a machine can transfer is
only less than that which it receives (from a source) to the extent that
friction intervenes. If it were possible to construct a machine without
friction, we could then say that work is a quantity that is never destroyed.

To provide a simple representation of the transmission of work in
the operation of machines, we can compare it with the motion of a fluid that distributes itself throughout several bodies by a transfer from
one body to another at their points of contact. This might happen by
the subdivision of the fluid into several currents, as in the case in
which one body pushes against several others. On the other hand, it
might take place by the reunion of several currents into one, as in the
case in which several bodies push against a single one. The fluid might
indeed accumulate in certain bodies and remain there in reserve until
new contacts permit a larger flow. The work held in reserve in this fluid analogy is what we have called *vis viva*. As we know, the latter depends on the velocities possessed by the bodies. According to this
analogy, a machine, in the meaning we ordinarily give to the term, is a
collection of bodies in motion arranged in such a way as to form a
kind of channel through which work can be transmitted most effectively to the places where it is needed. Once produced by the source

(e.g., an engine or motor), work passes in succession from one body to another. It can accumulate, divide itself into parts, and be later reunited. We shall see further on that work can be dissipated by friction or by striking against obstacles in the path. It may even distribute itself in the earth, where it no longer becomes evident to the senses.

Section 26. We shall now show that from the properties of machines with respect to work it results that work serves as the basis for the evaluation of the effectiveness of the motors (engines) of industry. It is the work that one should seek to economize, and it is to this same quantity that all questions of economy in the employment of motors principally relate.

Nothing necessary for human needs is produced save through the displacement of bodies and the change in their form. The only thing we can do on the earth's surface is the overcoming of resistance and the production of certain motions. The only useful faculty we have is the ability to produce displacement accompanied by force in the direction of the displacement, that is to say, the ability to produce the quantity we call work. Whether we acquire this from animals, from water or air in motion, from the combustion of coal, or from the fall of water, it is everywhere and at all times limited in amount. It is never created to our desire. Machines can only employ and economize work without being able to increase it.

If we did not have machines at our disposal, two different displacements would be two entities of totally distinct nature, admitting in general no mathematical basis for their calculation. Such displacements would then be like many otherwise useful things to which numerical values cannot be attached through mathematical calculations. But machines give the means of providing bases of evaluation for displacements analogous to those for different quantities of the same material.

When a machine, receiving its moving forces from a certain motor, is intended to bring about a certain useful effect, the result is that the points that act on bodies to displace or deform them are subject to resisting forces. But these forces are not in general the only ones that produce what we may properly call the "resistance" work. Frictional and various other resistances, which we cannot get rid of, add a resistance work to that associated with the useful effect. However, as there exists a rational possibility of having only those resisting forces that correspond to a useful effect, or since we can at any rate arrange matters to diminish considerably all other forces with respect to the latter, we can carry out our reasoning hypothetically, neglecting frictional forces. It thus becomes easy to see how we must modify in practice the conclusions drawn from this ideal assumption. Let us then, for the moment, assume that all the "resistance" work is that associated with the useful effect.

If we have the ability to produce a displacement in exerting a certain effort, we are then able with the aid of a machine (which can appropriately modify motion and force) to apply this ability to bring about a certain industrial process, as, for example, the grinding of grain in a mill or the spinning of thread. But it is clear that the grinding of each liter of grain or the spinning of each meter of thread, being in general associated with the same repeated circumstance, will demand that the points on which the machine acts will describe the same path and have applied to them the same force. Thus the grinding of the grain or the spinning of the thread will always involve the production of the same quantity of that which we have called "resistance" work. Consequently, the number of liters of grain ground or the number of meters of thread spun by means of the machine will be proportional to the "resistance" work produced in the machine by the grinding or spinning. But from the assumption we have just made, namely that we may at first neglect the resistances foreign to those associated with the useful effect, this resistance force is the only one exerted on the machine. The result is that, taking an interval of time that is not too small, the work transmitted by the machine is practically equal to the work done by the motor to which the machine is attached. The latter will then also be proportional to the quantity of grain that is ground or the amount of thread spun.

If we then wish to compare two motional processes it will suffice to assume that we have constructed machines by the aid of which we can apply these processes to the same industrial result. But it is clear that the comparative values of two grindings of grain will be measured by the number of liters of grain ground, and as the latter are essentially proportional to the quantities of motor work done on each machine, it follows that the two motional processes will have values proportional to the quantities of work that they can do on the machines.

At the present time we have the ability and expect to have it in even greater measure in the future to construct machines able to apply different motional processes, that is to say, different motors, and to bring about (transmit) with these machines work of the same kind. Because of this we can establish a mode of comparison between motors, through the quantities of the same kind of work they are capable of producing. The invention, improvement, and multiplication of such machines have vastly increased this method of evaluation, just as the invention and improvement of tools for the processing of materials (e.g., by cutting) have greatly enhanced the value of these implements in commerce, through the agency of the geometrical quantity we call volume.

[Editor's Note: The rest of this chapter of Coriolis' memoir is devoted largely to practical examples illustrating the role of work (energy) in the operation of machines.]

2

REFLECTIONS ON THE SCIENCE OF MACHINES

Jacques Lafitte

*This excerpt was translated expressly for this Benchmark
volume by R. Bruce Lindsay, Brown University, from
pp. 23–34 of* Reflexions sur la science des machines,
Paris: Libraire Blondet Gay, 1932, 122 pp.

PART ONE

I. Machines

The organized bodies constructed by man first attract our attention
by the differences manifested among them. It is these differences characterizing their particular organization that form the essential object of
scientific observation.

If in the vast aggregate of these bodies I compare several lineages
and if in a given lineage I compare the bodies having the most complicated organization, I am apparently fated to see only the differences
and to lose sight of the common characteristics of the various members of the aggregate; I am likely to consider as purely isolated the individuals and groups to which I am giving attention.

If for example I compare architectural creations with machine tools,
if I compare the primitive flint knife with the modern lathe, the hut
with the elaborate mansion of today, the simple abacus with the most
recent calculating machine, I shall assuredly have difficulty in discern-
ing, among the considerable obvious differences displayed by these
entities, their common characteristics.

Closer inspection, however, permits me to observe in all these entities their common characteristics, which belong essentially to them
and, which justify us in looking on them as a subaggregate of the
much larger domain of all natural bodies. To stress this distinction is
essentially to define the bodies in question. But at the same time it
also serves as the indication of the necessity of founding a new science.

The organized bodies or devices constructed by man display certain
differences from natural bodies.

There are indeed numerous natural arrangements possessing properties the observation of which has guided man in his primitive creations. The vertical support, the open or closed shelter, the dam, the
lever, the siphon are frequently found in primitive nature. As Reuleaux
has remarked, even the geyser can be considered analogous to certain
sophisticated machines. In short, it is wholly appropriate that these

natural systems should be compared with the systems resulting from man's creative activity and should therefore enter into the field of investigation of the mechanical and physical sciences. We can with advantage study the phenomena involved in them and investigate the conditions of their equilibrium, their motions, and the transformation that they bring about.

But each of these natural configurations formed as it were by chance maintains constant its form and has no capacity for developing any differences from the others beyond what it had to begin with. Thus these configurations have no capacity for evolution and in this sense distinguish themselves from the organized devices of which we are the creators.

Naturalists make us acquainted with other observations bearing on this matter. A chimpanzee uses a stone to break open a piece of fruit with a hard shell. A monkey uses a stick in the fashion of a lever to knock down hanging objects. Baboons roll stones on their enemies, birds make nests, and the beaver builds his dam. Thus animals (and not only those of advanced organization) not only employ natural shelters but also construct artificial shelters and use tools for this purpose.

In this primitive utilization of naturally existing configurations and in their construction by animals, it is reasonable to glimpse the future promise of the creative faculties in man. In observations of these phenomena one can always find precious indication of the origin and development of such faculties. But we always note that the configuration thus used or constructed remains a constant factor of the animal intelligence. In themselves they have no capacity to develop any differences greater than those originally observed. Based on instinct alone their organization has no chance of subsequent evolution.

I now observe, along with many other authors, that the ability to fashion a natural object with a view to obtaining definite results from its utilization is peculiar to man. I further observe that even our most primitive tools and constructed devices are different from natural objects and animal tools from the fact that the former are fashioned. I maintain that our primitive constructions and engines have given rise in the course of time to successively more complicated types. When I consider all these things I have ample reason to maintain that there exists the following definite difference between natural objects and the organized constructions of man: the former remain unchanged in the form that chance or animal instinct has given them, whereas the latter lead through the creative activity of man to organized forms capable of evolution.

Even if it is true that the difference in question has proved to be small in our primitive tool construction and some might say scarcely discernible, and if it is also the case that evolution in man's ability to make devices took millenia to accomplish, it still remains true that the primitive differences have become more pronounced in the passage of time and that the construction of complicated devices has progressed along with the development of man's creative intelligence.

If the organized devices constructed by us differ from the objects and nonliving things we encounter in nature, they also differ, though in a different way, from organized living things.

In some respects these natural objects are like living things in that some of their peculiar properties are due to the concentration of functions in specific organs. They also resemble living things in the fact that their operating competence increases with their degree of specialization, and finally, like living things, they can be studied for the mechanical and physical phenomena involved in them. But on the other hand they differ decidedly from living things in their inability to grow and reproduce themselves.

I am aware that certain contemporary authors, notably J. A. V. Butler* have

*[*Editor's Note:* See his "Science and Human Life" (1957) and also his earlier work "Man is a Microcosm."]

compared the genesis of machines to the phenomena exhibited in the growth and reproduction of human beings, and I shall have occasion to return later to this important consideration. I know also that other authors sometimes in fanciful and sometimes in serious fashion have projected for the future of machines very elaborate properties, which in the more sophisticated versions are those of living organisms. I realize that the bondage of man to the machine of which he is the creator has importance in the slow but already evident evolution of certain types of human kind into a type of what may properly be called a "Machine" man.

But if it is only in the future that I feel an obligation to attack such problems, which I have certainly no reason to consider devoid of interest, I can at any rate say now that if some day machines should acquire the power of growing and reproducing themselves in accordance with the same laws that govern the growth and reproduction of living things, they will then escape the control of man, their creator, and will cease to be true machines.

We can thus lay down certain characteristics that enable us to distinguish between the organized devices we create from natural objects and the living organisms found in nature. Similarly we should be able to distinguish these organized devices from the numerous products of our creative activity and the many organizations we encounter in human societies. If, for example, we consider the different social institutions that govern human relations, which regulate peoples' ways of living and working together, and communicating with each other; when we further consider the numerous substantive products that result from our continual technical activity, things like glass, alloys, and so on, we have to admit that between these and the bodies we inhabit there exists this difference: the one presents a certain plastic form not shared by the other.

Through the observation of slow and almost insensible modification, we can trace the development from the gross bodies of our experience through the materials that result from our technical activity to the first primitive engines and then the more sophisticated devices that have succeeded them. We are indeed able to indulge the hope of someday reproducing in our plastic creations some of the faculties of the living organisms of which we ourselves constitute the most highly perfected variety. We can then represent and for the sake of our investigation we ought to represent the organized entities constructed by us as forming in their aggregate a group distinguished among the natural entities by the specific characteristics common to all the individual members.

It is not true, of course, that there really exists such a sharply defined aggregate, since its boundaries are fuzzy and merge by insensible degrees in the larger organization of the universe as a whole. It results only from a classifying procedure, which supplments the gaps in our intelligence, that does not permit us to grasp at a single blow, so to speak, all the facts, causes, and laws of the world in which we live. Since our knowledge is only fragmentary and we can only distinguish the boundaries of the aggregate in question in a blurred fashion, it becomes convenient and helpful to study its existence as if it were real. It is also convenient that we should apply definite names to the bodies making up this aggregate and to the science describing their behavior.

In what follows I shall attach the name machine to all entities making up the aggregate I have just been considering. I shall adopt the term mechanology to designate the science of machines.

I have hesitated for a long time to propose the name machine for all organized devices constructed by man. In my point of view it encompasses the vast assemblage of engines, instruments, apparatus, tools, playthings, architectural constructions, in a word, all assemblages of organized bodies to which man has given a plastic form. I have therefore feared that the concisely narrow meaning of the term in current terminology might constitute an insurmountable obstacle to its employment in the very general sense I have attributed to it and that it originally had. For the aggregate I am considering I have sought for another term, which might be more convenient, reserving the name machine for just a part of the total aggregate in which the individual members are distinguished by special characteristics.

But then I reflected (and I shall return in due course to this point) that since every nomenclature is essentially an arbitrary one and results from a convention that we agree to set up for the expression of certain realities, it is always possible in the progress of science to replace any convention by one that seems ultimately to be a more satisfactory one. I have further reflected that the essential nature of machines has too often been neglected in the definitions that have hitherto been proposed. Since they are the product of human activity it is important, to understand them better, to make a comparative study of the forms, structures, and general organizations of all the devices constructed by man.

We may further bear in mind that contemporary discussions of the present state of the arts have accustomed us to see the strong connections connecting these various entities. For example, certain authors formally recognize the house as an inhabited machine, while certain others have recognized and emphasized the slow but steady evolution of the immobile forms of architecture to the mobile forms of mechanics. In my researches I have not found a more convenient term to express what I have in mind, nor could I feel like abandoning the term whose etymology, long history, and relation to my ideas seems to express correctly the reality I am seeking.

At the same time it appears to be not without interest to try, in my turn, to fix as precisely as possible the boundaries of the aggregate I am considering. To pretend to define what is meant by a machine amounts in a sense to supposing that the science of machines has already reached its final state or at any rate can do so soon in the near future. Otherwise what we are doing is to assign limits chimerically to the development of mechanical forms or to suppose at the outset a complete and perfect knowledge of the characteristics of all the individual mechanical devices past, present, and to come. For the perfection of a measuring instrument places all these devices in a defininte classification in accordance with the sum total of their characters. But this implies that the aggregate of these individual devices really constitutes a sharply defined aggregate with precise boundaries and having no connection with other devices.

In a universe in which everything happens through slow transformations, machines do not constitute an exceptional and completely isolated world. Every definition (of a machine) that can be proposed will only be an approximation without too great interest, subject to revision at every new step in the progress of science. Such a definition can be dangerous because it tends to crystallize in an immutable form the expression of what are essentially ever changing phenomena. Really one cannot define a machine any more than one can define a living being. For both, all one can do is to establish appropriate classifications of the groups they form. In this way we can grasp and measure the variable aspects of the bodies in question.

I am not indeed ignoring the definitions that people up to now have proposed for machines. There have been many of these. We find an excellent resume of them before 1870 in the book *Cinematique* by Reuleaux, published in France in 1866. In this book the author has investigated not less than eighteen different definitions and has shown that each presents definite shortcomings. Summarizing the critical study of the dominant ideas of that period and associating himself, for example, with the criticism of the ideas of Laboulaye, Reuleaux proceeded to give his own definition, a very penetrating one, which however does not seem to have attracted attention, at any rate in France.

If we consult, for the same period, the encyclopedic dictionary of Larousse in the edition of 1886, a work one naturally turns to if one wishes to learn about the ideas generally in vogue at that time, we find a definition analogous to that of Schrader, which was criticized by Reuleaux. Moreover the same work does not distinguish the machine from the measuring instrument or the engine by any well defined characteristic and leaves the reader uncertain about the meaning of these different terms.

Since the epoch we have just mentioned and during the contemporary period, some new definitions have come to light. Through the progress of the mechanical and physical sciences the earlier concept of a machine as a transformer of motion has evolved successively into a transformer of forces and energy.

It is easy to see that these various definitions rest wholly on certain phenomena involved in the operation of the machine and not on the consideration of the machine itself treated as a phenomenon. None of them are based on the observation of the characteristics that differentiate machines from other objects, since the enunciated characteristics, if they belong to certain machines, do not belong to all nor to those alone. Thus to take a very debatable hypothesis, a calculating machine does not possess any interest as a transformer of energy and the static transformer [*Editor's note*: machine like a lever?] is not a transformer of motion. To consider another example we can legitimately bring in the ox, which is clearly not a machine, and consider it analogous to an internal combustion engine from the standpoint of the transformation of energy, actually exhibited by both.

From this study we draw the conclusion that as far as machines are concerned each definition gives us only what we ourselves have put into it; that each definition results from a set of observations made directly on real things and not based on a *priori* concepts. It epitomizes the state of an already well developed science and follows in its various form the variations of this science. In order to define a machine it is first necessary to study the development of the science of machines through direct observation of what they do.

II. Mechanology, The Science of Machines

There exist two general types of sciences. The first, descriptive in character, undertake the rigorous description of observed phenomena. The second, normative in nature, have as their aim the investigation of the laws that govern the phenomena and the causes that produce them. This investigation has really only one goal: the explanation of the differences observed among the various phenomena.

The disciplines embraced in each of these types of science tend in time to become more and more specialized. Yet this specialization has not yet been completely accomplished and probably never will be. This combined with the fact the different disciplines are often involved at the same time in the same individual does not in any way prevent us from distinguishing between the two general types. Moreover, the highly perfected form assumed by the expressions for the differences in question in the most advanced sciences should not mislead us.

Thus the physical sciences through observation and measurement of natural and experimental phenomena endeavor to reduce to laws and causes all observed variations and differences. The natural sciences, in their turn, carry out the description and classification of observed natural phenomena in order to try to explain in terms of prior assumptions the differences they observe. Finally the social sciences through the description and classification of social phenomena provide the indispensable basis for the investigation of sociology.

The discipline of investigation, the descriptive procedures and methods of measurement vary from one science to another by reason of the differences in their objectives and to a less degree by reason of the state of sophistication of the sciences of more recent origin. The oldest sciences or those dealing with less complex phenomena have acquired, for example, remarkable power and precision in their procedures of observation and measurement as well as in their use of mathematics. But before attaining their present degree of perfection, these older sciences for a long time employed classificatory procedures that, seemingly the exclusive property of the natural sciences, are without doubt really only the primitive procedures of young and developing sciences. But cutting across this evaluation of methods and procedures the general division of scientific work is always characterized by the distinction between descriptive and normative types.

The science of machines could not hope to escape these essential distinctions, and examination of the facts associated with its slow formation indicate indeed that it has not actually eluded them.

The only aim of the science of machines or mechanology, a normative science, is the study and explanation of the differences that are observed among the various kinds of machines. And since science deals only with the real, its only object is the study of machines that really exist. It should shun all products of the imagination not applying to construction and use. It has the task of explaining the formation of all the various types of machine that come to our attention. In a word it is concerned with the problem of their existence.

Early science, in trying to understand complicated objects, necessarily displays a primitive organization and could not expect to succeed with considerable evaluation preliminary to a real descriptive science attaching more importance to the precision of its determinations than to the simple analysis of facts alone. In order to carry out its program, mechanology demands and will continue to demand even more strongly in the future that a science descriptive of machines or mechanography must modestly lay its own foundation. A very elaborate structure cannot be expected all at once. In certain domains of the world of machines the history, the description, and classification of the known facts will be sufficient. More detailed studies can follow.

Considering now the whole problem of machines, human interest in these devices appears to manifest itself under three distinct headings, corresponding to three different intellectual disciplines.

1. An art, antedating all science, and then developing parallel to science, drawing from science increasing power and developing into ever more varied forms. This art expresses in machines the creative aspirations of man, the needs he feels, and the possibilities he creates by the sustained application of his technical efforts.
2. A descriptive science of machines devoted to the history, description, and classification of existing machines. For this science I have proposed the name mechanography.
3. A normative science of machines that I denominate mechanology. This is a real science devoted to the study of the differences, and to the investigation of the causes and laws involved. Finally this science examines the ultimate problem of the reasons for the existence of machines.

The art of machine construction was followed by mechanography and mechanology in the course of time and will remain, I think, the general means by which we create, gain knowledge of, and explain machines.

But the very factual observations emphasized above allow us to perceive in each of the three great disciplines distinctions that become every day more evident. I feel impelled to emphasize this in all our activities with respect to the study of machines and to suggest the following order for this study, which seems to me indeed to be that actually exemplified in the historical development of the subject.

I.   The Art of Machine Construction
    a) Conception
    b) Material realization
    c) Assurance of proper functioning

Each of these main subdivisions of the art of machine construction is practiced by a specialist, e.g., architect, engineer, artisan, builder, entrepreneur, conductor, and so on.

II. Mechanography—the general descriptive science of machines
    a) Historical investigation: prehistory, history, archeology, ethnography, all relative to machines.
    b) Descriptive investigation of the elaboration and employment of various techniques: written presentation, graphical representations of forms and functions, symbolic representation.
    c) Taxonomic investigation of the increasing perfection of classification and measuring instruments of machine science.

Elaboration of general and particular distributions, of classification, nomenclature, and technology.

III. Mechanology—the normative science of machines
   a) Study of formal differences
   b) Study of structural differences
   c) Study of functional differences
   d) Study of general organizational differences
   e) Explanation of the origin of each type of machine

Each of these studies runs concurrently with the general problem of this science, namely the existence of machines.

[*Editor's Note:* The preceding material is the translation of an extract from the first section of a book of 122 pages. Later sections discuss in detail such subjects as earlier works on the history of machines, the distribution and classification of machines, problems and methods of mechanology, as well as social and philosophical considerations of the whole subject. Lafitte's position as a professional architect leads him to pay much attention to the relation between architecture and machines. This makes his conception of the nature of a machine rather more general than its restricted meaning in standard mechanics.]

3

Reprinted from *Gen. Mot. Eng. J.* **3**(3):34–39 (1956)

Operating Principles and Applications of the Fluid Coupling and Torque Converter to Automatic Transmissions

By OLIVER K. KELLEY
General Motors
Engineering Staff

In recent years the demands of modern traffic have brought about a rapid increase in the installation of automatic transmissions not only in passenger cars but also trucks, buses, and other vehicles. To date successful automatic transmission design has been based on the utilization of a hydro-dynamic drive component—either the fluid drive coupling or the torque converter—in conjunction with planetary gears and hydraulic controls. The operation of an overall automatic transmission is quite complex and can not be understood easily. The hydro-dynamic drive component, however, is basically a simple hydraulic mechanism. An easy approach to an understanding of how the fluid drive coupling and torque converter perform their intended function in an automatic transmission is by using the principle of a spinning flywheel.

Mechanics of hydro-dynamic

drive component based on

spinning flywheel principle

I N RECENT years no activity in the field of automotive engineering has been busier than that devoted to the development of automatic transmissions. By the end of 1955 General Motors had produced over 12 million automatic transmissions. Of this number approximately seven million were the well-known Hydra-Matic automatic transmission and the remainder were of the torque converter type. These figures represent passenger car installations only and do not include the thousands that have been installed on trucks, buses, ordnance vehicles, rail cars, and off-highway equipment. The automotive industry's production total of over 20 million automatic transmissions represents acceptance by the American motoring public of the convenience, ease, and safety of automatic drive.

To date successful automatic transmission design emphasis has been given to hydro-dynamic drive—either the fluid drive coupling or the torque converter. Each is basically a simple hydraulic driving mechanism, but despite their years of use and acceptance few people understand how they work.

The unique adoption and manner of combining the fluid drive coupling and the torque converter with full-power shifting gears can become complicated. In the same sense, the intricate designs of today's automatic transmissions keep the automatic transmission field very active. The basic mechanism of the hydro-dynamic drive component of an automatic transmission, however, is not as complicated as one might believe.

Operating Principles of a Fluid Drive Coupling

The simple mechanics of hydro-dynamic drive begins with the principle of a spinning flywheel. A spinning flywheel has stored-up energy and, when stopped, exerts a turning force on the mechanism stopping its rotation. Conversely, a turning force must be exerted against the flywheel to get it up to speed again after it has been stopped. A fluid drive coupling with an engine driving its input member and its output member stalled or stationary is the direct equivalent of the spinning flywheel principle.

Oil flowing radially outward through the input member of a fluid drive coupling, which is simply a centrifugal pump, leaves in the form of a spinning flywheel rim—a flywheel rim made of oil. The turning force exerted by a flywheel while it is being stopped is a function of the rate at which it is being stopped, its mean diameter, and its weight. Assume that the input member of

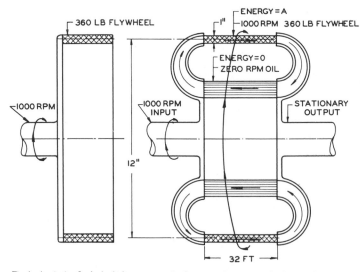

Fig. 1—A spinning flywheel, which exerts a turning force, or torque, on a mechanism stopping its rotation, parallels the operation of a fluid drive coupling which has its input member rotating and its output member stalled or stationary. If the input member of a fluid drive coupling extrudes out of its outlet, at a rate of 32 fps, a one-inch thick flywheel rim of oil rotating at 1,000 rpm, the turning force exerted each second on the output member to impart rotative motion will be 360 ft-lb.

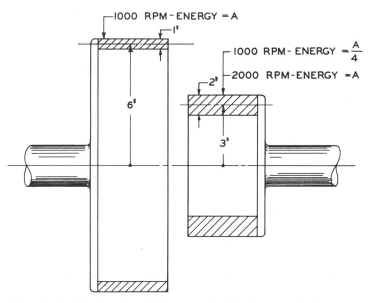

Fig. 2—Two flywheels, each rotating at 1,000 rpm and each weighing the same but with different mean diameters, contain specific energy characteristics. The energy contained in the larger diameter flywheel is four times greater than that contained in the flywheel of smaller diameter. Also, if both flywheels were to have the same energy, the flywheel of smaller diameter would have to rotate twice as fast as the larger diameter flywheel.

a fluid drive coupling extrudes out of its outlet, at a rate of 32 ft per sec, a flywheel rim of oil having a rotation of 1,000 rpm, a mean diameter of 12 in., and a rim section one inch in thickness (Fig. 1). The rotation of this spinning flywheel rim is being stopped by the stalled or stationary output member of the fluid drive coupling as soon as it emerges from

the input member at a rate of 32 fps. A 32-ft length of such a flywheel rim of oil will weigh close to 360 lb and will produce a turning force of 300 ft-lb when stopped in one second.

The foregoing events take place in a fluid drive coupling at stall conditions when an engine developing a torque of 300 ft-lb and running at full throttle is

trying to set a car in motion. The engine-driven input member is extruding a spinning fluid flywheel and pushing it into the blades of the stationary output member which stops the spin and in so doing receives a turning force on its blades amounting to a torque of 300 ft-lb.

Once the oil gets inside the blades of the stationary member, all resemblance to a spinning fluid flywheel is destroyed, and the oil merely comes out in a straight axial flow pattern and then returns again into the input member. There the oil is picked up from standstill and again re-created by the input member into a spinning fluid flywheel. An equal turning force, 300 ft-lb, must then be exerted by the engine against the oil while this is being done.

If the action of a fluid drive coupling running efficiently at higher speed is examined, another direct comparison with flywheels is recognized. Assume two flywheels, each running at 1,000 rpm and each weighing the same but with different mean diameters (Fig. 2). It is at once evident that the large diameter flywheel, while no heavier in total weight than the smaller one, is nevertheless more of a flywheel. If the diameter of the larger flywheel is twice as great as that of the smaller one, it is in reality four times the flywheel although no heavier in total weight. In other words, the energy contained in the larger diameter flywheel is four times greater than in the smaller

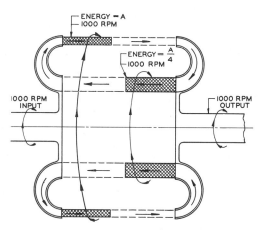
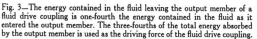

Fig. 3—The energy contained in the fluid leaving the output member of a fluid drive coupling is one-fourth the energy contained in the fluid as it entered the output member. The three-fourths of the total energy absorbed by the output member is used as the driving force of the fluid drive coupling.

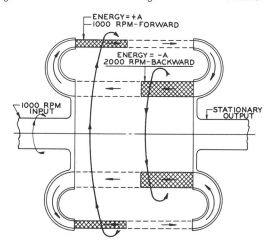

Fig. 4—If the straight blades of a fluid drive coupling's output member are replaced by blades that are strongly curved backward, the fluid leaving the output member will do so with a backward spinning motion which will cause a greater force to be exerted on the blades of the output member.

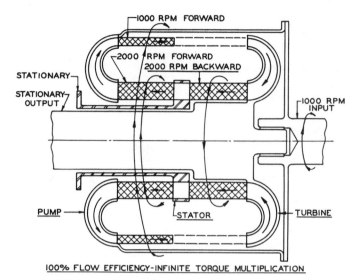

1000 RPM FORWARD

2000 RPM FORWARD
2000 RPM BACKWARD

STATIONARY

STATIONARY
OUTPUT

1000 RPM
INPUT

PUMP   STATOR   TURBINE

100% FLOW EFFICIENCY-INFINITE TORQUE MULTIPLICATION

Fig. 5—If the blades of a fluid drive coupling's output member are strongly curved backward, the fluid leaves the output member with a backward spinning motion opposite in direction to the rotation of the input member it is to enter. If the fluid were allowed to impinge directly on the blades of the input member, an appreciable amount of energy would be required from the engine to stop the backward spinning motion. To correct this situation a stationary member, called the stator, is installed between the input or pump member and the output or turbine member. The stator serves to reverse the backward spinning motion of the fluid leaving the turbine so that it has the same direction of rotation as the pump it enters. The reactive force on the stator establishes the amount of torque multiplication developed.

diameter flywheel even though both weigh the same and run at the same speed. Another basic fact also can be recognized. If both flywheels were to have the same energy, the smaller diameter one would have to run twice as fast as the larger one.

If the metal of the 1,000 rpm flywheel of larger diameter could be efficiently extruded down to the diameter and proportions of the smaller diameter 1,000 rpm flywheel, the process would give the energy difference between the two flywheels. Three-quarters of the original energy of the large diameter flywheel would be left to deal with while the smaller diameter flywheel would still be available, weighing the same and spinning at the same speed. Such an extrusion process would be difficult in metal but is easily accomplished with oil. This is exactly what goes on in a fluid drive coupling while running near 100 per cent efficiency.

Assume a fluid drive coupling having a circulation efficiency of 100 per cent (Fig. 3). Under this condition the input or driving member produces a continuous flywheel at a specific number of ft per sec. The coupling's output or driven member, acting as an extrusion die, squeezes the fluid flywheel down to a smaller diameter with a correspondingly heavier rim section. This smaller diameter flywheel leaves the exit of the driven member at the same rpm as when it entered but minus three-quarters of its original energy. This three-quarters of the original energy drives the car and also must be re-supplied by the engine when the flywheel diameter is again extruded into a larger diameter after it enters the input member.

The fluid drive coupling, in its efficient driving range, can be considered as a pair of rotating extrusion forms which first make a large diameter spinning fluid flywheel into one of smaller diameter of equal weight and then vice versa. Energy is absorbed by the output member as the flywheel diameter is decreased, and an equal amount is spent by the input member as the flywheel diameter is increased.

Simple Three-Element Torque Converter

The basic operating principles of the fluid drive coupling aid in establishing the operating principles of a simple, three-element torque converter.

Assume, as before, that the circulation efficiency of the fluid drive coupling is 100 per cent, that the input member is rotating at 1,000 rpm, and that the output member is stalled or stationary. However, replace the straight blades of the coupling's output member with blades that are strongly curved backward and which, by proper bending, can be made to receive the spinning oil without splash. By having the exit of the blades strongly curved backward the oil, as it is finally extruded out of the output member, will now be spinning backward (Fig. 4).

The backward speed of the spinning fluid flywheel depends upon the exit angle of the blades and the flow velocity of the oil. Assume that 32 fps is a high enough flow velocity and that the blades have a strong enough back bend so that the outcoming oil has a spinning speed of 2,000 rpm. The smaller diameter flywheel running at 2,000 rpm has the same energy as the larger diameter flywheel running at 1,000 rpm. If the output member is standing still and the blades have done their work on the oil 100 per cent efficiently, it would be possible to have this condition of equal energies in the input and output oils. However, the energies would be in opposite directions.

The turning force now felt by the output member, also referred to as the turbine, is twice as great as it was in the fluid drive coupling at stall conditions when the input member was sending out an equal fluid flywheel at 32 fps. The turbine feels the turning force of slowing down the 1,000 rpm fluid-flywheel rim section at a rate of 32 fps and also feels the reaction to the speeding up of the reversely spinning flywheel of equal energy. Obviously then, this turbine feels twice the turning force of the fluid drive coupling for equal flow velocity and equal input member conditions.

One condition is seriously wrong, however, in this arrangement—the direction of the backward spinning oil leaving the turbine. It would require the entire engine torque just to stop this backward rotation when it hits the entrance to the input member, or pump, and there would be no engine torque left to get the oil re-energized into a forwardly rotating fluid flywheel at the outlet of the input member.

This situation can be corrected by

installing a set of stationary curved blades, called the stator, between the turbine exit and the pump (Fig. 5). The addition of the stator changes the fluid drive coupling to a simple, single-phase, three-element torque converter. The curved stationary blades of the stator must be properly shaped to receive the backward rotation of the oil, bring it to a stop, and then direct it to a forward rotation. If this is properly accomplished without loss by having a very efficient blading arrangement in the stator, there will be no energy loss in the oil leaving the stator. The stator has merely received a fluid flywheel running backward at 2,000 rpm, brought it to a stop, and then let it convert itself into a forwardly rotating fluid flywheel at 2,000 rpm. The action is very similar to that which took place in the turbine.

It might be good to study the action of this conversion from one direction of rotation to the opposite direction of rotation. Consider first that the flywheel effect of any spinning body is in reality nothing but the total sum of the momenta of all of its particles due to their velocity and that the momentum is not lost if the velocity is not lost. From this it can be seen that as the oil is slowed down in its rotation by the stator blades and all of its rotation stopped, it is really accelerated into an axial flow without losing its velocity. This same absolute velocity, then, is directed by the stator blades into a new rotary direction, and as the oil leaves the blades, it is once again a fluid flywheel of the same speed but different direction of rotation.

Now consider what happens when this oil re-enters the pump. The oil re-entering the pump is in the form of a smaller diameter fluid flywheel rotating at 2,000 rpm with the same energy and direction of rotation as the larger diameter fluid flywheel coming out of the pump at 1,000 rpm. This means that the engine does not add any energy to the oil. The engine-driven pump could be designed with properly shaped blades to receive the 2,000 rpm smaller diameter flywheel and let it expand in a free whirl into the larger diameter flywheel. During expansion, the rotative speed of the flywheel would decrease while its radius increases, which is the natural behavior of liquids in free whirl, until it came out of the pump at 1,000 rpm with its original energy intact and essentially without exerting a force on the blades.

Theoretically, it would be possible to get a big driving torque on the turbine without having to spend any of the engine's torque. This would actually be the case if there were no flow losses. With 100 per cent flow efficiency perpetual motion would exist. However, friction in the fluid's flow makes it impossible to obtain such an ideal result because of the flow losses involved in taking the oil flywheel of larger diameter and extruding it to the smaller diameter in addition to extruding it into a backward spinning, smaller diameter flywheel.

The fluid drive coupling can never be 100 per cent. efficient. Part of the momentum of the 1,000 rpm larger diameter flywheel has to be used to overcome the friction of the oil flow. The same holds true for the simple, three-element torque converter. Part of the momentum of the larger diameter 1,000 rpm flywheel (and a greater part is now needed) must be used to accomplish the conversion through the turbine. The result is that it is difficult to obtain, even with accurately designed blades, a 1,900 rpm backward spinning flywheel in place of the 2,000 rpm flywheel.

The same kind of flow losses exist in the stator. The best that could be done would be to design the exits of the stator blades to produce a 1,800 rpm flywheel rotating in a forward direction from the 1,900 rpm backward spinning flywheel. This would give the energy equivalent

of a 900 rpm larger diameter flywheel. Even this 900 rpm fluid flywheel would not materialize in the larger diameter without some help from the pump. The pump blades can be slanted slightly backward, however, permitting the pump to run slightly faster (for example at 1,020 rpm) while the back bend of the blades aids the flow. This does not increase the torque on the pump, merely its speed. The pump senses a torque equal to the difference between the original 1,000 rpm flywheel and the 900 rpm flywheel equivalent of the small diameter flywheel entering the pump.

If the energy equivalents of the larger diameter flywheel are used, it can be said that the turbine has felt the turning force of the 1,000 rpm flywheel plus a 950 rpm flywheel, which is the equivalent of the small diameter 1,900 rpm backwardly rotating flywheel. This would give an index figure of 1,950 rpm as representing the turbine torque. The pump torque could be represented by a 100 rpm figure. The stator torque would be the equivalent of 1,850 rpm in the larger diameter flywheel—the equivalent sum of the 1,900 rpm and the 1,800 rpm small diameter flywheels.

It would be possible to make a torque converter where the relation of turbine torque to pump torque is 20 to 1 if the converter was designed for maximum flow efficiency at stall conditions when the turbine is standing still. If the turbine

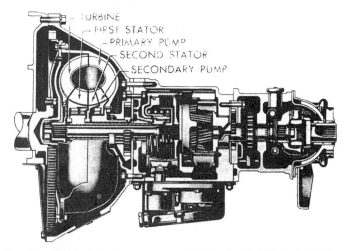

Fig. 6—The original Buick Dynaflow automatic transmission introduced in 1948 utilized a five-element polyphase torque converter in conjunction with a planetary gear set. The second pump and the two stators were free wheeling members which met varying conditions by providing a high oil-flow velocity at stall which rapidly diminished with an increase in turbine speed until a very nominal flow velocity transmitted the engine torque at cruising conditions.

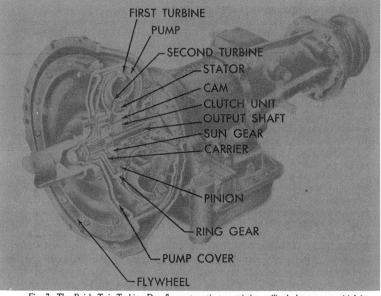

FIRST TURBINE
PUMP
SECOND TURBINE
STATOR
CAM
CLUTCH UNIT
OUTPUT SHAFT
SUN GEAR
CARRIER
PINION
RING GEAR
PUMP COVER
FLYWHEEL

Fig. 7—The Buick Twin-Turbine Dynaflow automatic transmission utilized the torque-multiplying characteristics of a planetary gear set in conjunction with the torque-multiplying ability of a fluid torque converter. At low speeds all of the power was transferred through the first turbine and its planetary gear set. As vehicle speed increased, the torque on the first turbine decreased and the second turbine took over the drive at cruising speeds.

is standing still, however, it is doing no work and the overall efficiency is zero.

A practical torque converter must be designed with blade entrance angles that receive the flow properly at some useful drive ratio, for example a 2 to 1 torque multiplication. If the same conditions of flow velocity and efficiency are assumed as before, it is again necessary to assign a loss through the turbine which is the equivalent of 50 rpm in terms of the larger diameter flywheel and also an equal loss for the stator. From this it can be seen that by the time the oil re-enters the pump there is a total equivalent of 100 rpm lost out of the original 1,000 rpm flywheel's energy for hydraulic flow losses through the turbine and stator.

There also is the turbine's rotation to contend with. If there is a 10 per cent loss of energy due to hydraulic flow through the turbine and stator, it becomes obvious then that, if the turbine torque is to be twice the pump torque, the turbine speed must be 10 per cent less than one-half the pump speed. This sets the turbine speed at 450 rpm and requires that the turbine blades be properly formed to receive the 1,000 rpm flywheel of oil when the turbine is running at 450 rpm. To achieve this action the entrance and exit angles of

the turbine blades must be accurately established for this condition.

The hydraulic loss assigned to the stator was the equivalent of 50 rpm of the 1,000 rpm initial flywheel energy, or five per cent. The rotational speed of the oil, therefore, before it enters the stator should be five per cent greater than when leaving the stator. The rpm of the oil entering the pump can be considered as an unknown quantity X and the rpm of the oil leaving the turbine as 105 per cent of X. The pump torque has been established as the difference in the flywheel values before and after passing through the pump, and the turbine torque is the sum of the fluid flywheel values before and after passing through the turbine. For 2 to 1 torque multiplication the turbine torque must be twice the pump torque. The following equation can be established and solved to determine the rpm of the oil entering the pump and leaving the turbine:

$$2(1{,}000 - X/2) = 1{,}000 + 1.05\ X/2$$
$$X = 656 \text{ rpm (oil entering pump)}$$

$$X(105 \text{ per cent}) = 688 \text{ rpm (oil leaving turbine)}.$$

The turbine blade exit angles must be designed to produce a net backward

spin to the oil of 688 rpm while the turbine has a forward speed of 450 rpm. This means that the turbine-blade exit angles must produce a fluid flywheel having a reverse spin of 1,138 rpm while the turbine is stationary, instead of the ideal stall-ratio torque converter's 1,900 rpm.

If a two per cent hydraulic loss is allowed for the pump member as before, along with providing a slight backward bend in the pump blades so that the initial 1,000 rpm energy flywheel leaves the pump rotating at 1,020 rpm, and if the entrance angle is set properly to receive 656 rpm oil from the stator, the basic design for the single-phase, three-element torque converter will be completed. The overall efficiency can be calculated by taking the pump speed times its torque and dividing by the turbine speed times its torque which gives an overall efficiency of 88 per cent. This is the so-called design point efficiency where ideal flow exists without entrance shock losses.

This single-phase, simple, three-element torque converter with stalled turbine cannot produce a 20 to 1 torque multiplication. It differs from an ideal torque converter designed for stall condition in two important respects. First, its members are not receiving oil properly when the turbine is stationary and large shock losses are encountered which increase the flow resistance. Second, its turbine and stator are not designed to accomplish the maximum possible turning of the oil as was done in the ideal stall-ratio torque converter. The turbine blades now produce a backspin of 1,138 rpm instead of 1,900 rpm, and the stator blades produce a forward spin of 656 rpm instead of 1,800 rpm. As a result, more oil must circulate through this milder path to produce the same forces in the turbine, and the pump must exert a greater torque on the oil because the stator no longer produces the maximum possible forward spin for the pump.

Both the shock losses and the simple circulation losses are proportional to the square of the flow velocity. This adds up to several times the losses in the ideal stall-ratio torque converter. Under the best of conditions it would be possible perhaps to obtain a 5 to 1 torque multiplication out of this single-phase converter at stall.

A similar situation exists at higher speed operation when the turbine runs faster than 450 rpm, the design speed. All of the entrance angles would be wrong in the opposite direction, and the shock losses would rapidly decrease the efficiency. Torque converters of the single-phase design have this shortcoming. Their efficiency at the design point can be good, and their torque multiplication at stall can be adequate. However, their efficiency at higher speed operation suffers so much that the single-phase design is not useable.

What the torque converter needs is a constantly changing blade entrance angle for its members so that they can accommodate themselves to the ever changing conditions of turbine speed. This presents quite a complex design problem for which there seems to be no apparent solution. A part of the solution, however, can be obtained by permitting the stator member to free wheel when the direction of oil flow has changed sufficiently to actually exert a forward rather than a backward reaction on the stator. If the stator free wheels at this point, there will be no additional shock losses.

A second way to obtain part of the solution, and one which has now been made practical in the Buick Variable-Pitch Dynaflow, is to change the angle of the stator blades to better satisfy varying car speed conditions.

Buick Dynaflow Automatic Transmission

The Buick Dynaflow torque converter automatic transmission is an example of progress made in the application of hydro-dynamic drive. The original Dynaflow was introduced in 1948. The design, which adapted a basic torque converter into a five-element polyphase converter, consisted of a primary pump, a secondary pump mounted on a free wheeling clutch, a turbine, and two stators also mounted on free wheeling clutches (Fig. 6). The free wheeling members met the varying conditions by providing a high oil-flow velocity at stall which rapidly diminished with increased turbine speed until a very nominal flow velocity transmitted the engine torque at cruising conditions.

In 1952 Buick introduced the Twin-Turbine Dynaflow which greatly improved acceleration and torque converter efficiency and reduced the engine speed for a given car speed. This design consisted of a pump, a first and second turbine, and a free wheeling stator (Fig. 7). The first turbine was connected to

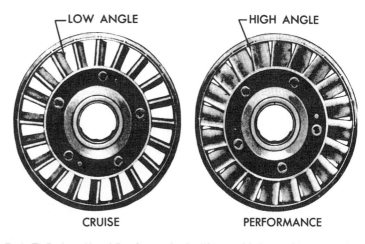

Fig. 8—The Buick variable pitch Dynaflow, introduced in 1954, retained the features of the twin-turbine Dynaflow but also added a stator having variable or controllable pitch blades. The 20 blades were mounted on individual crank pins so that the pitch could be changed approximately 75°. The stator blades were actuated by a stator piston linked to the throttle linkage through a control valve. Oil pressure, created by throttle movement, moved the stator piston and the blades to low angle for cruising conditions or high angle for high performance conditions.

the converter's output shaft through a torque-multiplying planetary gear set. The second turbine was directly connected to the converter's output shaft. At low speeds, when top performance is required, all of the power was transferred through the first turbine and its planetary gear set. As vehicle speed increased, the torque gradually and smoothly diminished on the first turbine and increased on the second turbine until it completely took over the drive at cruising speeds. Meanwhile, the first turbine was rotating freely whereby it could re-enter the drive whenever its high torque-producing ability was required.

In 1954 Buick introduced Variable-Pitch Dynaflow. This design retained the twin-turbine feature but also added a stator having variable or controllable pitch blades (Fig. 8). The automatic control of the stator-blade pitch provided the best angle for cruising and economy driving, as well as the best angle for performance and fast acceleration. With the fixed stator-blade angle of the previous Twin-Turbine Dynaflow design a compromise angle which would give the best results for both conditions had to be used.

In 1955 Buick introduced a still further step in automatic transmission design improvement. This new design contains an additional fixed-blade stator located between the first and second turbines. This additional double regenerative stator provides additional performance at lower speed where it is most needed.

Developmental steps of the type described here have constantly improved the Buick Dynaflow transmission and reflect the ability of the basic torque converter, when properly and uniquely adapted, to provide absolutely smooth, fast responding, automatic drive without the need of shifting gears.

Conclusion

The ultimate in automatic drive has not been reached, and activity in this field of automotive engineering continues. The results obtained thus far in the development of automatic transmissions, however, have proven to be highly successful, and the percentage of cars equipped with automatic drive has steadily increased. In 1955 alone over 65 per cent of all passenger cars manufactured were equipped with automatic drive.

What the future will bring will depend on the abilities of engineers to develop new or better ways for providing automatic power transmission from the engine to the driving wheels.

35

4

Reprinted from pp. 47–51 of *Acoustics: A Text on Theory and Applications*,
G. W. Stewart and R. B. Lindsay, D. Van Nostrand Co., Inc., New York,
1930, 358 pp.

COMBINATION OF ACOUSTIC ELEMENTS

G. W. Stewart and R. B. Lindsay

[*Editor's Note:* In the original, material precedes this excerpt.]

2·3. Helmholtz Resonator and Acoustic Impedance.—The Helmholtz resonator is an enclosure communicating with the external medium through an opening of small area (see Fig. 2·2). The opening may be flat, as in the figure, or it may be in the form of a neck. In either case it is a simple matter to separate the resonance elements. Inside the resonator there is a volume of gas of magnitude V which is alternately compressed and expanded by the movement of the gas in the opening. It thus provides, so to speak, the *stiffness* element of the system. The gas in the opening moves as a whole and provides the mass or *inertia* element. At the opening, moreover, there is a radiation of sound into the surrounding medium leading to the dissipation of acoustic energy and providing the *dissipation*

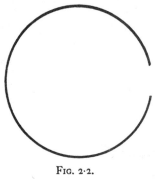

Fig. 2·2.

element. To write down the equation of motion of the gas in the resonator we must estimate the magnitudes of the above elements.

If the opening has a neck of length l small in comparison with the wave length, with a cross sectional area S, the mass of gas in the opening is $\rho_0 S l$. It is customary and convenient to write this in another form by introducing the quantity

$$c_0 = S/l, \qquad (2\cdot13)$$

which is called the acoustic *conductivity* of the opening. A more thorough discussion of the conductivity will be given in Section 2·4. For the present we merely substitute it into the mass expression, whence the latter becomes $\rho_0 S^2/c_0$. It is to be noted that this representation is possible whether the opening is in the form of a neck or is flat.

To get the expression for the dissipative force we need to calculate the amount of acoustical energy radiated from a hemispherical source of sound in a fluid. We shall give a rigorous derivation of this quantity in a later section (Sec. 3·2). It will suffice here to note that the final result for the dissipative force is

$$\frac{\rho_0 \omega k}{2\pi} S^2 \dot{\xi},$$

where $k = 2\pi/\lambda$ as usual. The reader should note the dependence on the velocity.

Finally, we must compute the stiffness coefficient. For this it is necessary to calculate the force acting on the area S of the opening. If the volume V of the resonator is decreased adiabatically by the amount dV, the excess pressure is (see eq. (1·14))

$$p' = \rho_0 c^2 s = -\rho_0 c^2 \frac{dV}{V},$$

for by definition, since the mass is constant, $d(V\rho) = 0$ and

$$s = \frac{\delta\rho}{\rho_0} = -\frac{dV}{V}.$$

Now $dV = -S\xi$, if the displacement producing the volume change is ξ. Therefore the force acting on area S is

$$\frac{\rho_0 c^2 S^2}{V} \xi.$$

We are now ready to write the equation of motion. Putting $S\xi = X$ and supposing that the resonator is driven by an external force producing a pressure p, we have

$$\frac{\rho_0}{c_0}\ddot{X} + \frac{\rho_0\omega k}{2\pi}\dot{X} + \frac{\rho_0 c^2}{V}X = p. \qquad (2\cdot14)$$

Mathematically, if p is a harmonic function of the time, $(2\cdot14)$ is precisely similar to $(2\cdot1)$. For the steady state its solution therefore is

$$\dot{X} = \frac{p}{\dfrac{\rho_0\omega k}{2\pi} + i\left(\dfrac{\rho_0\omega}{c_0} - \dfrac{\rho_0 c^2}{V\omega}\right)}, \qquad (2\cdot15)$$

the real part of which written in trigonometric form becomes

$$\dot{X}_{\text{real}} = \frac{p_1 \cos(\omega t - \alpha)}{\rho_0\sqrt{\left(\dfrac{\omega k}{2\pi}\right)^2 + \left(\dfrac{\omega}{c_0} - \dfrac{c^2}{V\omega}\right)^2}}, \qquad (2\cdot16)$$

where, as usual,

$$\tan\alpha = \frac{2\pi\left(\dfrac{\omega}{c_0} - \dfrac{c^2}{\omega V}\right)}{\omega k}. \qquad (2\cdot17)$$

The maximum value of $\dot{X}$ occurs approximately when $\omega/c_0 = c^2/V\omega$, that is, the approximate resonance value of ω is

$$\omega = \omega_0 = c\sqrt{\frac{c_0}{V}}. \qquad (2\cdot18)$$

The more accurate expression for the resonance frequency is

$$\omega_0 = c\sqrt{\frac{c_0}{V}} \cdot \frac{1}{\sqrt{1 + \dfrac{k^2 c_0^2}{8\pi^2}}},$$

but the term $k^2 c_0^2/8\pi^2$ is usually negligible, for $c_0 < < \lambda$, as a rule.

These theoretical conclusions are confirmed by experiments on resonators.

We now define *acoustic impedance* analogously to the mechanical impedance of the preceding section.

Thus, we write

$$Z = \frac{\text{pressure}}{\text{rate of volume displacement}} = \frac{\text{pressure}}{\text{volume current}}$$

$$= \frac{p}{\dot{X}}$$

$$= Z_1 + iZ_2,$$

where

$$Z_1 = acoustic\ resistance = \frac{\rho_0 \omega k}{2\pi} = \frac{\rho_0 c k^2}{2\pi}, \qquad (2\cdot19)$$

and

$$Z_2 = acoustic\ reactance = \rho_0 \left(\frac{\omega}{c_0} - \frac{c^2}{\omega V} \right). \qquad (2\cdot20)$$

It is customary to call the quantity ρ_0/c_0 the *inertance*, while the quantity $V/\rho_0 c^2$ is the *acoustic capacitance*. It is thus usual to write for the reactance $M\omega - 1/\omega C$ with $M = \rho_0/c_0$ and $C = V/\rho_0 c^2$. The latter is seen from eq. (2·14) to be the ratio of the volume displacement to the pressure for the case of static displacement. With regard to the former, it should be noted that the *inertance* is *not* the mass of the system. Rather we have the relation

$$\text{inertance} = \frac{\text{mass}}{S^2}.$$

As might be expected from the previous section the maximum $\dot{X}$ and the maximum pressure (or displacement) do not occur at quite the same frequency. The resonance frequency is, of course, given to a close approximation by (2·18), while the maximum displacement occurs for the value of ω which makes

$$\frac{\omega^4 k^2}{4\pi^2} + \left(\frac{\omega^2}{c_0} - \frac{c^2}{V} \right)^2$$

a minimum. This comes out to be

$$\omega_1 = c \sqrt{\frac{c_0}{V}} \cdot \frac{1}{\sqrt{1 + \frac{k^2 c_0^2}{4\pi^2}}}. \qquad (2\cdot21)$$

This is but slightly less than ω_0, as a rule.

The amplification constant of the resonator is the ratio of the squares of the maximum excess pressure in the resonator and the maximum external operating pressure. For the former we have at once the expression,

$$p_{max} = \rho_0 c^2 s_{max} = \frac{\rho_0 c^2 X_{max}}{V},$$

while the latter is simply p_1. Now we have

$$X_{max} = \frac{p_1}{\rho_0 \omega \sqrt{\left(\dfrac{\omega k}{2\pi}\right)^2 + \left(\dfrac{\omega}{c_0} - \dfrac{c^2}{V\omega}\right)^2}}.$$

Hence after some reduction we arrive at

$$Amplification = \frac{p^2_{max}}{p_1^2} = \frac{1}{\left(\dfrac{k^3 V}{2\pi}\right)^2 + \left(1 - \dfrac{k^2 V}{c_0}\right)^2}, \qquad (2\cdot22)$$

which for the resonance case reduces simply to

$$\frac{4\pi^2}{k^6 V^2}.$$

[*Editor's Note:* Material has been omitted at this point.]

5

Reprinted from *J. Appl. Phys.* **9**:612–622 (Oct. 1938)

The Filtration of Sound, I

By R. B. Lindsay

Brown University, Providence, Rhode Island

1. Introduction

ACOUSTIC filtration is only one aspect of the general problem connected with the transmission of energy through a medium possessing periodic nonhomogeneity of structure. Many illustrations come to mind, of which we note only a few. Thus a perfectly flexible homogeneous stretched string when loaded with equally concentrated and equally spaced mass particles constitutes such a medium from the standpoint of the propagation of transverse elastic waves. Similarly an electric transmission line containing an iterated combination of inductances, resistances and capacitances is a periodic nonhomogeneous medium for the passage of long electromagnetic waves. Finally a crystal metal lattice is a medium of similar nature for the transmission of electrons.

The mathematical problem of determining the transmission through the various structures just mentioned is fundamentally the same, and it is found that in each case certain frequency ranges (or energy ranges in the case of the electrons) are transmitted while others do not pass. From the purely mathematical point of view there would be a decided gain in developing the subject of energy filtration from the most general standpoint. Here it becomes, however, simply a problem in mathematics, namely the solution of second order differential equations subject to iterated boundary conditions. In this article on the other hand we are primarily interested in the physical phenomena associated with transmission and filtration.

2. Compressional Waves in a Confined Fluid Medium

Let us first examine the propagation of compressional elastic waves in a fluid medium, e.g., a gas like air, confined in a tube. Such situations are still of more than academic interest, as witness the stethoscope, speaking tubes on shipboard, air-conditioning conduits, etc. A very interesting and significant feature of such transmission is its frequency selectivity with respect to alterations in the mode of confinement of the medium and with respect to side attachments.

As an initial illustration consider sound passing through a tube with a constriction or an expansion. (See Fig. 1 (a), (b).) In each case it is as-

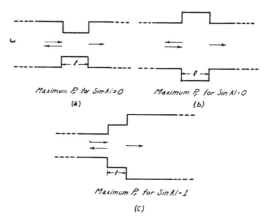

Maximum P_r for $\sin kl = 0$ Maximum P_r for $\sin kl = 0$

(a)   (b)

Maximum P_r for $\sin kl = 1$

(c)

Fig. 1. Constrictions in tubes in which the sound enters from the left and travels to the right.

sumed that the sound wave comes in from the left and goes off to the right with no reflection from a terminus at the right (i.e., ideally speaking the tube extends to ∞ on the right). The ratio of the transmitted energy flow (e.g. joules per second or watts) at the right of the constriction or expansion to the incident energy flow at the left of the constriction or expansion is called the *power transmission ratio* and is denoted by P_r. The mathematical treatment of the problem shows that P_r is a function of S_2/S_1, i.e., the ratio of the areas of the main line and constriction or expansion respectively, and $\sin^2 kl$, where l is the length of the constriction or expansion and $k = 2\pi\nu/c$, where ν is the frequency and c the velocity of the sound. This means that for a given mode of confinement, P_r is periodic in frequency with a period equal to $c/2l$. In particular it turns out that when sin $kl = 0$, P_r is a maximum, falling to a minimum (not zero, of course) for sin $kl = 1$. This is interesting since sin $kl = 0$, or $kl = n\pi$ (n integral) is the condition for resonance of a single tube of length l, either open at both ends or closed at both ends. We may look upon the former situation as corresponding to Fig. 1(a), while the latter corresponds to Fig. 1(b). Interestingly enough, Fig. 1(c), where the area of cross section does not revert to its original value, corresponds, as seems quite reasonable, to the case of a tube closed at one end and open at the other, so the condition for maximum P_r is sin $kl = 1$. Some actual transmission curves for cases (a) and (b) are indicated in Fig. 2.

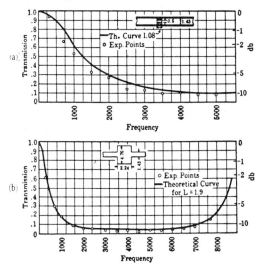

FIG. 2. Transmission curves for two types of tubes.*

A second illustration is provided by a tube with a Helmholtz resonator inserted as a branch. The situation is shown diagrammatically in Fig. 3(a), where the resonator consists of a closed chamber at the side of the tube with a small orifice as the entrance. It now develops, as one might expect, that P_r (i.e., the power transmission ratio across the branch) is no longer periodic in frequency. Rather, as Fig. 4 indicates, P_r has a minimum at the resonance frequency of

* Figs. 2, 4, 5, 6, 10 and 12 are taken with permission from Stewart and Lindsay, *Acoustics* (D. Van Nostrand Co., N. Y.)

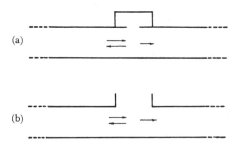

FIG. 3. Further examples of tubes used for transmitting sound.

the resonator and a relatively small value for a considerable frequency range about the resonance point. We should expect this since it is just at the resonance frequency that the resonator will absorb energy most readily from the incident sound wave.

For a third illustration we take the case of sound passing through a tube with a side orifice open to the air (Fig. 3(b)). Here the orifice acts as a branch and P_r measures again the transmission ratio across the branch. The physical description of the behavior of this arrangement is most simply given by noting that the incident sound sets the air in the orifice in motion. Some of the energy of this motion is radiated away into the air. The resultant effect is to reflect some of the incident sound energy and dissipate some of it through radiation. The rest gets by in transmission across the branch. Both absorption effects are most marked at low frequencies. Hence P_r is smaller at low frequencies and rises to nearly unity at high frequencies, as Fig. 5 indicates. This is not really inconsistent with the well-known fact that for given displacement

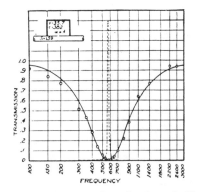

FIG. 4. Transmission curve for tube shown in Fig. 3(a).

velocity of a sound source radiation takes place more efficiently at high frequencies than at low. The point in the present case is that at high frequencies it is harder to get the air in the orifice to vibrate, hence the maximum displacement velocity of the air there is less and there is actually *less* radiation, in spite of the fact that the radiation efficiency is higher. Indeed at sufficiently high frequencies the transmission through the tube behaves as if the orifice were not there at all.

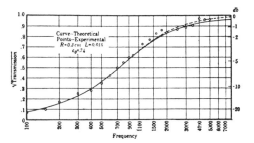

FIG. 5. Transmission curve for tube shown in Fig. 3(b).

3. Acoustic Filtration in Air. Physical Considerations

The discussion of the previous section suggests the possibility of using the selectivity of the structures there considered for the purpose of constructing actual acoustic filters. Consider, for example, the tube with a side orifice. The indication is that if a single orifice decreases the transmission at low frequencies, a succession of orifices will cut P_r for this frequency range nearly to zero. This proves to be true experimentally, as the accompanying figure (Fig. 6)

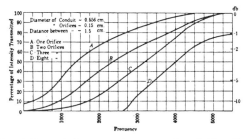

FIG. 6. Transmission curves for a succession of side orifices in a tube.

shows. We should expect that whereas for low frequencies P_r is negligible, when the fundamental frequency of the equivalent open tube

between orifices is reached, the transmission will have risen to unity. For higher frequencies the orifice effect will usually have become negligible and large P_r is thus assured independently of tube resonance. This neglects the mathematical refinements we shall presently introduce. However, these purely physical considerations imply that the structure under consideration acts like a *high pass filter* with a transition frequency between the nontransmission or attenuation band and the transmission band which cannot be higher than the first resonance frequency of the main line considered as an open tube. Actually the latter may be considerably above the transition or cut-off frequency.

Continuing the descriptive review of acoustic filtration in air, consider a tube with equally spaced Helmholtz resonators as side branches (Fig. 7). From physical considerations we do not

FIG. 7. A tube with equally spaced Helmholtz resonators as side tubes.

expect the resonators to influence P_r for frequencies far away from the resonance frequency of the resonators. Hence we should expect this structure to behave as a low pass filter. As the frequency rises close to the resonance frequency, the resonators go into action, P_r falls and attenuation sets in. However, as the frequency increases still further the effectiveness of the resonators decreases and large P_r again becomes the rule except possibly for sin $kl=1$, where l is the distance between resonators. The exact cut-off frequency will depend on the interaction of the main line resonance and the resonator resonance. This can only be found by a closer study of the problem.

4. Transmission Theory of Acoustic Filtration. Ideal Infinite Case

There are two principal theories of acoustic filtration. The first, due to G. W. Stewart,[1] is based on the electrical filter analogy and considers the acoustic line as made up of concentrated or "lumped" elements. This is only a first approximation since acoustical elements are

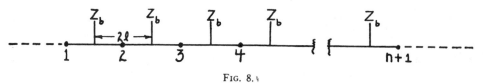

FIG. 8.

rarely lumped in the strict sense. Nevertheless it has been used with considerable success. Since it is very fully described in the book just mentioned we shall not go into it here. Rather we shall confine our attention to the more general transmission theory developed by W. P. Mason.[2] It is somewhat difficult to assign priority since the fundamental idea of the propagation of elastic waves in a medium possessing periodic nonhomogeneity appears to be an old one.[3] Consider the system indicated schematically in Fig. 8. This consists of an infinite acoustic line of cross-sectional area S with equally spaced branches (separation distance $2l$) in which the branch impedance at the junction is Z_b, a pure reactance (i.e., a pure imaginary impedance). Though we shall begin by thinking of the acoustic medium in the line and branches as a gas like air, the method is also applicable to liquid and solid media and to transverse as well as longitudinal waves. For the moment we shall confine the discussion to gases and describe the plane harmonic wave transmission in the line by means of the excess pressure p and the volume current $\dot{X}$ (equal to the product of cross-sectional area and particle velocity). The application of the appropriate boundary conditions satisfied by p and $\dot{X}$ at any branch (in the case of a fluid medium, these are simple hydrodynamical continuity conditions; in the case of solids, the situation is a little more elaborate, as will be seen below) yields the following equations connecting p and $\dot{X}$ for two successive mid-section points j and $j+1$ (e.g., 2, 3, etc. in Fig. 8).

$$p_{j+1} = A p_j - iBZ\dot{X}_j, \qquad (1)$$

$$\dot{X}_{j+1} = A\dot{X}_j - iC/Z \cdot p_j, \qquad (2)$$

where

$$\begin{aligned} A &= \cos 2kl + iZ/2Z_b \cdot \sin 2kl, \\ B &= \sin 2kl + iZ/Z_b \cdot \sin^2 kl, \\ C &= \sin 2kl - iZ/Z_b \cdot \cos^2 kl. \end{aligned} \qquad (3)$$

In these expressions $Z = \rho_0 c/S$ is the acoustical impedance of a plane harmonic progressive wave. The mean density of the fluid is ρ_0 and the velocity of a compressional wave in the fluid is c. As usual $k = 2\pi\nu/c$, where ν is the frequency. Examination of (3) discloses that

$$A^2 + BC = 1. \qquad (4)$$

If we introduce the angle W with

$$A = \cos W, \qquad (5)$$

we can write $B = (B/C)^{\frac{1}{2}} \cdot \sin W$ and $C = (C/B)^{\frac{1}{2}} \cdot \sin W$. Eqs. (1) and (2) may now be written

$$p_{j+1} = p_j \cos W - iZ\dot{X}_j (B/C)^{\frac{1}{2}} \cdot \sin W, \qquad (6)$$

$$\dot{X}_{j+1} = \dot{X}_j \cos W - i/Z \cdot p_j (C/B)^{\frac{1}{2}} \cdot \sin W. \qquad (7)$$

Now $p_j/\dot{X}_j$ is by definition[4] the acoustic impedance at the point j. Its value for given frequency depends entirely on the structure of the acoustic line to the right (or left) of j. But since the line is, by definition, infinite in extent, the impedance must be the same at all mid-section points. Let us call this the *characteristic impedance* of the line and denote it by Z_0. We have then

$$p_j/\dot{X}_j = p_{j+1}/\dot{X}_{j+1} = \cdots = Z_0. \qquad (8)$$

The evaluation of Z_0 follows at once by combining (6), (7) and (8). We find

$$Z_0 = Z(B/C)^{\frac{1}{2}}. \qquad (9)$$

But by resubstituting into (6) and (7) this yields

$$p_{j+1}/p_j = \dot{X}_{j+1}/\dot{X}_j = e^{-iW}. \qquad (10)$$

Now when W is *real*, p_{j+1} and p_j can differ only in phase but not in magnitude and the same is of course true of $\dot{X}_{j+1}$ and $\dot{X}_j$. Therefore the structure transmits those frequencies for which W is real. On the other hand when W is complex, p_{j+1} and p_j differ in magnitude and so do $\dot{X}_{j+1}$ and $\dot{X}_j$. Hence for frequencies for which W is complex, a given excess pressure maintained at j will correspond to a smaller one at $j+1$ and so on along the line, so that effectively there is no transmission. Now cos W is real if Z_b is a pure reactance. Hence the condition for transmission through the infinite structure is

$$|\cos W| \leq 1 \qquad (11)$$

while the condition for attenuation is

$$|\cos W| > 1. \qquad (12)$$

Reference to the characteristic impedance (9) shows that when $|\cos W| \leq 1$, sin W is real and hence Z_0 is real. However when $|\cos W| > 1$, sin W is pure imaginary and the same is therefore true of $(B/C)^{\frac{1}{2}}$. Hence in this case, Z_0 is pure imaginary. We may therefore characterize the properties of the structure in terms of Z_0: for frequencies such that the characteristic impedance is real, one gets transmission; for frequencies such that Z_0 is imaginary, attenuation results.

If Z_b contains both real and imaginary components, cos W is complex and conditions (11) and (12) no longer possess meaning. In this case, as might be expected, there is no complete transmission for any frequency range. For then W is always complex and p_{i+1} and p_i will always differ in magnitude as well as in phase. We shall not deal with this case in the present paper.

5. Illustrations of the Application of the Transmission Theory to Air Filters

Consider first the low pass filter mentioned in Section 3, consisting of a tube with equally spaced Helmholtz resonators as branches. The relevant data follow:

ρ_0 = equilibrium density of air at 20°C, 76 cm pressure = 12.05×10^{-4} gram/cm³,
c = velocity of sound in air (20°C) = 3.44×10^4 cm/sec.,
S = area of main line = $9\pi/16$ cm²,
$2l$ = 1.67 cm,
$Z_b = i(\omega M - 1/\omega C')$, where $M = \rho_0/c_0$, $C' = V/\rho_0 c^2$, and c_0 = acoustic conductivity of orifice into resonator[5] = 2.26 cm,
V = volume of resonator chamber = 4.36 cm³,
n = number of sections = 4.

The transmission characteristics of this structure are indicated in the accompanying plot (Fig. 9) of cos $W = \cos 2kl + Z \sin 2kl/2(\omega M - 1/\omega C')$ as a function of frequency. We see that there is a

transmission band extending from $\nu = 0$ to $\nu = 3100$ cycles. The cut-off frequency is the first nonvanishing root of the equation

$$\cos W = -1. \qquad (13)$$

The attenuation band extends from $\nu = 3100$ cycles to the frequency corresponding to the first nonvanishing root of

$$\cos W = +1 \qquad (14)$$

which in this case is 5600 cycles. Thereafter transmission again sets in and except for very narrow attenuation bands where sin $2kl = 0$, is the rule for all higher frequencies. The transmission through this filter has been measured experimentally[6] and the results are indicated in Fig. 10, which also indicates the appearance of the filter. The sharpness of the cut-off is to be noted as well as the approximate agreement with the theoretically calculated value. Considering the nature of the measurement and the fact that the filter is by no means "infinite," consisting as it does of only 4 sections, the agreement must be considered very good. Of course the value of P_r is by no means unity in the transmission region as the above idealized theory predicts. This is due partly to the neglect to take account of dissipation and partly to the finiteness of the structure. (Cf. Section 6 for the treatment of the finite case.) The method of measurement was hardly sensitive enough to do more than just indicate the presence of the second transmission band at 5600 cycles.

Inspection of the expression for cos W in this case reveals that the following general statements may be made concerning the low pass type of filter just discussed: (1) lowering the resonance

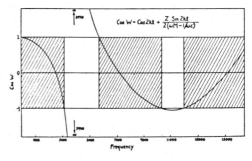

FIG. 9. Variation of cos W with frequency.

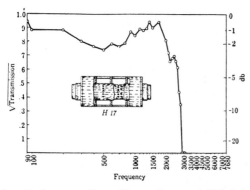

FIG. 10. Low pass filter with equally spaced Helmholtz resonators as branches

frequency of the resonators lowers the cut-off frequency; (2) lowering the cut-off frequency makes the first attenuation region narrower, while raising the cut-off frequency widens this attenuation band; (3) other things remaining the same, increase in the cross sectional area S decreases the cut-off frequency.

As a second illustration of an acoustic filter in air, consider a tube with equally spaced side orifices. The dimensions and other data follow:

$2l = 5$ cm; $n = 12$,
$Z = 56.09$ gram/cm^4 sec.,
$Z_b = i\rho_0\omega/c_0 = 0.0896i\nu$ (c_0 = conductivity of orifices = 0.085 cm),
$Z/2Z_b = 313/i\nu$.

The plot of $\cos W = \cos 2kl + (c_0Z/2\rho_0\omega) \sin 2kl$ in Fig. 11 shows the transmission characteristics

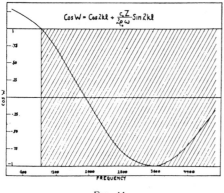

FIG. 11.

of the structure. It is a high pass filter with a theoretical transition frequency at $\nu = 800$ cycles,

the first root of $\cos W = +1$. The transmission band which begins there continues indefinitely. The curve of the experimentally measured transmission is shown in Fig. 12. The agreement between theory and experiment is seen to be of the same order as that observed in the low pass type filter. We can make the additional statements that other things remaining the same, increase in the acoustical conductivity, c_0, of the orifices increases the transition frequency. Other things being equal, on the other hand, increase in the cross-sectional area decreases the transition frequency.

6. Finite Filters

The theory presented in Section 4 takes no account of the finite length of the filter structure.

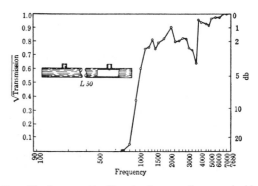

FIG. 12. An acoustic filter having equally spaced side orifices.

In this section we shall introduce the appropriate modification of the theory for this purpose. Referring once more to Fig. 8, we now suppose that the system there indicated consists of n sections beginning at 1 and ending at $n+1$. The whole structure is considered to be inserted in an infinite homogeneous acoustic line with cross-sectional area S_i to the left and S_t to the right of the filter. The main line of the filter itself is assumed to have cross-sectional area S. Acoustic radiation in the form of a plane harmonic wave proceeding from the left is incident on the system at 1 and leaves it at $n+1$ in the form of a plane wave moving solely to the right, the infinite line ruling out the possibility of a wave to the left in the region beyond $n+1$. Eqs. (1) to (7) inclusive still apply. But now from (6) and (7) we can show by mathematical induction that we also have

$$p_{n+1} = p_1 \cos nW - iZ\dot{X}_1(B/C)^{\frac{1}{2}} \cdot \sin nW, \quad (15)$$

$$\dot{X}_{n+1} = \dot{X}_1 \cos nW - i/Z \cdot p_1(C/B)^{\frac{1}{2}} \cdot \sin nW. \quad (16)$$

From the boundary conditions at $n+1$ and from what has just been said about that point, it follows that $\dot{X}_{n+1}$ and p_{n+1} are the volume current and excess pressure respectively in the wave traveling out of the filter at its terminus. We shall denote the pressure and volume current in the main line to the left of the filter by

$$p = p_0 e^{-ikz} + p_0' e^{ikz}, \quad (17)$$

$$\dot{X} = S_i/\rho_0 c \cdot (p_0 e^{-ikz} - p_0' e^{ikz}). \quad (18)$$

Here p_0 is the excess pressure in the incident wave

at the point 1 (taken as $x=0$) while p_0' is the excess pressure in the reflected wave at the same point. In (17) and (18) we are leaving out the common harmonic time factor $e^{i\omega t}$ for the sake of convenience. The power transmission ratio of the filter is now defined to be

$$P_r = Z_i |p_{n+1}|^2 / Z_t |p_0|^2, \quad (19)$$

where $Z_i = \rho_0 c / S_i$ and $Z_t = \rho_0 c / S_t$. This is the ratio of the average flow of energy per second out of the filter to the average flow of energy per second into the filter. Utilizing the boundary conditions at 1, $viz.$

$$p_1 = p_0 + p_0'; \quad \dot{X}_1 = S_i/\rho_0 c \cdot (p_0 - p_0'), \quad (20)$$

we are finally led to

$$P_r = \frac{4Z_i/Z_t}{(1+Z_i/Z_t)^2 \cos^2 nW + [Z_i/Z \cdot (C/B)^{\frac{1}{2}} + Z/Z_t \cdot (B/C)^{\frac{1}{2}}]^2 \sin^2 nW}. \quad (21)$$

The simplest experimental arrangement is that in which $Z_i = Z = Z_t$ by the choice of a common cross-sectional area. In this case (21) reduces to the form

$$P_r = \frac{1}{1 + Z^2/|Z_b|^2 \cdot (\sin^2 nW)/4 \sin^2 W}. \quad (22)$$

The connection between P_r for the finite filter and the transmission characteristics of the corresponding infinite filter given by $\cos W$ in (11) and (12) is now very readily seen. When $|\cos W| > 1$, i.e., when the infinite structure has an attenuation band, $W = W_r + iW_i$, where due to the real character of $\cos W$, $W_r = n'\pi$ (n' integral). Hence in this case $\sin W = \pm i \sinh W_i$ and $\sin nW = \pm i \sinh nW_i$. Consequently as n increases, $\sin^2 nW / \sin^2 W$ will increase monotonically and therefore P_r will become negligibly small: this corresponds to attenuation. On the other hand, when $|\cos W| \leq 1$, W is real and both $|\sin nW|$ and $|\sin W|$ are less than or equal to unity. Moreover in this case (cf. Eq. (3)) $|Z_b|$ cannot be very small compared with Z unless $\sin 2kl$ is also very small. The indications therefore are that for $|\cos W| \leq 1$, P_r can become of the order of magnitude of unity and will not become zero even if n is very large.. This corresponds to the transmission band.

We are now ready to consider some examples. Here it will be interesting for comparison purposes to investigate the low and high pass filters already discussed in Section 5.

Figure 13 shows the plot of P_r for the low pass filter described in Section 5 together with the measured values as taken from Fig. 10. It is of interest to note that the general trend of the two curves is the same, though the theoretical P_r is in every case higher than that measured, as we should expect from the neglect of irreversible dissipation. The agreement of the cut-off is particularly good. We have already mentioned that the second transmission band could hardly be more than just barely detected by the experimental arrangement employed. It is interesting to note that the resonance frequency of the Helmholtz resonator used as a side branch is about 4000 cycles and thus comes at about the center of the attenuation band.

The three peaks in the theoretical curve are readily attributed to the values of W between 0 and π (i.e., in the transmission band, cf. Fig. 9) for which $\sin nW$ ($viz.$ $\sin 4W$ in this case) vanishes. These are, of course, $W = \pi/4$, $\pi/2$ and $3\pi/4$. Note that the frequency is plotted on a logarithmic scale. The last peak does not appear

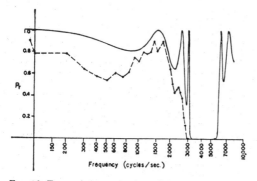

FIG. 13. Transmission curve for a finite low pass filter.

to have been detected experimentally. This is hardly surprising since it is a very sharp one.

The P_r curves calculated and observed for the finite high pass filter of Section 5 are shown in Fig. 14. The general remarks concerning the low pass filter also apply here. We note, however, in addition, the following interesting points.[7] The noticeable dip in the experimental P_r curve at about 3500 cycles is faithfully mirrored in the theoretical curve. This received no explanation at the time when the measurements were originally made. The ripples in the theoretical curve, which are also roughly followed by the experimental results, are due to the $n-1$ (i.e., in this case 11) zeroes of $\sin nW$ as W goes from 0 to π in the transmission region. Another similar series occurs as W goes from π to 2π, etc. The dip comes at the frequency for which $W = \pi$, since here the indeterminate form $\sin^2 nW/\sin^2 W$ has the limiting value n^2. It will be noted that there is but one attenuation band and after the transition frequency, transmission occurs for all higher frequencies. This is in agreement with all experimental results so far obtained.

7. Alternative Definition of Transmission Ratio for Finite Filters

While the definition of the power transmission ratio P_r in Eq. (19) is the one commonly meant by this term in acoustics it is well to point out that it is not always the one most simply realized in experimental measurement. To check it experimentally it is necessary to measure the sound intensity in the line directly after the filter and, replacing the filter by a section of the original acoustical line equal in length to the

filter, to repeat the measurement at the same place. The ratio of the former to the latter measurement is P_r. Obviously it would be simpler from an experimental point of view to place detecting devices in the line *before* and *after* the filter and make the measurements simultaneously. The ratio in this case, however, will not be P_r. If we assume that the device measures something proportional to the square of the displacement velocity, we can define the new ratio as

$$P_r' = |\xi_{n+1}|^2/|\xi_1|^2. \tag{23}$$

The principal difference between this and P_r given in Eq. (19) lies in the fact that the denominator refers to the *resultant* displacement velocity at the beginning of the filter rather than that in the wave incident on the filter from the left. P_r' follows from Eqs. (15) and (16) by eliminating p_1 between them and replacing p_{n+1} by $Z_t \dot{X}_{n+1}$. We then get

$$P_r' = S_i^2/S_t^2$$

$$\frac{1}{\cos^2 nW + Z_t^2/Z^2 \cdot C^2 \cdot \sin^2 nW/\sin^2 W}. \tag{24}$$

And if we take the conventional situation where $S_i = S_t$, $Z_t \doteq Z$, this reduces to

$$P_r' = \frac{1}{1 - i\dot{Z}/Z_b \cdot C \cdot \sin^2 nW/\sin^2 W}. \tag{25}$$

It must be emphasized that P_r' is not a genuine transmission coefficient. In particular it can exceed unity. Nevertheless it can be used as a measure of transmission, since by inspection we see that for $|\cos W| \leq 1$, P_r' will in general be large, while for $|\cos W| > 1$, the circular functions become hyperbolic and for n large, P_r'

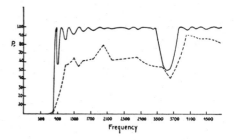

FIG. 14. Transmission curve for a finite high pass filter.

becomes vanishingly small. It is worth pointing out that although P_r' can greatly exceed unity, it cannot become infinite for any frequency. The analytical reason for this is clear from inspection of (24), since infinite P_r' would involve the simultaneous vanishing of both $\cos nW$ and $Z_t/Z \cdot C \cdot \sin nW/\sin W$. This contradicts the restrictions placed on these quantities. Physically, we see that $P_r' = \infty$ would imply a vanishing $\dot{X}_1$ for nonvanishing $\dot{X}_{n+1}$, i.e., complete reflection of the wave incident on the filter with simultaneous nonvanishing wave disturbance in the line beyond the filter. Since the latter is assumed to be infinite, this situation is self-contradictory. Moreover, if P_r' is measured experimentally, P_r can be estimated as

$$P_r = P_r'\left(\frac{1 - iZ/Z_b \cdot C \cdot \sin^2 nW/\sin^2 W}{1 + Z^2/|Z_b|^2 \cdot \sin^2 nW/4 \sin^2 W}\right) \quad (26)$$

Since a graphical comparison of P_r and P_r' may be of some interest, Figs. 15 and 16 present

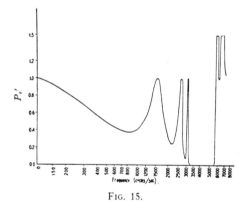

Fig. 15.

the plot of P_r' for the low pass and high pass filters discussed in the last section and for which the P_r curves are shown in Figs. 13 and 14. It will be noted that both P_r and P_r' follow the same general trend. It follows that P_r' can serve as a very useful measure of the transmission through the filter structure.

8. Finite Filter with a Finite Termination

In our discussion of finite filters so far we have supposed the acoustic line in which the filter is inserted to be infinite to the right. In any actual filter arrangement this part of the line will be

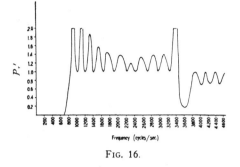

Fig. 16.

finite. It may be made very long and appropriately treated (e.g. lined with felt) so as to cut down the reflected wave toward the left, but the latter will always be present to some extent. We can get some indication of its effect by considering the general case of a finite filter terminated by a finite line. Let its length be l' and the impedance of a plane progressive wave in it be Z'. Since we are still discussing air filters it will be simplest to consider it a cylindrical tube. If its area of cross section is S', we have $Z' = \rho_0 c/S'$. The impedance at the far end of the terminal tube, $Z_{l'}$ is given in terms of the terminal impedance of the filter Z_t, by means of the relation[8]

$$Z_{l'} = \frac{Z_t \cos kl' - iZ' \sin kl'}{\cos kl' - iZ_t/Z' \cdot \sin kl'}. \quad (27)$$

When this is solved for Z_t we get

$$Z_t = p_{n+1}/\dot{X}_{n+1} = \frac{Z_{l'} \cdot \cos kl' + iZ' \cdot \sin kl'}{\cos kl' + iZ_{l'}/Z' \cdot \sin kl'}. \quad (28)$$

Ideally, if the far end of the terminal tube is open we shall have (neglecting the motion of the air in the opening and the radiation) $Z_{l'} = 0$ and thus Z_t becomes

$$Z_{t0} = iZ' \tan kl'. \quad (29)$$

On the other hand if the far end of the tube is rigidly closed, $Z_{l'} = \infty$ and Z_t becomes

$$Z_{tc} = -iZ' \cot kl'. \quad (30)$$

Going back to (15) and (16) we recall that

$$\xi_{n+1}/\xi_1 = S_i/S'$$

$$\cdot \frac{1}{\cos nW + iZ_t/Z \cdot C \cdot \sin nW/\sin W}. \quad (31)$$

Hence the P_r' expressions for the cases of the open end and closed end respectively are

$$P_r' = S_i^2/S'^2$$

$$\cdot \frac{1}{(\cos nW - Z'/Z \cdot \tan kl' \cdot C \sin nW/\sin W)^2} \quad (32)$$

$$P_r' = S_i^2/S'^2$$

$$\cdot \frac{1}{(\cos nW + Z'/Z \cdot \cot kl' \cdot C \sin nW/\sin W)^2}. \quad (33)$$

Inspection of (32) and (33) indicates that for $|\cos W| \leq 1$, they can become rather small, leading to large P_r' and therefore a transmission band. For $|\cos W| > 1$ the circular functions become hyperbolic and for n large, the denominators become very small, leading to negligible P_r'. Hence we have again the succession of transmission and attenuation bands as before. The situation in the transmission bands is indeed rather complicated due to the resonance effects of the terminating tube. The resultant P_r' values are due to the interaction of the tube resonance and that of the finite filter structure itself. Hence relative dimensions play an important role. We can at any rate make general statements about two limiting cases. Thus if: (1) the length l' of the terminating tube is so small that its fundamental frequency is above the frequency range of the transmission band being considered, the plot of P_r' in this band as W goes from 0 to π will show a definite series of n infinite peaks, corresponding to the n changes of sign of $\cos nW$ in the band. The position of the peaks does not depend markedly on the terminating tube. The infinite value of P_r' at the peak frequencies *is*, of course, an effect of the finite termination and the confinement of the sound energy in a finite system with neglect of dissipation through friction and radiation. On the other hand, if: (2) l' is so large that there are many harmonics of the terminating tube in the frequency range of the transmission band, one will expect subsidiary peaks due to tube resonance. Of course, at the anti-resonance tube frequencies, e.g. those for which $\tan kl' \to \infty$, P_r' will drop to zero even in a transmission band.

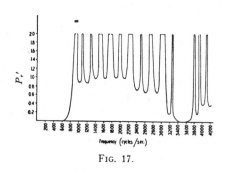

Fig. 17.

By way of illustration we present in Fig. 17 the plot of P_r' for the case of the high pass filter already discussed terminated by a closed tube 5 cm long. For convenience the infinite peaks are terminated at $P_r' = 2$. This situation corresponds more or less to limiting case (1) above. However, certain modifications are to be noticed. In the first place, due to the fact that the first anti-resonance frequency of the terminating tube falls at about 3500 cycles, the transmission there is cut to zero and the dip in the P_r' curve shown in Fig. 15 now becomes effectively a small attenuation band in what is really the transmission band of the equivalent infinite filter. This band extends approximately from 3350 to 3750 cycles. As the anti-resonance frequency is passed the succession of transmission peaks again recurs. It will also be noticed that in the transmission region from 800 cycles to 3350 cycles there are 11 peaks instead of the 12 which would properly correspond to the actual number of sections. The 12th change of sign of $\cos nW$ is masked completely by the large values of $\cot kl'$ and hence the corresponding peak does not appear. It is interesting to observe that the terminating tube has no influence on the transition frequency from the low frequency attenuation band to the transmission band.

We can readily handle the case in which the impedance at the open end of the terminal tube is not taken to be zero but given its actual value in terms of the inertance of the opening as an orifice and the radiation resistance. Thus for

$$Z_{l'} = (\rho_0 \omega k/2\pi) + i\rho_0 \omega/c_0, \quad (34)$$

where c_0 is the acoustic conductivity of the opening[9] we get

50

$$Z_{10}' = \frac{\rho_0\omega k/2\pi + i(\rho_0\omega/c_0 + Z' \tan kl')}{1 - \rho_0\omega/c_0 Z' \cdot \tan kl' + i(\rho_0\omega k/2\pi Z') \cdot \tan kl'}. \tag{35}$$

We shall not trouble to substitute this into (31) but merely note than in doing so we are led to a complex denominator for ξ_{n+1}/ξ_1. This means that P_r' can never become infinite: there is now a definite loss of energy from the system through radiation in addition to the usual frictional dissipation.

Bibliography

1. Stewart and Lindsay, *Acoustics* (New York, 1930) p. 159 ff.
2. W. P. Mason, Bell System Tech. J. **6**, 258 (1927). Phys. Rev. **31**, 283 (1928).
3. Cf. Horace Lamb, Mem. and Proc. Manchester Lit. and Phil. Soc. **42**, 1 (1897–98). Also cf. N. Kasterin, Amst. Ak. Versl. **6**, 460 (1898) and Arch. Néerland. **5**, 506 (1900).
4. Cf. reference 1, p. 50, 55.
5. For the physical meaning of the symbols M, C', c_0 the reader may consult reference 1, p. 47 ff. We are using C' instead of the conventional C to avoid confusion with (3).
6. For the method of measurement, cf. reference 1, p. 197.
7. The rather poor agreement between theory and experiment noted in the paper "Finite Acoustic Filters" (J. Acous. Soc. Am. **8**, 211 (1937)) appears to have been due to an erroneous assignment of dimensions. The experimental curve shown in Fig. 3 of that paper appears actually to correspond to $2l = 5$ cm and $n = 12$. This was kindly called to the author's attention by A. Belov of Leningrad. Figures 11, 12 and 14 of the present paper are based on the revised assignment.
8. Cf. reference 1, p. 138.
9. Reference 1, p. 125.

6

Reprinted from *Phys. Rev.* **57**:921–924 (May 15, 1940)

Fresnel Formulae Applied to the Phenomena of Nonreflecting Films

KATHARINE B. BLODGETT

Research Laboratory, General Electric Company, Schenectady, New York

(Received March 18, 1940)

The Fresnel formulae are applied to the calculations of the amplitudes of rays reflected and transmitted by nonreflecting films of transparent isotropic substances. Two cases are considered. In the first, the film of refractive index n_1 is bounded on both sides by a medium of refractive index n_0. In the second, the film is bounded by media of refractive indexes n_0 and n_2. It is shown how in the first case these formula lead to zero reflection when the thickness t of the film is given by $n_1 t \cos r_1 = (2n+2)\lambda/4$. In the second case for zero reflection the thickness must be given by $n_1 t \cos r_1 = (2n+1)\lambda/4$.

SEVERAL workers[1-3] have coated glass with films for the purpose of diminishing or extinguishing the reflection of light from the glass. Their published papers have given the following analysis of the requirements which must be met in order that a film shall reflect no light of wave-length λ. A film which meets these requirements is called a nonreflecting film.

There are two types of nonreflecting films made of transparent substances: *Type (a).* A film having a refractive index n_1 bounded on both sides by a medium of refractive index n_0. (Example: wall of a soap-bubble.) *Type (b).* A film having a refractive index n_1, bounded by media of refractive indices n_0 and n_2, where $n_0 < n_1 < n_2$. (Example: film of fluorite on glass.) The requirements for the extinction of reflected light are given by the following equations.

[1] J. Strong, J. Opt. Soc. Am. **26**, 73 (1936).
[2] K. B. Blodgett, Phys. Rev. **55**, 391 (1939).
[3] C. H. Cartwright and A. F. Turner, Phys. Rev. **55**, 595(A) (1939).

THICKNESS

The film must have a thickness t which is given by the equations

$$n_1 t \cos r_1 = (2n+2)\lambda/4 \text{ for Type (a),} \quad (1)$$

$$n_1 t \cos r_1 = (2n+1)\lambda/4 \text{ for Type (b),} \quad (2)$$

where r_1 is the angle of refraction of light in the film, and n has the values 0, 1, 2, $\cdots$.

REFRACTIVE INDEX OF FILM

Type (a). The film may have any value of refractive index.

Type (b). Case (1). Film viewed by perpendicular light, $i = 0$. The film must have a refractive index which satisfies the equation

$$n_1^2 = n_0 n_2. \quad (3)$$

Case (2). Film viewed at an angle of incidence $i \neq 0$. The refractive index must satisfy one of

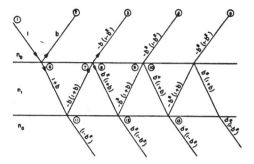

FIG. 1. Paths of a ray of light polarized perpendicular to the plane of incidence. The film of refractive index n_1 is bounded on both sides by a medium of refractive index n_0.

the following equations:

For the R_s ray $\tan i \tan r_2 = \tan^2 r_1$,   (4)

For the R_p ray $\sin^2 r_1 \cos^2 r_1$

$$= \sin i \cos i \sin r_2 \cos r_2, \quad (5)$$

where R_s and R_p are the polarized rays having the plane of polarization perpendicular and parallel, respectively, to the incident plane; and r_1 and r_2 are the angles of refraction of light in the media of refractive indices n_1 and n_2.

The present paper will apply the Fresnel formulae to calculations of the amplitudes of the rays reflected and transmitted by nonreflecting films of transparent isotropic substances. The Fresnel formulae are[4]

$$R_s = -E_s \frac{\sin (i-r)}{\sin (i+r)}, \quad (6)$$

$$D_s = E_s \frac{2 \cos i \sin r}{\sin (i+r)}, \quad (7)$$

$$R_p = -E_p \frac{\tan (i-r)}{\tan (i+r)}, \quad (8)$$

$$D_p = E_p \frac{2 \cos i \sin r}{\sin (i+r) \cos (i-r)}, \quad (9)$$

where E_s, R_s, D_s represent the amplitudes of the incident ray, reflected ray, and transmitted ray, respectively, for R_s-polarized light (as defined above), and E_p, R_p, D_p have corresponding values for R_p-polarized light.

Equations (6), (7), (8) and (9) are in agree-

---

[4] *Handbuch der Physik*, Vol. 20, p. 211.

ment with the requirement imposed by energy relationships between the three rays. The energy of a light wave per unit volume is $na^2/8\pi$ where a is the amplitude of the wave. Therefore the energy per unit time of a wave having a wave front of cross section A and traveling with the velocity c is $Acna^2/8\pi$. From this we obtain the following value for the energy of the ray D,

$$D^2 = (E^2 - R^2)A_0 n_0 / A_1 n_1, \quad (10)$$

where A_0 and A_1 represent the cross sections of the waves in the media in which the light is reflected and refracted, respectively. This equation may be written

$$D^2 = (E^2 - R^2) \cos i \sin r / (\cos r \sin i). \quad (11)$$

The values of D_s, E_s and R_s from Eqs. (6) and (7) satisfy Eq. (11); also the values of D_p, E_p and R_p from Eqs. (8) and (9).

From Eqs. (6) and (7) we obtain the result

$$E_s + R_s = D_s. \quad (12)$$

That is, the sum of the amplitudes of the incident and reflected rays on one side of a boundary is equal to the amplitude of the refracted ray on the opposite side of the boundary.

From Eqs. (8) and (9) we obtain

$$E_p - R_p = D_p \sin i / \sin r, \quad (13)$$

which can be written $E_p - R_p = n_1 D_p / n_0$ where $n_1/n_0 = \sin i / \sin r$.

Figure 1 represents the paths which a ray of R_s-polarized light takes when it strikes a film of Type (a). The amplitude of the incident ray is taken as unity. The diagram shows the amplitude

TABLE I. *Amplitudes of reflected and transmitted rays for incident light polarized both perpendicular (R_s) and parallel (R_p) to the plane of incidence. The film of refractive index n_1 is bounded on both sides by a medium of refractive index n_0.*

RAY	R_s-POLARIZED	R_p-POLARIZED
1	1	1
2	b	d
3	$-b(1-b^2)$	$-d(1-d^2)$
4	$-b^3(1-b^2)$	$-d^3(1-d^2)$
5	$-b^5(1-b^2)$	$-d^5(1-d^2)$
6	$1+b$	$(1-d)n_0/n_1$
7	$-b(1+b)$	$-d(1-d)r_0/n_1$
8	$b^2(1+b)$	$d^2(1-d)n_0/n_1$
9	$-b^3(1+b)$	$-d^3(1-d)n_0/n_1$
10	$b^5(1+b)$	$d^4(1-d)n_0/n_1$
11	$1-b^3$	$1-d^2$
12	$b^3(1-b^2)$	$d^3(1-d^2)$
13	$b^5(1-b^2)$	$d^4(1-d^2)$

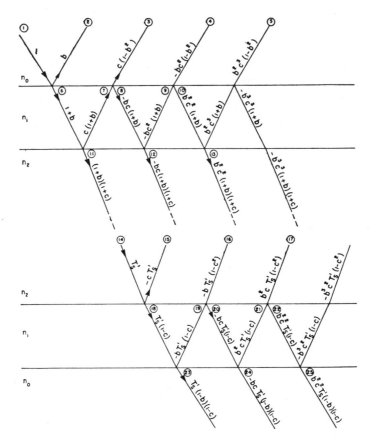

FIG. 2. Paths of a ray of light polarized perpendicular to the plane of incidence. The film of refractive index n_1 is bounded by media of refractive indexes n_0 and n_2.

of the fraction of the light which follows each path. Ray (1) is split into the following components.

(A) Ray (2) reflected from the upper surface of the film. The amplitude of (2) is $b = \sin (i-r) / \sin (i+r)$ from Eq. (6).

(B) Ray (6) refracted in the film. From Eq. (12) ray (6) has the value $1+b$.

(C) Ray (7) reflected from the second surface of the film, with the value $-b(1+b)$, since the values of R_s given by Eq. (6) are equal in value and opposite in sign for rays striking the inner and outer surfaces of a film of Type (a).

(D) Ray (8), and other successively reflected rays (9), (10) $\cdots$ having the values given on the diagram.

(E) Ray (3) refracted with the value $-b(1-b^2)$ calculated from rays (7) and (8) by Eq. (12). Other refracted rays are (4), (5) $\cdots$.

(F) Transmitted ray (11) calculated from rays (6) and (7) by Eq. (12). Other transmitted rays are (12), (13) $\cdots$.

When the thickness of a film of Type (a) is given by Eq. (1), the path difference between the successive reflected rays (2), (3), (4) $\cdots$ is λ, and also between the successive transmitted rays (11), (12), (13) $\cdots$. Therefore the amplitude of the sum of the rays is

$$R = b - b(1-b^2)(1+b^2+n^4\cdots)$$
$$= b - b(1-b^2)/(1-b^2)$$
$$= 0. \qquad (15)$$

The amplitude of the sum of the transmitted rays is

$$T = (1-b^2)(1+b^2+b^4\cdots)$$
$$= (1-b^2)/(1-b^2)$$
$$= 1. \qquad (16)$$

TABLE II. *Amplitudes of reflected and transmitted rays for incident light polarized both perpendicular (R_s) and parallel (R_p) to the plane of incidence. The film of refractive index n_1 is bounded by media of refractive indexes n_0 and n_2.*

RAY	R_s-POLARIZED	R_p-POLARIZED
1	1	1
2	b	d
3	$c(1-b^2)$	$e(1-d^2)$
4	$-bc^2(1-b^2)$	$-de^2(1-d^2)$
5	$b^2c^3(1-b^2)$	$d^2e^3(1-d^2)$
6	$1+b$	$(1-d)n_0/n_1$
7	$c(1+b)$	$e(1-d)n_0/n_1$
8	$-bc(1+b)$	$-de(1-d)n_0/n_1$
9	$-bc^2(1+b)$	$-de^2(1-d)n_0/n_1$
10	$b^2c^3(1+b)$	$d^2e^3(1-d)n_0/n_1$
11	$(1+b)(1+c)$	$(1-d)(1-e)n_0/n_2$
12	$-bc(1+b)(1+c)$	$-de(1-d)(1-e)n_0/n_2$
13	$b^2c^3(1+b)(1+c)$	$d^2e^3(1-d)(1-e)n_0/n_2$
14	$T_s{}'$	$T_p{}'$
15	$-cT_s{}'$	$-eT_p{}'$
16	$-bT_s{}'(1-c^2)$	$-dT_p{}'(1-e^2)$
17	$b^2cT_s{}'(1-c^2)$	$d^2eT_p{}'(1-e^2)$
18	$T_s{}'(1-c)$	$T_p{}'(1+e)n_0/n_1$
19	$-bT_s{}'(1-c)$	$-dT_p{}'(1+e)n_0/n_1$
20	$-bcT_s{}'(1-c)$	$-deT_p{}'(1+e)n_0/n_1$
21	$b^2cT_s{}'(1-c)$	$d^2eT_p{}'(1+e)n_0/n_1$
22	$b^2c^2T_s{}'(1-c)$	$d^2e^2T_p{}'(1+e)n_0/n_1$
23	$T_s{}'(1-b)(1-c)$	$T_p{}'(1+d)(1+e)$
24	$-bcT_s{}'(1-b)(1-c)$	$-deT_p{}'(1+d)(1+e)$
25	$b^2c^2T_s{}'(1-b)(1-c)$	$d^2e^2T_p{}'(1+d)(1+e)$

This means that when parallel rays of light of unit amplitude strike the surface of the film, the rays which leave the film also have unit amplitude since each ray is the sum of a series of rays, the sum being unity. The successive members of the series usually diminish very rapidly in value. For example, when light strikes a film of refractive index $n=1.50$ at an angle $i=30°$, the value of b calculated by Eq. (6) is 0.240. Therefore the rays (11), (12), and (13) $\cdots$ have the values 0.9422, 0.0545, 0.0031, 0.0002, $\cdots$. Table I gives the corresponding values of the amplitudes of the rays for R_p-polarized light. The value of d can be calculated from Eq. (8). In this case also the sum of the reflected rays is zero, and of the transmitted rays is unity.

Figure 2 gives the paths of the rays and their formulae for a film of Type (b). The formulae are listed in Table II. The values of b and d in the formulae can be calculated by means of Eqs. (6) and (8) for the upper boundary of the film which

separates the media of refractive indices n_0 and n_1, and c and e for the lower boundary between n_1 and n_2.

When a film satisfies Eq. (2) the path difference between the successive reflected rays is $\lambda/2$. Therefore in writing the sum of the reflected rays (2), (3), (4), (5) $\cdots$, alternate members of the series must be written with the sign opposite to that which appears in Fig. 2 and in Table II. The sum R is

$$R=b-c(1-b^2)(1+bc+b^2c^2+\cdots)$$
$$=b-c(1-b^2)/(1-bc). \qquad (18)$$

Eqs. (4) and (5) were derived from Eqs. (6), (7), (8), (9) for the cases $b=c$ and $d=e$. When Eq. (4) is satisfied, Eq. (18) becomes

$$R=0. \qquad (19)$$

The path difference between the rays (11), (12), (13) transmitted in the medium n_2 is also $\lambda/2$. Therefore the sum $T_s{}'$ of these rays is

$$T_s{}'=(1+b)(1+c)(1+bc+b^2c^2\cdots)$$
$$=(1+b)(1+c)/(1-bc)$$
$$=(1+b)/(1-b) \text{ when } b=c. \qquad (20)$$

For R_p-polarized light, the sum is

$$T_p{}'=n_0(1-b)/n_2(1+b). \qquad (21)$$

Figure 2 shows the paths of the rays into which a ray $T_s{}'$ [ray (14)] is split when it strikes the second boundary of the medium n_2. The sum of the rays (15), (16), (17) $\cdots$ is zero. The sum T of the rays (23), (24), (25) $\cdots$ is

$$T=T_s{}'(1-b)(1-c)/(1-bc).$$

When $b=c$ and $T_s{}'=(1+b)/(1-b)$ from Eq. (20), we have the result

$$T=1.$$

Similarly the equations for the case of R_p-polarized light give the result $T=1$.

The writer is indebted to Dr. F. Seitz for assistance in deriving some of the formulae in this paper.

7

Reprinted from *Nature* **187**(4736):493–494 (1960)

STIMULATED OPTICAL RADIATION IN RUBY

T. H. Maiman

*Hughes Research Laboratories
A Division of Hughes Aircraft Co.,
Malibu, California*

Schawlow and Townes[1] have proposed a technique for the generation of very monochromatic radiation in the infra-red optical region of the spectrum using an alkali vapour as the active medium. Javan[2] and Sanders[3] have discussed proposals involving electron-excited gaseous systems. In this laboratory an optical pumping technique has been successfully applied to a fluorescent solid resulting in the attainment of negative temperatures and stimulated optical emission at a wave-length of 6943 Å.; the active material used was ruby (chromium in corundum).

A simplified energy-level diagram for triply ionized chromium in this crystal is shown in Fig. 1. When this material is irradiated with energy at a wave-length of about 5500 Å., chromium ions are excited to the $^4F_2$ state and then quickly lose some of their excitation energy through non-radiative transitions to the ^2E state[4]. This state then slowly decays by spontaneously emitting a sharp doublet the components of which at 300° K. are at 6943 Å. and 6929 Å. (Fig. 2a). Under very intense excitation the population of this metastable state (^2E) can become greater than that of the ground-state; this is the condition for negative temperatures and consequently amplification via stimulated emission.

To demonstrate the above effect a ruby crystal of 1-cm. dimensions coated on two parallel faces with silver was irradiated by a high-power flash lamp;

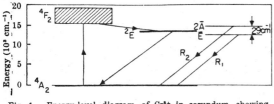

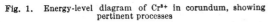

Fig. 1.   Energy-level diagram of Cr^{3+} in corundum, showing pertinent processes

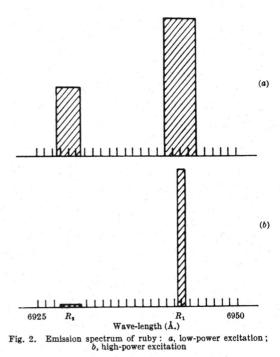

the emission spectrum obtained under these conditions is shown in Fig. 2b. These results can be explained on the basis that negative temperatures were produced and regenerative amplification ensued.   I expect, in principle, a considerably greater ($\sim 10^8$) reduction in line width when mode selection techniques are used[1].

I gratefully acknowledge helpful discussions with G. Birnbaum, R. W. Hellwarth, L. C. Levitt, and R. A. Satten and am indebted to I. J. D'Haenens and C. K. Asawa for technical assistance in obtaining the measurements.

[1] Schawlow, A. L., and Townes, C. H., *Phys. Rev.*, 112, 1940 (1958).
[2] Javan, A., *Phys. Rev. Letters*, 3, 87 (1959).
[3] Sanders, J. H., *Phys. Rev. Letters*, 3, 86 (1959).
[4] Maiman, T. H., *Phys. Rev. Letters*, 4, 564 (1960).

Fig. 2.   Emission spectrum of ruby :   *a*, low-power excitation ; *b*, high-power excitation

8

Reprinted from *Electronics* **17**:110–113 (Sept. 1944)

PHOTOTUBE CONTROL OF FLUID FLOW

Robert C. McNickle

Brooke Engineering Co., Inc., Philadelphia, Pa.

ROTAMETERS are simple instruments which accurately indicate the rate of flow of liquid or gas in a pipe line. They are widely used in chemical process, power, and other industries. The instrument, shown in Fig. 1, usually consists of a vertical, transparent, tapered, glass tube and a metering element inserted inside the tube. The small end of the tube is at the lower portion, and the metering element, variously shaped depending upon fluid requirements, is free to move up or down along the axis of the tube. The position assumed by this element directly indicates flow rate. The gas or liquid being metered flows from the bottom to the top of the tube.

Theory underlying the operation of this type of flow indicator is based on the flow equation, $Q = CA\sqrt{2gh}$, where h is a constant by virtue of the constant net weight of the metering element (commonly called the float or rotor), and A is a variable due to the taper of the tube. The force counter to the float weight is the head differential induced by the fluid flowing through the annular aperture between the outside diameter of the float, and the inside diameter of the tube. Thus the rate of flow varies directly as annular area and the calibration of the instrument is linear.

In other words, forces on the float are in balance; weight of the float minus buoyancy equals area of the top of the float, times differential pressure. If the flow rate increases, the differential across the float will increase, and the float will rise to a new position to maintain a fixed differential pressure across it. Rate of flow is accurately measured

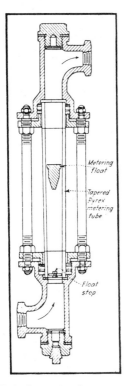

FIG. 1—Construction of a rotameter is shown in this sectional drawing. Position of the float depends on differential pressure which is automatically maintained constant by variations in the annular opening between float and tube as the float rises and falls. Flow-rate indication is linear

Metering float

Tapered Pyrex metering tube

Float stop

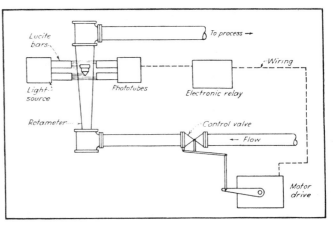

FIG. 2—Block diagram shows general operation of the flow-control unit. Provisions are incorporated to cancel out effects of line-voltage fluctuation and changes in color of the controlled fluid

Lucite bars

To process →

Wiring

Light source

Phototubes

Electronic relay

Rotameter

Control valve

← Flow

Motor drive

FIG. 3—The complete unit. Elevating handwheels are below and hand-automatic controls are on the front panel of the controller. Plastic light-transmission bars project through slots on either side of the rotameter tubes

Close-up of electronic control unit. Covers have been removed from chassis in the foreground to reveal plastic bars which transmit light from the source to the rotameter glass and back to the phototubes

throughout the entire tube range due to the variable orifice (fixed differential) principle, and is not limited to a narrow range such as is the case in the fixed orifice type of flow meters.

Self-supervised Operation

For the past ten years, there has been a demand for automatic control applied to rotameters. Brooke Engineering Co. developed a control for this type of meter several years ago. The control operated on an induction principle such as is commonly used in other types of instruments. The float had an iron rod which hung down inside a center-tapped coil, mounted below the rotameter. This coil was electrically balanced against a similar, remote coil having an iron rod that could be positioned by hand.

Differences in balance of the two coils when the float moved out of its preset balance, were fed into an electronic relay which caused a motor to operate. This, in turn, opened or closed a valve to correct the flow rate. This type of control was not

entirely satisfactory, due to the fact that its sensitivity was only plus or minus 1 percent, and phase shifts caused unstable operation if the control point was not near the center of the coil. Also, it could not be used on small sizes of rotameters, because the magnetic effect of the coil on the iron rod attached to the float was so great as to affect the sensitivity and accuracy of the rotameter.

About a year ago, a large refinery had a process in which they wished to maintain the flow rate of a fluid to ±¼ percent and it was decided to use a fully compensated electronic relay, receiving its signals from phototubes. This system had been in successful operation for many years, measuring the smoke density of large industrial boilers and adusting the air supply to maintain a fixed smoke color at the boiler outlet.

For the rotameter application, it was decided to use two light beams, one shining across the top of the float, and the other across the bottom of the float. Each light beam

was applied to a phototube, and the outputs of the phototube amplifiers were electrically balanced against each other to cancel out the effects of voltage changes and color deviation of the liquid.

If the float in the rotameter moved due to a change in flow, one phototube received more light and the other, less light. This unbalance caused a thyratron to operate a motor with compensation and with full torque at all times. This motor changed a control valve in the correct direction to restore the float to the set value within ±¼ mm (0.0098 in.)

Disposition of Parts

Arrangement of the various pieces of equipment is shown in Fig. 2. The rotameter is mounted on a stand, in front of a panel as in Fig. 3. The electronic relay, with its light source common to both beams, and its phototubes, amplifiers, thyratrons, and allied equipment, is mounted behind the panel, as in Fig. 4. Lucite or Plexiglass bars are used to transmit the light

FIG. 4—In a rear view, motor drives and control valves appear below the electronic relays. Mercoid units mounted on the valve shafts include limit switches and extras which can be used to operate relays on terminal panel for automatic shut-down feature

out to the rotameter and back to the phototubes. The electronic relay is mounted on an elevator assembly which permits the operator to raise or lower the entire assembly to change the rotameter control setting. The unit is also supplied with a switch which permits the operator to remove the control from "automatic", and to raise or lower the flow to any desired value.

In addition to the compensated controlling action, described later, other features may be included in the electrical circuit when required. For example, automatic shut-down if any of the fluids in the process cease flowing or go beyond predetermined values, or if tubes or light source fail.

In most cases, a lock-in circuit must be employed to return the float within the range of the light beam if the characteristics of flow are such that an occasional surge in the fluid will raise or lower the float out of the light beam. This lock-in feature is most important. Essentially, it consists of electromagnetic relays which are actuated by the phototube amplifiers just before the float leaves the light beam. The lock-in feature discerns which way the float moved and operates the motor at full speed to return the float to the light beams where the thyratron can come into operation.

Referring to the circuit diagram in Fig. 5, when the output of V_4 goes beyond a predetermined limit, as set by R_{13}, a type 2050 thyratron in the lock-in circuit fires to energize a relay. When energized, this relay disconnects the cathode of an opposing 2050 so that it cannot operate, and puts a high positive voltage on the grid of the thyratron V_6 to run the motor at full speed. The lock-in circuit does not come into play until the float is about ready to leave the light beam.

The electronic relay employed converts impulses from a pair of phototubes into signals sufficiently large to run a reversible motor. Net movement of the motor is proportional to the signal and the motor has full torque at all times. The electronic relay provides electrical compensation by means of a circuit which varies the thyratron grid voltage by charging and discharging a capacitor at variable rates. Such compensation is similar in results to mechanical throttling and reset but is accomplished electrically without the use of relays, open contacts, or moving parts.

The light source is a 120 v bulb directed through two pieces of Lucite or Plexiglass to form two light beams shining across the rotameter. They are spaced so that when the float is in balance, one half of each light beam is blocked out and the other half of the beam passes

through the fluid and metering tube to other bars of Lucite, and thence to the phototubes. The phototubes are of the high-vacuum type and are not appreciably affected by voltage changes.

Circuit Details

Direct-current bias for the phototubes, amplifiers, and thyratrons is furnished by the 6H6 rectifier. One half of the tube acts as an ordinary high-voltage half-wave rectifier, taking voltage drop across a load resistor, while the other half utilizes the drop across the tube, giving low voltage. Both plate outputs are filtered. Resistors R_1 and R_2 provide grid bias for V_3 and V_4 while C_1 and C_2 provide a-c grid-to-cathode return.

Phototubes V_1 and V_2 are so connected to their amplifiers V_3 and V_4, that the more light received by the phototubes, the lower the amplifier plate output. With this arrangement, it is possible to keep a nearly constant plate output of the amplifiers with a variable voltage. For instance, if the voltage falls, the light source dims, plate voltage of V_3 and V_4 drops, and the phototubes decrease the negative grid voltage of the amplifiers, thereby increasing plate output to approximately the same value as at the higher voltage. This will take care of line voltage changes in the order of ±5 v at 115 v. For greater changes of voltage the plate outputs of V_3 and V_4 will vary up or down together, but since these plate outputs are balanced against each other in T, it will not affect the control. With the amplifier plates balanced against each other, color changes of the liquid in the rotameter will also cancel out.

The plate of V_3 is connected to one half of the primary of push-pull transformer T, and the plate of V_4 is connected to the other half. Assuming the float is in neutral, light on V_1 equals light on V_2, plate outputs of V_3 and V_4 will be equal and cancel each other in the primary of T, and there will be no voltage developed in the secondary. If the float moves down from the neutral position, V_1 receives more light, and V_2 less light, therefore the grid of V_3 will become more negative, and the grid of V_4 less negative. The plate output of V_4 will become greater and that of V_3 less, causing a voltage to be developed in the secondary of T. Since the primary voltage is pulsating half wave, an a-c voltage will be developed in the secondary.

A fixed, negative d-c bias of approximately 10 v is maintained on the grids of V_6 and V_7 by the rectifier when the float is in balance. If the float falls, as assumed above, the positive half of the a-c voltage developed in one half of the secondary of T reduces the negative grid voltage to approximately 2 v and V fires to run the motor in a direction which changes the valve to restore the float to neutral. Note that the emission of tubes V_3 and V_4 is 180 deg out of phase with the emission of V_6 and V_7.

Anti-Hunt Compensation

It can readily be seen that if the motor were to run at full speed until the float reached its neutral position, hunting would occur. To introduce compensation C_3 is used in the following manner. When V_7 fires to run the motor, its plate becomes approximately 70 v negative. This negative voltage, in addition to running the the motor, charges C_3 through R_{11}, R_{10}, and R_9, to put a

high negative voltage on the grid of V_7 to stop it from firing. Then, C_3 discharges through R_9, R_{10}, R_{11}, and the motor. Thyratron V_7 cannot fire again until C_3 discharges.

While the motor is not running, the float has a chance to come to rest. If its point of rest is not the neutral position, a signal from T will again cause the motor to run. If the float moves a great distance from neutral, a strong signal will be developed by T, and the motor will have to run a long time before C_3 is charged with a high enough voltage to overcome the effect of the secondary voltage of T. The motor will then have a long time delay. If the signal from T is weak, the motor will only be energized a short time to charge C_3 a small amount, and the time of delay will be short. In other words, the net rate of movement of the control valve will be proportional to the amount the float moves from neutral.

Capacitor and resistor values are so chosen that if the float moves a great distance, the motor will run at full speed without interruptions. As the float approaches neutral, the motor begins to step at a speed proportional to the distance from neu-

tral. Capacitor-resistor combinations C_4-R_5 and C_5-R_7 also have a somewhat compensating effect but on a relatively small scale.

If the float had moved in a direction to fire V_6, a negative feedback voltage would have been induced in the motor field not being used. This induced voltage would be equal to the voltage running the motor, and would cause compensation as described above.

By this system the motor always has full torque, regardless of how close the float is to balance. This is extremely important for highly sensitive controls,—otherwise, should the valve stick, the torque delivered by the motor near the neutral point would not be sufficient to restore the float to its original balance point. This is often a limiting factor in many control devices.

Successful field operation of 64 of these units for several months has demonstrated that an extremely accurate control can be applied to devices where it is necessary to maintain a movable body in a fixed position. If greater sensitivity is required, concentrated light beams can be employed. By this method, a slight change in float movement will cause a greater signal change.

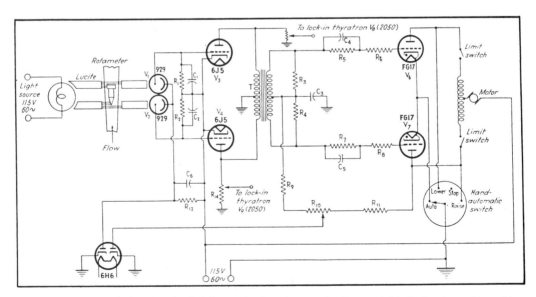

FIG. 5—Basic circuit of electronic relay includes a pair of FG17 thyratrons which can be replaced with FG57 type where a large motor is required for valve operation. Light source, shown as directly connected to the line, is sometimes put in series with the filaments of the rectifier and amplifier tubes for long operating life at reduced voltage

9

Reprinted from *Van Nostrand's Eng. Mag.* **32**:495–501 (1885)

THE DISTRIBUTION OF ELECTRICAL ENERGY BY SECONDARY GENERATORS.

By J. DIXON GIBBS.

From "Iron."

THE remarkable results obtained during the last few years from the production of electrical energy and its application to lighting purposes as well as to the transmission of mechanical power, have naturally brought into prominence the great problem of the distribution of electricity. In a complete system of electrical distribution, it is necessary that electrical energy in all its forms should be at the disposal of individual householders, whatever the service they may require it to perform; that is to say, if light is desired, currents should be available at will for feeding every type of lamp, whether arc or incandescent. If mechanical power is required, motors should supply it; whilst currents suitable for electro-chemical purposes should be obtainable with equal facility. The distribution should be over a large area, central stations being preferably situated in the outskirts of towns at a distance from the area to be supplied. Where water power exists it should be utilized, or, if steam power is used, a site should be chosen in proximity with water so as to secure the economy in fuel effected by the employment of condensing apparatus.

It is scarcely necessary to remark that nothing hitherto done in the way of street lighting or of lighting large establishments, such as theaters, hotels, and public buildings, by means of machinery on or near the premises, can be claimed to constitute a distribution of electrical energy.

The same may be said of the transmission of a given force to a single point. This is not a distribution of mechanical power. Gas and water companies do not set up separate works for the supply of special consumers, but from central stations distribute to all consumers in conformity with their several requirements. In order to arrive at the results just described as necessary to a complete system of electrical distribution, the following conditions are essential :—

1. Every receiving apparatus must be supplied with its proportion of electrical energy, so that it may act independently of the others and without affecting them.

2. The regulation must be automatic, instantaneous in its action, and require no attention.

3. The regulation must be of such a nature that the generating dynamo machine shall produce each moment the exact amount of electricity necessary to supply all the apparatus in action.

The chief difficulty in realizing these conditions has been that the intensity and E. M. F. of an electric current being always exactly determined, the uses to which the current can be put are necessarily limited to feeding apparatus of a given resistance and of an electrical capacity, in harmony with the quantity of available electricity, from which it results, that by means of a given current it is only possible to employ apparatus of consumption of identical construction, that is to say, connected together under certain conditions of resistance that must be maintained constant.

The author need not review the various systems, more or less ingenious, which have been invented during the last few years with the result, not of solving, but of going round, the difficulty, for in the interesting lectures recently delivered by Professor Forbes before the Society of Arts, the mechanism of all these combinations has been ably explained. It seems, however, to have resulted from the facts adduced in these lectures :—

1. That the future of electrical distribution lies in the direction of the employment of currents of high potential and small quantity, requiring conductors of small diameter.

2. That the employment of an apparatus, by means of which the factors of the initial energy can be transformed to suit the requirements of each consumer, is essential to the solution of the problem which we are discussing this evening, and the secondary generators under review are such transformers. They are known as the Gaulard-Gibbs secondary generators. The phenomena of induction which have immortalized the name of Faraday are utilized in these instruments. Numerous predecessors have certainly conceived the idea of utilizing secondary currents localized and of different kinds, but coming later in this path of research, the inventors of the secondary generators have labored under more propitious circumstances, because results already arrived at have enabled them to produce these phenomena under such conditions that their employment has been rendered absolutely practical and economical. This consideration certainly inspired them with the courage to pursue with perseverance those researches to which their predecessors had given but passing attention.

However this may be, the present inventors have regarded the employment of these phenomena from a purely industrial point of view. Their first thought after having experimentally verified the actual transformation of primary electrical energy into electrical currents of different kinds, and capable of being applied to every practical purpose, was to make a careful analysis of the phenomena observed. They were thus able to determine the special conditions under which the primary and secondary circuits would yield the highest effective and most economical return for the energy expended ; they arrived at the conclusion that the two circuits, inducing and induced, must have the same mass of metal and a position absolutely symmetrical with the common magnetic field. Since it is upon the determination of these conditions that the invention of the secondary generator is based, it may be interesting to know that these conditions, which are a *sina qua non* of an economical return, have never been previously determined.

But it was not sufficient to determine philosophically what the industrial conditions should be, it was also necessary to realize them practically. The inventors accordingly constructed their apparatus with a sufficient number of spirals to produce the required practical E. M. F by means of a cable formed of an inducing circuit of low resistance surrounded parallelly to its axis by forty-eight wires composing the induced circuit, the sum of the sections of which was equal to the section of the inducing circuit. By means of this arrangement the theoretical conditions already alluded to were approximately fulfilled, since the mean distances of the induced circuits from the magnetic field were equal to the distance of the inducing circuit from the same magnetic field. Further, it was easy to group the extremities of the secondary wires so as to give to the factors E. I. of energy the different values required according to the work to be done. These apparatus served, during five consecutive months, without interruption, to light with arc and incandescent lamps, five stations of the Metropolitan Railway, one apparatus being placed at each station. The primary circuit in which they were placed was composed of a single wire 15 miles in length and $\frac{1}{5}$ of an inch in diameter. This primary circuit was metallically closed throughout its entire length with the terminals of the dynamo machine at Edgware Road. The results as regards effective work formed the subject of a report by Dr. Hopkinson, the conclusions of which were perfectly satisfactory, and are too well known to require repetition.

The anticipations of an economical return having been thus fulfilled, the next step was to seek the most simple and practical methods of applying economically the principle upon which the construction of the apparatus reposed. These researches led to the formation of the inducing and induced circuits by means of copper discs superposed and furnished with ear pieces for the purpose of connecting them together. This arrangement, which allows of the juxtaposition of the two circuits, has also the advantage of permitting the employment of any insulating material that may be found to give the best results. The simplicity of this method of construction is obvious; the weight and size of the apparatus are remarkably small in relation to the work it is capable of performing.

The apparatus is identical in form with the generators exhibited at Turin.

It is worthy of remark that in these apparatus the actual resistances of the inducing and induced circuits are kept as low as possible, so as to render the work expended in the interior circuits very small in proportion to the work available in the exterior circuits. As an example, take the instrument we have before us: it is intended to supply in the exterior circuit an effective work of 750 Watts under the influence of a primary current of 12 amperes—the total resistance of its two circuits induced and inducing is $\frac{3}{10}$ of an ohm, so that $12^2 \times \frac{3}{10}$ Watts represents the work absorbed by the apparatus, and, consequently, useless; thus the theoretical loss of energy resulting from the interposition of these apparatus is $\frac{43.2}{750}$, or $5\frac{1}{2}$ per cent. Nevertheless, when the inventors announced an effective return of 90 per cent., many doubts were expressed in consequence of the unfavorable results hitherto obtained from researches in the same direction—these doubts were really testimonies to the novelty of the results—it remained only to demonstrate conclusively the truth of these results. The authority of an eminent electrician had been insufficient to carry conviction to every mind. An exceptional circumstance, however, enabled the inventors to determine definitely, without room for further question, the accuracy of their assertion.

In the month of January, 1884, the Italian government offered a grand prize of 10,000 fr., to be competed for internationally for the most important advance made in the transport of electrical energy to a distance, and invited other governments to name representatives who should constitute a jury for deciding the question. The Italian government was probably influenced in taking this step by the conviction that the industrial development of Italy would be largely aided by the prompt utilization of the vast natural forces in which the country abounds. The jury was composed of: M. Tresca, membre de l'Institut de France, honorary president; Professor Ferraris, acting president; M. Wattmann, rector of the University of Geneva; Professor Voit, of Munich; Professor Webber, of Zurich; Professor Roiti, of Florence, member of the International Commission for the determination of the ohm; Professor Kittler, of Darmstadt; Professor Cossa, of the School of Engineers at Turin; Professor Farini, of Milan, &c. Professors Voit and Kittler were the gentlemen deputed to take electrical measurements at the Vienna and Munich Electrical Exhibitions.

Practical experiments of the distribution of electrical energy by means of the secondary generators were made under conditions which M. Tresca, in the name of the international jury of the Turin Exhibition, communicated to the Académie des Sciences de Paris, in terms of which the author will read a translation. The original was published in the *Lumiere Électrique* of October 18, 1884. It runs as follows: "An International Electrical Exhibition is now being held at Turin, in connection with which an important prize is offered by the Italian government and the town. I am charged by my colleagues of the jury of this exhibition to bring to the notice of the Académie the following facts: Messrs. Gaulard and Gibbs have established at the exhibition, the station of Lanzo, and the intermediate stations, a circuit whose length, including return, is 80 kilometers by means of a bronze chrome wire 3.7 millimeters in diameter without covering. This wire carries an alternating current produced by a Siemens electro-dynamic machine of the 60 horse-power type in such a way that the current can be simultaneously utilized for different modes of lighting, whether at the exhibition, or at the Turin station, or at the Lanzo station, or at the intermediate stations, by its transformation at each point of the two factors constituting its energy by means of the secondary generators of the new type, shown by Messrs. Gaulard & Gibbs. On September 25, we verified the simultaneous regular working.

"1. At the exhibition of the following apparatus, which had to be necessarily supplied with very different potentials— nine Bernstein lamps, one Sun lamp, one Siemens lamp, nine Swan Lamps, and five other Bernstein lamps situated at a small distance.

"2. At the Turin Lanzo station, 10 kilometers away, thirty-four Edison lamps of sixteen candles, forty-eight of eight candles, and a Siemens arc lamp.

"On September 29, the experiments were still more conclusive, the system being extended to the Lanzo station, 40 kilometers distant, by the perfectly regular action of twenty-four Swan lamps of 100 volts. The numerous transformations required by the variety of these different methods of lighting are effected with accuracy, and, although we are not able to give the exact figures, it is perfectly demonstrated that the secondary generators may be considered, at all events within certain limits, as transformers giving a relatively large return of the energy of alternating currents. The actions of lighting and extinction are effected without any disturbance (of the other lights) and by means of simple commutators. The principal object of this communication is limited, however, to testifying to the complete success of a distribution of different modes of lighting over an (effective) distance of 40 kilometers. The importance of the realized fact alone demands that it should be fixed by a precise date, but it should be borne in mind that we are not dealing here with the transport of mechanical power."

More than 300 Italian engineers and architects who had witnessed the experiments, assembled for the 1884 congress, passed a resolution of which the author will read a translation. "The fifth congress of Italian Engineers and Architects cannot ignore the great importance of the experiments now being made at Turin by means of the Gaulard and Gibbs secondary generator, and, having examined the working of such a system of distribution, record the hope that the government, the corporations of towns, and manufacturers will patronize this system, and that the expectations which five months of trial on the Metropolitan Railway, the experiments at Turin, and the sound scientific conceptions upon which the system is based have raised in the field of science and industry, may thus be realized."

The measurements taken by means of the electrometer of Mascart of the effective return of the secondary generators are shown by the curves on the diagram, fig. 2. In taking for abscissæ the resistances introduced in the secondary circuit, and for ordinates the primary and secondary work, it will be seen that the progression, at first increasing, arrives at its maximum between the resistances of 6 and 10 ohms, which are the resistances under which the apparatus works normally, but these measurements having been considered by some members of the jury not absolutely free from theoretical objections, a commission was appointed by the jury, composed of Professors Webber, Voit, Roiti, and Ferraris, to prepare a calorimeter, by means of which Professor Ferraris carried on his experiments during seven consecutive days, and arrived at the conclusions which formed the subject of a special report, which is very voluminous. The conclusions of this report are condensed in the table of which the following is a copy, and which gives an average practical return (column (N)) of 94 per cent. when the apparatus worked under the conditions for which they were constructed.

In the annexed table are shown the theoretical and practical co-efficients of the return from a secondary generator coupled in tension, calculated in the following manner, for a series of resistances of the secondary circuit. In column R the values of the total resistances of the secondary circuit vary from 0.28 to 40.0 ohms. In column M are shown the theoretical values of the coefficient of the total return. In column N are shown the values of the coefficient of the exterior theoretical return, and in column (N) the values of the coefficient of the exterior practical or effective return.

R.	M.	N.	(N.)
0.28	0.500	0 000	0.00
2	0.876	0.753	0.74
4	0.933	0.867	0.86
6	0.956	0.911	0.90
8	0.962	0.928	0.92
10	0 967	0.940	0.93
12	0.971	0.948	0.94
14	0.973	0.954	0.94
16	0.974	0.957	0.95
18	0.975	0.959	0.95
20	0.975	0.961	0.95
22	0.976	0.963	0.95
24	0.976	0.964	0.95
26	0.975	0.964	0.95
28	0.975	0.965	0.95
30	0.975	0.966	0.96
32	0.974	0.966	0.96
84	0.973	0.965	0.95
36	0.973	0.965	0.95
38	0.972	0.965	0.95
40	0.971	0.965	0.95

Now that the secondary generator has been absolutely demonstrated to be a perfect transformer of the energy of alternating currents, it only remains to

the author to examine whether the conditions under which these instruments act are in perfect accord with the conditions laid down at the commencement of this paper for the solution of the problem. In order to thoroughly understand that this is so, let us suppose that we have to distribute 10,000 glow lamps, 200 arc lamps, and 200 mechanical horse-power in varying proportions, over a circuit 15 miles in length. The author adopts these figures because they represent somewhere about the average requirements of the future. With the aid of the secondary generators this distribution would be effected in the following manner :—

The initial electrical work would be produced by four alternating-current dynamo machines, of the Siemens model, for example, supplying 100 amperes and 3,000 volts each. This work would be distributed over four distinct circuits, metallically closed with the terminals of each dynamo, and formed of a cable having a diameter of one centimeter only, connected with the secondary generators, one of which would be placed in the house of every consumer. The form and size of these secondary generators would necessarily be proportioned to the quantity of work required of each one respectively. It may be remarked here that the secondary generators, fed by an electrical quantity which is constant, develop on the current which feeds them a counter electro-motive force which is proportionate to the work they develop in their external or secondary circuits. From this it follows that the quantity or ampere value of the primary current must remain fixed, whatever may be the number of secondary generators to be fed, and that the E. M. F. only of the primary current will vary according to the sum of the resistances set up by the number, more or less important, of the generators in action on the circuit. This result is automatically obtained by means of a regulator of intensity, which, placed in the primary circuit, acts upon the derivation of the exciting machine, by introducing variable resistances, so as to proportion the intensity of the magnetic field of the generating dynamo machine to the E. M. F., which it must develop in order to overcome the resistance opposed by the secondary generators in action.

It is necessary to remember that the work developed by the secondary generators depends absolutely upon the number of spirals of which each is composed, and the form of energy developed depends on the manner of grouping these spirals. From this it follows that each consumer may, as it pleases him, put his apparatus in action, and cause it to produce the special form of energy he wishes to employ without troubling himself about his neighbors. The most absolute independence of each apparatus, and an automatic proportioning of the work produced to the work expended, permit the realization by this system of the essential conditions already indicated for enabling a distribution of every alternating form of electrical energy in currents resulting from the phenomena of induction, which produce always, and necessarily, the alternating form of current.

The old ideas, attributing special danger to the employment of alternating currents, have been ably corrected by Professor Forbes, Dr. Hopkinson, and others; but it is worthy of remark that in an installation under the system the author is discussing—whatever may be the E. M. F. of the primary current, which, it must be remembered, circulates always in a closed circuit—the difference of potential between the terminals of the secondary generators will never be greater than that necessary for the lamps fed by them—that is to say 100 volts or 50 volts, as the case may be. Nothing, therefore, short of culpable carelessness could possibly give rise to a condition of things presenting any danger whatever to the public.

If consumers required electrical energy only for producing light, a solution brought to this point would be as complete as possible; but the applications of continuous currents are too numerous not to render their distribution also desirable, and that with the same facility which, it has been shown, attends the distribution of alternating currents.

It has been already pointed out that the *sine qua non* of the practical and economical transport and distribution of electrical energy to a great distance is to give to the energy to be transported the form of small quantity and high tension, or E. M. F. But although the known types of alternating-current dynamos adapt themselves with the greatest facility to the production of currents of the highest E. M. F., this is, unfortunately, not the case in regard to the collection of continuous currents of a higher E. M. F. than 2,000 volts. On this account the inventors of the secondary generators sought and found a means of redressing the alternating currents produced by their secondary generators. These currents, it will be remembered, have already, by transformation, a low E. M. F., that is to say, they are in the form most readily utilizable. On November 16 last, Professor Ferraris, president of the International Jury of the Turin Exhibition, witnessed the perfect redressing of a current produced by a secondary generator—this current had 16,000 changes of direction per minute.

The instrument for redressing an alternating current is composed of several electro magnets coupled in series and fixed on a cast iron frame, a similar number of electro magnets attached to a movable frame turning on its axis, a redressing commutator fixed on the same axis, having as many changes of polarity as bobbins, and lastly of collectors receiving the currents. The alternating current enters by the fixed bobbins traversing them in series; the point where the current leaves is attached to the brush which communicates with the commutator; the opposite pole communicates directly with the other brush. These brushes are so arranged that they can never come into contact with the same metallic pieces. If the apparatus is at rest, the movable electro magnets are also traversed by the alternating currents; but as soon as the apparatus begins to work, the commutator inverts the poles of 'the movable electro magnets. When the speed has reached the synchronism of the alternations, the current becomes continuous in the movable bobbins, and maintains the synchronism. It is necessary only to take a derivation on the collectors to have a continuous current.

This instrument has, then, the property of taking. under the influence of a very small alternating current, a speed which is synchronic with the changes of direction of the current which feeds it. Thus, then, until it has been found possible to do important work with continuous currents of high E. M. F., the secondary generators will enable the distribution in all its forms, whether alternating or direct, whether of high or low potential of electrical energy, over distances sufficiently great to enable the practical utilization of natural forces. The value of the progress thus realized has been officially recognized in Italy by a grand prize of 10,000 francs.

Besides installations in Italy and elsewhere, there .is one now in course of preparation, of which the central station is situated in New Bond Street, which will doubtless be studied with interest. This installation will consist of the distribution of more than 5,000 lamps of various systems—the steam engines now being placed in position are of over 600 horse-power indicated, and have been manufactured by Messrs. Marshall & Company, of Gainsborough. The dynamos to be used are by Messrs. Siemens Brothers, and are the largest yet made by that firm; they have already been tested, and give most satisfactory results. An excavation of 6,000 square feet has been made under the Grosvenor Gallery, and in this space the machinery is being placed.

It may be interesting to mention that, pending the laying down of the large engines, a temporary engine of 30 horse-power nominal is driving two Siemens W′ dynamos coupled in parallel, which give a current of 24 amperes and 800 volts. This current traverses sixteen secondary generators of 2 horse-power each, which supply currents for 300 glow lamps distributed in the library and club at the Grosvenor Gallery, and in two adjoining establishments in Bond Street. When the satisfactory working of the permanent installation shall have been demonstrated, it is not unreasonable to expect that a wide extension of the application of electricity for the purpose of house-to-house lighting upon the principle described by the author will take place.

10

Reprinted from *R. Soc. (London) Proc.* **74**:476–487 (Jan. 24, 1905)

ON THE CONVERSION OF ELECTRIC OSCILLATIONS INTO CONTINUOUS CURRENTS BY MEANS OF A VACUUM VALVE

J. A. Fleming

An electric oscillation being an alternating current of very high frequency, cannot directly affect an ordinary movable coil or movable needle galvanometer.

Appliances generally used for detecting electric waves or electric oscillations are, therefore, in fact, alternating current instruments, and must depend for their action upon some property which is independent of the direction of the current, such as the heating effect or magnetizing force. The coherer used in Hertzian wave research is not metrical, since the action is merely catastrophic or accidental, and bears no very definite relation to the energy of the oscillation which starts it. Even the demagnetising action of electric oscillations, though more definite in operation than the contact action at loose joints, is far from being all that is required for quantitative research. It is obvious it would be an advantage if we could utilise the direct current mirror galvanometer for the detection and measurement of feeble electric oscillations. This can be done if we can discover a medium with perfect unilateral conductivity.

Some time ago, I considered the use of the aluminium-carbon electrolytic cell with this object. It is well known that a cell containing a plate of aluminium and carbon, immersed in some electrolyte which yields oxygen, such as dilute sulphuric acid or an aqueous solution of any caustic alkali, or salt yielding oxygen, has a unilateral conductivity within limits. An electric current under a certain electromotive force can pass through the cell from the carbon to the aluminium, but not in the reverse direction.

This action has been much studied and is the basis of many technical devices, such as the Nodon electric valve.

The electrochemical action by which this unilateral conductivity is produced involves, however, a time element, and after much experimenting I found that it did not operate with high frequency currents. My thoughts then turned to an old observation made by me in 1889, communicated to the Royal Society, amongst other facts, in a Paper in 1889, and also exhibited experimentally at the Royal Institution in 1890.* This was the discovery : that if a carbon filament electric

---

* See ' Roy. Soc. Proc.,' vol. 47, p. 122, 1890, " On Electric Discharge between Electrodes at different Temperatures in Air and High Vacua," by J. A. Fleming, communicated December 16, 1889; see also ' Proceedings of the Royal Institution,'

glow lamp contains a pair of carbon filaments or a single filament and a metallic plate sealed into the bulb, the vacuous space between possesses a unilateral conductivity of a particular kind when the carbon filament, or one of the two filaments, is made incandescent. I have quite lately returned to this matter, and have found that this unilateral conductivity exists even with alternating currents of high frequency and is independent of the frequency. Hence, in a suitable form, it seemed possible that such a device would provide us with a means of rectifying electric oscillations and making them measurable on an ordinary galvanometer. The following experiments were, therefore, tried :—

Into a glass bulb, made like an incandescent lamp, are sealed in the ordinary way two carbon filaments, or there may be many filaments. On the other hand, one carbon filament may be used and a platinum wire may be sealed into the bulb terminating in a plate or cylinder of platinum, aluminium or other metal surrounding the filament. It is preferable to use a metal plate carried on a platinum wire sealed into the glass bulb, the plate being bent into a cylinder which surrounds both the legs of the carbon loop. The diagrams in fig. 1 show various forms of the arrangement. Diagram *a* shows a bulb with a single carbon filament surrounded by a metal cylinder, *b* shows one with two carbon filaments, and *c* a carbon filament and two insulated metal plates. The ends of the carbon filament which is rendered incandescent are marked + and − and the terminal of the other electrode of the valve is marked *t*. The bulb must be highly exhausted to about the pressure usual in the case of carbon filament incandescent lamps, and the metal cylinder or plate must be freed from occluded air.

Suppose that we employ such a bulb containing one carbon filament surrounded by a metal cylinder (see *a*, fig. 1). The filament may be of any voltage, but I find it most convenient to employ filaments of such a length and section that they are brought to bright incandescence by an E.M.F. of 12 volts. The voltage and section of the filament should be so arranged that the temperature of the filament corresponds with an "efficiency," as a lamp-maker would say, of 2·75 or 3 watts per candle. The filament is conveniently brought to incandescence by a small insulated battery of secondary cells. A circuit is then completed through the vacuous space in the bulb between the cylinder and the filament by another wire which joins the external terminal *t* of the metal cylinder and that terminal of the carbon filament which is in connection with the negative pole of the heating battery. In this last circuit is placed a sensitive mirror galvanometer of the movable needle or movable coil type, and also a coil which may

vol. 13, Part LXXXIV, p. 45, Friday evening discourse on February 14, 1890, "Problems on the Physics of an Electric Lamp," when this unilateral conductivity was experimentally shown.

be the secondary circuit of an air core transformer in which electric oscillations are set up. As is now well known, the vacuous space in the bulb permits negative electricity to move in it from the hot filament or cathode through the vacuous space to the cylinder or anode and back through the galvanometer and coil, but not in the reverse direction, as long as the cylinder is cool and the carbon filament not at a temperature much above the melting point of platinum. To illustrate the action of the bulb as an electrical valve, the following experiments can be shown :—

Electric oscillations are set up in a metal wire circuit by the discharge of a Leyden jar, as usual. This circuit takes the form of a thick wire of one or more turns, bent into the form of a circle or square. Some distance from this, we place another wire, of several, say eight

Fig. 1.

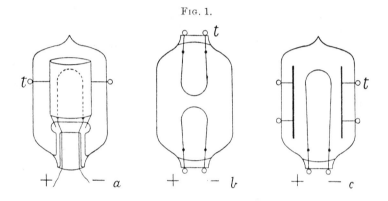

or ten turns, also bent into the form of a circle, and connect this last wire into the circuit of a galvanometer and vacuum bulb made as described, so that it is a circuit having unilateral conductivity. On exciting the oscillations in the primary circuit by an induction coil we have an alternating high frequency magnetic field produced, which affects the secondary circuit at a distance. The oscillations in this last are, however, able to flow only in one direction. Hence, the galvanometer is acted upon by a series of intermittent but unidirectional electro-motive forces, and its needle or coil deflects. Since the field is a high frequency field, we can show the screening effect of a sheet of tin foil or silver paper in a very simple and effective manner by the effect it produces in cutting down the galvanometer deflection when the metal sheet is interposed between the primary secondary circuits. Also, if we move the secondary coil away from the primary coil or turn the two coils with their planes at right angles to one another, then the galvanometer deflection diminishes or falls to zero because the induc-tion is decreased. Accordingly, we have in this vacuum valve and

71

associated mirror galvanometer a means of detecting feeble alternating electric currents or oscillations. Another method is to employ a differential galvanometer and two vacuum valves. These must then be arranged, as shown in fig. 2, one circuit G_1 of the differential galvanometer is in series with one valve V_1 and the other circuit G_2 with the other valve V_2, but so joined up that currents flowing through the valves in opposite directions pass round the two galvanometer wires in the same direction as regards the needle and, therefore, their effects are added together on the galvanometer needle. Each valve must then have its own separate insulated battery to ignite the filament. Also, it is necessary that the connection with the oscillatory circuit must be made in both cases to the hot filament by that terminal which is in

FIG. 2.

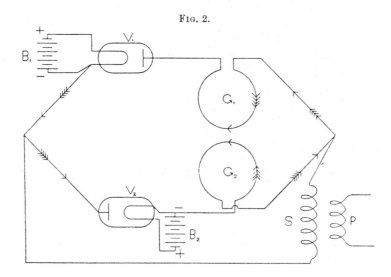

connection with the negative pole of the local battery used to ignite the filament (see fig. 2).

This arrangement of a differential galvanometer and two valves transforms, of course, more of the alternating oscillation into direct current than when one valve alone is used. It provides us with a means of detecting electrical oscillations not merely in closed circuits but in open electrical circuits.

When so using it, it is necessary to associate with the oscillation valve and galvanometer an oscillation transformer for raising the voltage. The resistance of these valves, when in operation, may be anything from a few hundred ohms up to some megohms, depending on the state of incandescence of the filament and upon the electromotive force employed to drive the current through the vacuous space, as well as upon the size of the filament and the plate. This resistance does not obey Ohm's law, but the current increases to a maximum and then slightly decreases as the voltage progressively increases. The form of oscillation transformer employed with the device is as follows : A small air-core induction coil has a primary circuit, which consists of 52 turns of gutta-percha covered wire, wound in a helical groove cut on an ebonite rod 0·5 inch in diameter and 6 inches in length. The primary circuit is made of a No. 20 or No. 22 S.W.G. copper wire. The secondary circuit consists of 36,000 turns of fine silk-covered wire, No. 36, wound in six coils, each having about 6000 turns, and all joined in series. This secondary circuit has one terminal connected to one common terminal of the galvanometer and the other to the common terminal of the two oscillation valves (see fig. 4). The primary coil of this oscillation transformer has one terminal connected to earth and the other to a long insulated rod which acts as an aerial or electric wave collector. To prevent the direct action of the transmitter upon the secondary coil by simple electromagnetic induction, it is best to wind the secondary coil in two equal parts in opposite directions and to wind the primary in a corresponding manner.

If an electric wave sent out from a similarly earthed transmitter falls upon the rod, then an electrical oscillation is set up in the receiving circuit and therefore in the primary coil of the oscillation transformer inserted in series with it. This oscillation is raised in voltage by the secondary coil of the transformer, and by reason of the unilateral conductivity of a vacuum valve, placed in series with the coil, one part of the oscillation, viz., the positive or the negative current, passes round the galvanometer coils and affects it.

If we employ a sensitive dead beat galvanometer of the type called by cable engineers a "Speaking Galvanometer," then intelligible signals can be sent by making small and larger deflections of the galvanometer corresponding to the dot and dash of the Morse alphabet; anyone who can "read mirror" can read off the signals as quickly as they can be sent on an ordinary short submarine cable with this arrangement.

The arrangement, although not as sensitive as a coherer or magnetic detector, is much more simple to use. Also it has one great advantage, viz., that it enables us to examine the behaviour of any particular form of oscillation producer. By means of it we can detect changes in the wave-making power or uniformity of operation of the transmitting arrangement, by the variation of the deflection of the galvanometer. Thus, for instance, if a spark-ball transmitter is being employed and the deflection of the galvanometer in association with the receiving aerial is steady, if we put the slightest touch of oil upon the spark-balls of the transmitter, their wave-making power is increased and the deflection of the galvanometer at once increases. Since the current through the galvanometer is the result of the groups of oscillations which are created in the receiving circuit, and since in the ordinary transmitter these oscillation groups are separated by wide intervals of silence, it is obvious that we can increase the sensitiveness of the above described arrangement by employing a very rapid break or interruptor with the induction coil. If, for instance, we employ a Wehnelt break with the induction coil or a high speed mercury break or alternating current transformer, we get a far better result as indicated by the deflection of the galvanometer than when employing the ordinary low frequency spring or hammer break.

The point of scientific interest in connection with the device, however, is the question how far such unilateral conductivity as is possessed by the vacuous space is complete. The electrical properties of these vacuum valves have accordingly been studied.

A bulb containing a 12-volt carbon filament rendered brightly incandescent by a current of about 2·7 to 3·7 ampères was employed. The filament was surrounded by an aluminium cylinder. The length of the carbon filament was 4·5 cm., its diameter 0·5 mm., and surface 70 square mm.

The aluminium cylinder had a diameter of 2 cms., a height of 2 cms., and a surface of 12·5 square cms. The filament was shaped like a horse-shoe, the distance between the legs being 5 mm. This filament was rendered incandescent to various degrees by applying to its terminals 8, 9, 10, and 11 volts respectively. Another insulated battery of secondary cells was employed to send a current through the vacuous space from the cylinder to the filament, connection being made with the negative terminal of the latter. The current through the vacuous space and the potential difference of the cylinder and negative end of the hot carbon filament were measured by a potentiometer. The effective resistance of the vacuous space is then taken to be the ratio of the so observed potential difference (valve P.D.) to the current (valve current) through the vacuum.

The following table records the observations. The column headed P.D. gives the potential difference between the hot filament and the cylinder, that headed A gives the current through the vacuous space in milliampères, that headed R the resistance of the space in ohms, and that headed $K10^5$ is 100,000 times the conductivity.

The result is to show that the vacuous space does not possess a constant resistance, but its conductivity increases rapidly up to a maximum and then decreases as the valve potential difference progressively increases. If we plot the current values as ordinates and potential difference of the valve electrodes as abscissæ, we find that the current curve quickly rises to a maximum value and then falls again slightly as the potential difference increases steadily. The conductivity curve also rises to a maximum and then decreases (see fig. 3).

The facts so exhibited are well-known characteristics of gaseous

Table I.—Variation of Current through, and Conductivity of, a Vacuum Valve with varying Electromotive Force, the Electrodes being an Incandescent Carbon Cathode and Cool Aluminium Anode.

Carbon filament at 11 volts, 3·77 amp., 41·47 watts.				Carbon filament at 10 volts, 3·44 amps., 34·43 watts.			
Vacuum Space.				*Vacuum Space.*			
P.D.	A.	R.	K10⁵.	P.D.	A.	R.	K10⁵.
0·6	0·024	25,000	4·0	0·7	0·014	50,000	2·0
5·4	0·264	20,550	4·86	2·8	0·073	38,360	2·6
8·8	0·480	18,330	5·45	8·2	0·392	20,920	4·76
18·2	3·880	4,691	21·4	12·8	0·824	15,530	6·56
22·9	26·790	855	118·1	16·2	1·739	9,316	10·70
29·1	28·02	1,038	96·1	20·1	5·352	3,756	26·6
37·1	28·426	1,305	76·6	23·3	9·68	2,407	41·4
49·0	26·50	1,719	58·0	35·9	10·037	3,577	28·0
70·2	26·87	2,613	38·3	49·7	9·794	5,075	20·0
100·0	24·36	4,105	25·0	71·6	8·920	8,027	12·5
				100·08	8·331	12,010	8·32

Carbon filament at 9 volts, 3·112 amps., 28·0 watts.							
Vacuum Space.							
P.D.	A.	R.	K10⁵.	P.D.	A.	R.	K10⁵.
0·5	0·005	100,000	1·0	24·2	2·389	10,130	10·0
2·5	0·049	50,020	2·0	28·2	2·437	11,650	8·6
5·2	0·128	40,625	2·46	36·6	2·508	14,590	6·86
8·3	0·324	25,620	4·0	48·6	2·535	19,170	5·0
8·8	0·361	24,380	4·1	58·5	2·374	24,640	4·0
12·6	0·70	17,970	5·5	72·5	2·253	32,180	3·0
16·4	1·735	9,452	10·5	102·0	2·067	49,350	2·0
20·4	2·351	8,677	11·2				

conduction in rarified gases.* It may be noted that there is in these current-voltage and voltage-conduction curves a general resemblance to the magnetisation and permeability curves of iron.

To examine further the nature of this conduction, the following experiments were made. If a vacuum bulb, as described, is joined up in series with a galvanometer and an electrodynamometer and an alternating electromotive force applied to the circuit, the two instruments will both be affected. The galvanometer is, however, affected only by

* See J. J. Thomson, 'Conduction of Electricity through Gases,' Chap. VIII.

Fig. 3.

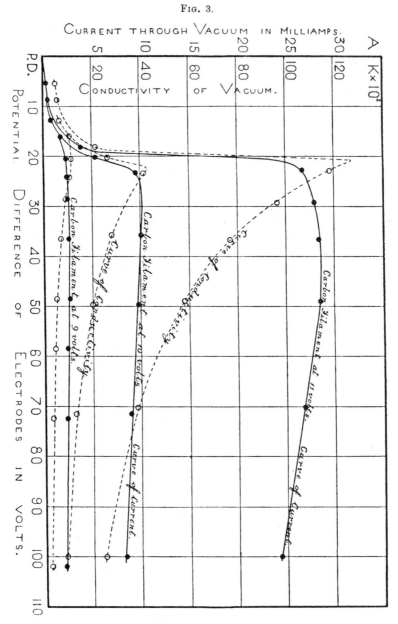

the resultant flux of electricity in one direction. It measures the unidirectional current. The dynamometer is affected by the bilateral flux of electricity and it measures the total or alternating current. If, therefore, the vacuous space is totally non-conducting in one direction, one half of the alternating current will be cut out. The galvanometer will read the true mean (T.M.) value of the remanent unidirectional current, and the dynamometer will read the root-mean-square (R.M.S.) value. If the conductivity in one direction is not zero, then the galvanometer will read the T.M. value of the difference of the positive and negative currents, but the dynamometer will read the R.M.S. value of their sum.*

In the last case, the current through the valve may be considered to be a continuous current superimposed upon an alternating current.

If we call I the maximum value of the nearly sinoidal current in one direction, and I′ the maximum in the opposite direction, then we may say that the dynamometer reading (D) expressed in true current value is equal to g (I + I′) where g is the *amplitude factor*, and also that the galvanometer reading (G) in true current value is equal to g/f (I − I′) where f is the *form factor* of the current.†   Hence—

$$\frac{D}{G} = f\frac{I+I'}{I-I'}, \text{ or } \frac{D/G+f}{2f} = \frac{I}{I-I'}.$$

The fraction $\dfrac{2f}{D/G + f}$, say β, expressed as a percentage may be called the *rectifying power* of the valve, for it expresses the percentage which the actual unilateral electric flow or continuous current through the valve is of that continuous current which would flow if the unilateral conductivity were perfect.

Perfect rectifying power, however, does not exist. There is not an infinite resistance to movement of negative electricity from the metal cylinder to the hot filament through the vacuum, although this resistance is immensely greater than that which opposes the movement of negative electricity in the opposite direction. This point was examined, as follows : A very sensitive electrodynamometer was skilfully constructed by my assistant, Mr. G. B. Dyke, the fixed coil having 2000 turns of No. 47 silk-covered copper wire and the movable coil 1000 turns. The suspension of the movable coil was by a fine flat phosphor-bronze wire at top and bottom. The deflection was observed by a mirror and scale.

---

* If i is the instantaneous value of a periodically varying current with maximum value I and periodic time T, then the root-mean-square value (R.M.S. value) of i is defined to be $\left(\dfrac{1}{T}\displaystyle\int_0^T i^2 dt\right)^{\frac{1}{2}}$ and the true mean value (T.M. value) of i is defined to be $\dfrac{2}{T}\displaystyle\int_0^{\frac{1}{2}T} i\,dt.$

† The *form factor f* and *amplitude factor g* are the names given by the author (see 'Alternating Current Transformer,' J. A. Fleming, vol. 1, p. 585, 3rd edit.) to the ratio of the R.M.S. to the T.M. value of the ordinates of a single valued periodic curve, and to the ratio of the R.M.S. value of the ordinates to the maximum value during the period.

This dynamometer was placed in series with a shunted movable coil galvanometer of Holden-Pitkin pattern, and the two together placed in series with a variable section of an inductionless coil through which an alternating current was passing. A vacuum valve as above described was in series also with the galvanometer and dynamometer. The alternating current was derived from an alternator giving a nearly true sinoidal electromotive force. The form factor of the electromotive force curve of this alternator was determined and found to be 1·115, that for a true sine curve being 1·111.

The vacuum valve sifted out the alternating current flow and allowed the currents in one direction to pass, but nearly stopped those in the opposite direction. The indications of the electrodynamometer were proportional to the root-mean-square (R.M.S.) value of the sum of the two opposite currents, and that of the galvanometer to the true mean value (T.M.) of their difference. The galvanometer and dynamometer were both calibrated by a potentiometer by means of continuous current, and curves constructed to convert their scale readings to milliampères. Then with various alternating current electromotive forces, their readings were taken when in series with a vacuum valve and recorded in the following tables. The letter D denotes current in milliampères as read by the so calibrated dynamometer and G that read by the galvanometer. The ratio D/G is denoted by α, and the rectifying power, viz., $2f/\alpha + f$ by β.

The table shows that the value of α is not constant, but for each state of incandescence of the filament reaches a maximum which, however, does not greatly differ from the mean value for the range of currents used. If we set out the mean values of β in a curve (see fig. 4), in terms of the power expended in heating the carbon filament, we see that the rectification is less complete in proportion as the temperature of the carbon filament increases. This is probably due to the fact that as the filament gets hotter, it heats the enclosing cylinder to a higher temperature and enables negative electricity to escape from the latter.

Hence, I feel convinced that if the metal cylinder could be kept quite cool by water circulation the rectification would reach 100 per cent. or be complete.

An ideal and perfect rectifier for electric oscillations may, therefore, be found by enclosing a hot carbon filament and a perfectly cold metal anode in a very perfect vacuum. With a bulb such as that used for the above experiments all we can say is that the current passed through the vacuum is from 80 to 90 per cent. continuous, 100 per cent. implying that the vacuum is perfectly non-conducting in one direction and permits the flow of negative electricity only from the hot to the cold electrode. The necessity for keeping the cathode cold is shown by the following experiment :—An alternating-current arc was

Table II.—Ratio of Electrodynamometer (D) to Galvanometer (G) Readings in Milliampères. Form Factor of E.M.F. Curve = 1·115 = *f*.

Carbon filament at 11 volts, 3·77 amps., 41·7 watts.

D.	G.	D/G = α.	2f/α+f=β.	
0·85	0·57	1·49	0·86	
1·33	0·85	1·56	0·83	
1·87	1·16	1·61	0·82	
2·30	1·40	1·64	0·81	Mean = 0·82.
3·20	1·88	1·73	0·78	
3·52	2·10	1·68	0·80	
4·54	2·81	1·62	0·82	

Carbon filament at 10 volts, 3·44 amps., 34·43 watts.

0·50	0·34	1·47	0·86	
1·34	0·86	1·56	0·83	
2·28	1·48	1·54	0·84	
2·72	1·68	1·62	0·82	
2·78	1·71	1·63	0·81	Mean = 0·83.
3·02	1·87	1·62	0·82	
3·53	2·17	1·63	0·81	
4·30	2·92	1·47	0·86	
4·25	2·88	1·48	0·86	

Carbon filament at 9 volts, 3·112 amps., 28·0 watts.

0·40	0·31	1·29	0·93	
0·73	0·50	1·46	0·87	
1·28	0·83	1·54	0·84	
1·65	1·15	1·43	0·88	
1·82	1·26	1·44	0·87	
1·78	1·26	1·41	0·88	Mean = 0·89.
1·93	1·35	1·43	0·88	
1·94	1·41	1·38	0·89	
1·87	1·41	1·38	0·91	
1·83	1·39	1·32	0·92	
1·73	1·37	1·26	0·94	

formed between carbon rods, and an iron rod was placed so that its end dipped into the arc. An ammeter was connected in between either carbon and the iron rod, and indicated a continuous current of negative electricity flowing through the ammeter from the iron rod to the carbon pole. This current was, however, greatly increased by making the iron rod of a piece of iron pipe closed at the end and

79

kept cool by a jet of water playing in the interior. In this manner I have been able to draw off a continuous current of 3 or 4 ampères from an alternating-current arc using 15 alternating-current ampères.

Returning, then, to the vacuum valve, we may note that the curves in fig. 3 show that the vacuous space possesses a maximum conductivity corresponding to a potential difference of about 20 volts between the electrodes, for the particular valve used. The interpretation of this fact may, perhaps, be as follows:—In the incandescent carbon there is a continual production of electrons or negative ions by atomic dissociation. Corresponding to every temperature there is a certain electronic tension or percentage of free electrons. If the carbon is

Fig. 4.

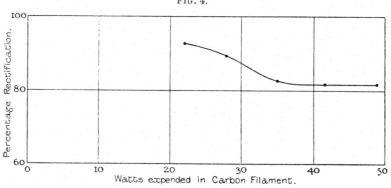

made the negative electrode in a high vacuum these negative ions are expelled from it, but they cannot be expelled at a greater rate than they are produced. Therefore, there is a maximum value for the outgoing current and a maximum value for the ratio of current to electromotive force, that is for the conductivity.

This fact, therefore, fixes a limit to the utility of the device. The current through the vacuous space is, to a very large extent, independent of the electromotive force creating it, and is at no stage proportional to it, or at least only within a narrow range of electromotive force near to the maximum conductivity.

Whilst, therefore, the device is useful as a simple means of detecting electric oscillations, it has not that uniformity of conductivity which would make it useful as a strictly metrical device for measuring them. It can, however, perform the useful service of showing us how far any device for producing electric oscillations or electric waves produces a uniform or very irregular train of electric oscillations, and what changes conduce to an improvement or reduction in the efficiency of the transmitting device.

11

Reprinted from *Franklin Inst. J.* **190**(1):1–38 (1920)

THE AUDION—ITS ACTION AND SOME RECENT APPLICATIONS.*

BY

LEE DE FOREST, Ph.D., Sc.D.

ANALOGIES are apt to be interesting, and in scientific matters frequently instructive and clarifying. The title of to-night's paper, "The Audion," suggestive of *Sound,* prompts the consideration of an analogy in the realm of *Sight*—the microscope. The audion, in a measure, is to the sense of sound what the microscope is to that of sight. But it is more than a magnifier of minute sounds, electrically translated; the audion magnifies and translates into sensation electric energies whose very existence as well as form and frequency, would but for it remain utterly unknown. As the microscope has opened to man new worlds of revelation, studies of structure and life manifestations of natural processes and chemical reactions whose knowledge has proven of inestimable value through the past three generations, so the audion, like the lens exploring a region of electro-magnetic vibrations but of a very different order of wave-length, has during the scant thirteen years of its history opened fields of research, wrought lines of useful achievement, which may not un-

---

* Presented at a joint meeting of the Section of Physics and Chemistry and the Philadelphia Section, American Institute of Electrical Engineers, held Thursday, January 15, 1920.

[Note.—The Franklin Institute is not responsible for the statements and opinions advanced by contributors to the JOURNAL.]

fairly be compared with the benefits from that older prototype and magnifier of light waves. But when the first steps were taken in the work, which eventually resulted in the audion of to-day, I no more foresaw the future possibilities than did the ancient who first observed magnification through a drop of water realize the present application of the high-power microscope to bacteriology.

In 1900, while experimenting with an electrolytic detector for wireless signals, it was my luck to be working by the light of a Welsbach burner. That light dimmed and brightened again as my little spark transmitter was operated. The elation over this startling discovery outlasted my disappointment when I proved that the unusual effect observed was merely acoustic and not electric. The illusion had served its purpose. I had become convinced that in gases enveloping an incandescent electrode resided latent forces, or unrealized phenomena, which could be utilized in a detector of hertzien oscillations far more delicate and sensative than any known form of detecting device.

The first " commercial " audion, as it originally appeared in 1906, was therefore no accident, or sudden inspiration. For failing to find in an incandescent mantel the genuine effect of response to electrical vibrations I next explored the bunsen burner flame, using two platinum electrodes held close together in the flame, with an outside circuit containing a battery of some 18 volts and a telephone receiver. See Fig. 1—the form used in 1903. Now when one electrode was connected to the upright antenna and the other to the earth, I was able to clearly hear in the telephone receiver the signals from a distant wireless telegraph transmitter. The resistance of this new "flame detector" was decreased when the flame was enriched by a salt. Next the incandescent gases of an electric arc were considered; and likewise the action in the more attenuated gases of an ordinary lamp bulb, surrounding an incandescent filament or filaments.

But during these early years I was afforded little time to concentrate on this laboratory problem, and it was not until 1905 that I had opportunity and facilities for putting to actual proof my conviction that the same detector action which had been found in the neighborhood of an incandescent platinum wire, or carbon filament, in a gas flame existed also in the more attenuated gas surrounding the filament of an incandescent lamp. In one case the burning gases heated the electrodes; in the other the electrodes heated the remanant gases. But in both it was *first* the electrons from the hot electrodes, and *second* ionization of the gases which these electrons produced, which established an electrically conducting state, extraordinarily sensitive to any sudden change in electrical potential produced on the electrodes from some outside source.

Considering therefore this actual genesis of the audion it will be seen that it was never, strictly speaking, a *rectifying* device. True both electrodes were seldom alike and a " polarization " was always had from the outside battery, but any rectification of the alternating currents impressed on the detector was merely inci-

FIG. 1.

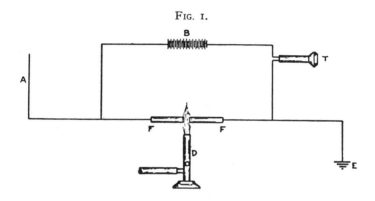

dental and played no vital part in the action of the audion. From the beginning I was obsessed with the idea of finding a *relay* detector, in which local electric energy should be controlled by the incoming waves—and not a mere manifestation of the electrical energy of the waves itself. Hence it was that the external battery as a source of local energy was always employed, when the incandescent filament was utilized as source of the electric carriers through the gas. The battery for lighting the filament I was styled the "A" battery and, as distinguished from this, the other battery was named the "B" battery. This nomenclature has been retained, and is to-day commonly accepted, even by the many who for various reasons refuse to recognize the name " audion."

At the period now under consideration, 1903–05, I was familiar with the Edison effect and with many of the investigations thereof carried on by scientists, Prof. Fleming among others. In 1904 I had outlined a plan of using a gas heated by an incandescent carbon filament in a partially exhausted gas vessel as a wireless detector, in place of the open flame. But here the rectification effect between hot filament and a cold electrode was not considered. Two filaments, heated from separate batteries would give the desired detector effect equally well. What I had already found in the flame detector, and now sought in a more stable and practical form, was a constant passage of electric carriers in a medium of extraordinary sensitiveness, or tenuity, which carriers could be in any conceivable manner affected to a marked degree by exceedingly weak electrical

FIG. 2.

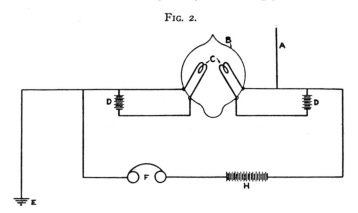

impulses, delivered to the medium, indirectly or through the hot electrodes. Fig. 2.

The ordinary small incandescent lamp of that epoch supplied admirably the conditions I required, merely by the introduction of a second electrode. That added electrode could be either hot or cold. Obviously therefore, use it cold, avoiding thus the unnecessary battery. Then obviously, too, I must so connect my telephone "B" battery as to make this cold electrode positive, for otherwise no local current could flow through the gaseous space in the lamp between the unlike electrodes. Fig. 3.

The high frequency impulses to be detected were, as in the earlier flame type, originally applied directly to these same two electrodes. That these alternating electric currents were thereby rectified was merely incidental. A glance at the plate-current, plate-voltage curve (Fig. 4), shows why, even were both anode and cathode hot, the receipt of a train of high-frequency current waves would produce a resultant change in the normal telephone current and result in a signal.   This typical curve, taken from a "gasey" lamp, such as I first employed, is curvilinear over two

FIG. 3.

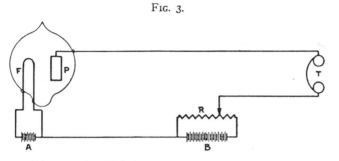

portions.   If now the "B" battery potential was so adjusted that the detector was operating on either knee of this curve the increase to this locally applied voltage, resulting from the positive halves of the wave-train, would produce a greater (or less)

FIG. 4.

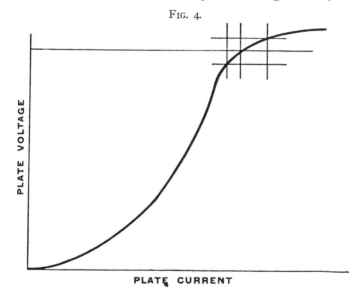

PLATE VOLTAGE

PLATE CURRENT

increase in the local current flowing across the gap than would the negative halves of the wave-train produce a decrease (or increase, as the case might be).

In other words, the responsive action of this two-electrode

audion was due to the asymmetry of its characteristic curve, rather than to its rectifying property. This latter property could be made to aid, to increase the intensity of the signal produced originally and mainly by the so-called "trigger," or genuine *relay* action of the device, which was always *controlling* the local energy by means of a much smaller income energy.

In others words, then, the two-electrode audion, with A and B batteries, was not primarily a "valve." And I have always objected to this misapplication of the name *valve* to the audion; a name which our British friends have from the first persisted. with a stubbornness worthy of a better cause, in misapplying!

Long before the two-electrode relay audion of 1905 had a

FIG. 5.

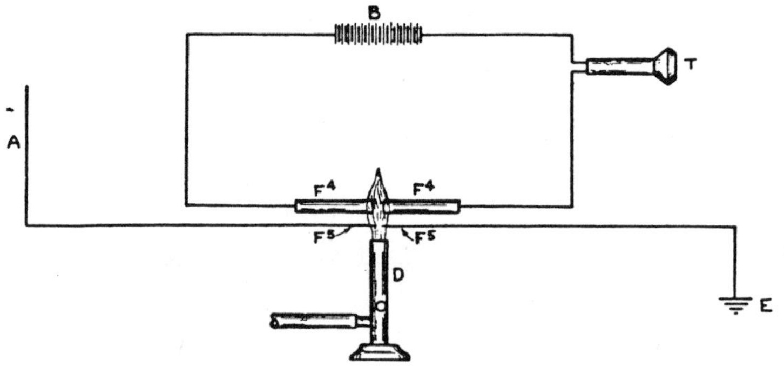

chance to prove its worth in commercial wireless service I had found that the influence of the high-frequency impulses could be impressed to better advantage on the conducting medium from a third electrode. In its first inception the third electrode also dates back to the flame detector of 1903. Fig. 5, taken from the earliest patent of the audion group, shows the original idea of keeping the high-frequency current path distinct from that of the local telephone current. Consequently when in 1906, having secured the maximum efficiency from the two-electrode vacuum type, I cast about for further means of improvement, it was but natural to revert to this plan of separating the two circuits. The new electrode connected to the high-frequency secondary circuit was at first applied to the outside of the cylindrical lamp vessel; the other terminal of the secondary circuit was led to one terminal of the lamp filament. Fig. 6, of a 1906 patent, shows this pro-

genitor of the third electrode. This simple arrangement proving
a step in advance, I concluded that if this auxiliary electrode
were placed within the lamp the weak charges thereto applied
would be yet more effective in controlling the electron-ionic current passing between the filament and plate.

FIG. 6.

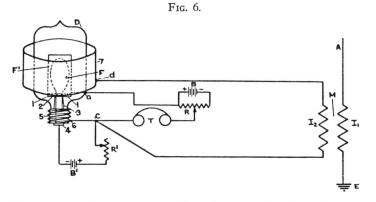

Fig. 7, taken from a patent filed two months after the preceding one, illustrates the next arrangement tried. Here I used two
plates, one either side of the filament—one in the telephone
circuit, the other in the high-frequency circuit associated
with the antenna. It will be noted also that here for

FIG. 7.

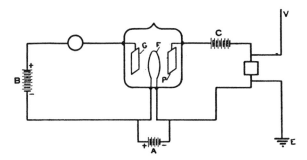

the first time was shown the third, or "C" battery, in the
input circuit, so much employed of late, notably when the audion
is used as amplifier of telephonic currents. This two-plate device proved another decided step forward, and I realized then
that if this third electrode were placed directly in the path of the
carriers between the filament and plate anode I would obtain the

maximum effect of the incoming impulses upon the local current flow. But obviously another electrode thus placed directly in the stream must not be a plate—it must be perforated to permit the carriers to reach the anode. A wire bent back and forth in form of a grid should answer admirably. Fig. 8, taken from the patent filed in January, 1907, the so-called "Grid Audion" patent, illustrates the preferred form which the idea promptly assumed.

In surveying the wide field of electric communication to-day one cannot look back at that little figure, of the first grid electrode, without a sense of wonder at the enormous changes which it has wrought. It has made possible *commercial* trans-oceanic radio telegraphy. It has realized trans-Continental telephony:

FIG. 8.

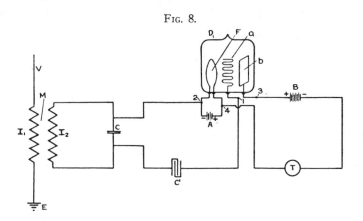

it has made reception of wireless signals half-way around the globe an everyday occurrence. The uncanny accuracy of millions of shells from the Allies' guns, the clock-like precision of advancing barrages, would have been impossible save for the effectiveness of their trench and airplane radio service, in which the grid audion was the essential heart. To-day this little grid controls and modulates an ever increasing kilowattage of radio telephone energy, which as early as 1915 conveyed the spoken voice from Arlington to Honolulu, and more recently from New Brunswick to the transport *George Washington,* in the harbor of Brest. It has already placed twenty simultaneous telephone messages upon a single pair of wires. A few ounces of grid wire make possible the saving of hundreds of tons of copper in long distance telephone conductors. It has given to the physicist a tool for the exploration of unprobed fields of research; and to the electrical engineer a generator, without moving parts, of alternating currents of any desired frequency, from one to ten million period per second—a machine absolutely constant and reliable in its silent work.

Let us consider briefly the explanation of all this radical advance, the theory of the invisible mechanism whereby this astonishing control of powerful energies by minute impulses is effected. Lacking a very concrete conception of just what electrons are and just how electric charges residing on a grid can so effectively dam back the flood of electrons expelled at enormous velocities from a hot cathode at the urge of high potentials—our minds must be content with pictures of characteristic curves and mathematical formulæ, at best but crudely interpretive.

The fundamental operating characteristic of the audion is that expressing the current flowing from filament to plate in terms of the potentials supplied to the grid. Fig. 9 expresses this relation graphically. Here we see that a moderate negative potential (10 volts) applied to the grid completely cuts off the current between filament and plate. As this negative grid charge is reduced to O the plate current rapidly increases. As the grid potential becomes positive this plate current continues to increase up to a point S, after which it rapidly approaches a saturation value, above which the plate current will not rise, regardless of how high is the positive potential applied to the grid. This curve was taken with a given fixed potential difference applied between filament and plate, and for a given filament temperature. This sample characteristic was taken from a "hard" audion, from which the gas has been sufficiently well exhausted to show no irregularity in its curve, due to ionization. It will be observed that over the straight line portion of this curve, between points Q and S, when the grid potential varies a small but equal amount on right and left of zero, the amplitude of variation in the plate current is directly dependent on the variation of the applied grid potential. In other words, there will be over this range no distortion between the wave form of the incoming alternating potential impressed on the grid and that of the current fluctuations produced in the plate, or output, circuit. Obviously therefore the straight line portion of the audion characteristic is the one to utilize in an audion amplifier, or repeater.

whether for amplifying radio or voice-frequency currents. Consequently we find telephone engineers going to extreme lengths to so design their audions and circuits, and to so regulate the potentials applied to the grid, as to operate entirely within this straight line characteristic. The result is a perfect reproduction of voice currents, but magnified to any extent desired by the use of two or more such amplifiers connected in cascade—from ten to twenty thousand times or more.

But when the audion is used as a simple detector of damped radio signals, where it is desired to obtain the maximum possible integrated effect from a train of incoming high-frequency waves

FIG. 9.

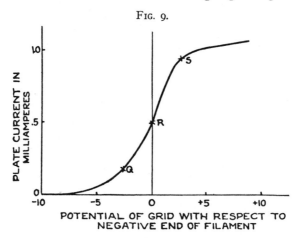

PLATE CURRENT IN MILLIAMPERES

POTENTIAL OF GRID WITH RESPECT TO NEGATIVE END OF FILAMENT

upon the direct current in the plate circuit, it is desirable to operate at the lower, or upper, knee of this characteristic curve, for the reason already explained. Here we have taken advantage of the asymmetry of the curve, so that the sum of the decreases in the plate current as the grid potentials decrease, greatly outvalues the sum of the increases in the plate current, when the grid potentials increase. This results in an integrated decrease in the current through the telephone receiver, which may represent much greater energy than that of the incoming wave-train.[1] It is thus that the audion can operate as a true *relay* device, possessing a sensitiveness far greater than that of the most perfect crystal rectifier, or of any valve. Consideration of the advan-

---

[1] Where a " C " battery is employed for keeping the grid always negative, the audion is operated on the lower knee of its characteristic curve.

tage of thus working on the asymmetric portion of the audion curve would lead us to expect an increase in sensitiveness as a detector of spark signals if this asymmetry be further emphasized by the introduction of a small amount of gas into the bulb, thus producing a very appreciable amount of ionization. Such has long been known to be a fact—no high vacuum audion to-day equals, as a radio detector, the "soft" bulbs which were in very general use a few years ago. The presence of such an amount of gas is usually evidenced by a blue haze seen around the anode, when high potentials (say from 60 to 100 volts) are applied across plate and filament. Ionization phenomena always introduce certain irregularities in the operation of the audion, and kinks, or cusps, in its characteristic curves—even where the gas pressure does not greatly exceed one-ten-thousandth of a millimetre of mercury.

Our knowledge of electrons is of comparatively recent date. In 1899 J. J. Thompson showed that negative electricity is given off from a heated carbon filament in the form of electrons having a mass of 1/1800 that of a hydrogen atom. These electrons may be considered as atoms of electricity. Richardson, in 1903, first applied the electron theory of metallic conduction to emission from heated conductors. He assumed that electrons are ordinarily held bound within the metal by an electric force at the surface, by a tension similar to the surface tension of liquids. But if the velocity of an electron be made sufficiently high, as by applied heat, it is able to overcome this surface force and escape. The number of electrons, therefore, which attain the necessary critical velocity to escape will increase very rapidly with the temperature. The laws are similar to those governing the increase in vapor tension of a liquid with increasing temperature. Richardson thus concluded that the electronic emission from an incandescent metal should increase according to a similar equation:

$$i = a\sqrt{T}\ e^{\frac{-k}{T}}$$

where i is the current per square centimetre at temperature T, and k is a constant dependent on the latent heat of evaporation of the electrons. But actual investigations of the Richardson law, notably by Dr. Langmuir, showed that as the heat of a cathode filament was increased the thermionic current increased first in accord with Richardson's equation, but that beyond a certain point further increase in temperature produced no further increase in thermionic current.

A family of saturation curves, each one corresponding to a certain applied fixed potential between cathode and anode, results, as shown in Fig. 10, where the first parts of the several curves combine to form a single curve following Richardson's law. These curves [2] show that the thermionic current does not continue to increase as expected, because the space surrounding the hot filament is capable of carrying only a certain current for a given potential difference. The explanation offered is that the electrons surrounding the filament soon set up a " space charge " which repels new electrons escaping from the filament, causing some to return to the filament.

From a study of the curve family of Fig. 10, Langmuir has evolved a formula introducing the factor of plate potential, in the case of a filament coaxial with a cylindrical anode. Here the current in amperes

$$i = 14.65 \times 10^{-6} \cdot \frac{V^{\frac{3}{2}}}{r}$$

where r is the radius of the cylinder in centimetres. But extremely minute amounts of gas vitiate the correctness of this formula by neutralizing more or less the space charge.

Therefore for a power oscillator a certain definite amount of gas in the tube may prove a distinct advantage. The filament, if of tungsten, has a tending to absorb gas, so that if a small amount only is left in the tube on exhaustion the audion shows a tendency to grow " harder " with use. A perfect vacuum is never attained; spectrum analysis shows traces of residual gases always present, even in the " hardest " of tubes. If powerful bombardment of anode plates is long continued, gases will be thereby driven out from the metal and the tube rapidly become too soft to be of use—unless the gases have first been thoroughly exhausted from all metal parts within the oscillion, by methods well known to the X-ray tube art. Moreover, too much positive ionization tends to disintegrate the cathode filament. With negligible ionization there appears to be no disintegration of the filament by the electronic discharge, and its life is as great as though no electronic current flowed.

---

[2] Due to Dr. Langmuir.

We have just seen, then, how the space charge surrounding a hot cathode can, in the absence of sufficient ionization, produce a saturated condition of the plate current. This current, or number of electrons emitted, is fixed by the cathode temperature for a given applied voltage, while the velocity of the electrons is dependent on this applied voltage. These electrons, escaping from the cathode and producing the choking space charge, can equally well charge up a third electrode located in this space to a

FIG. 10.

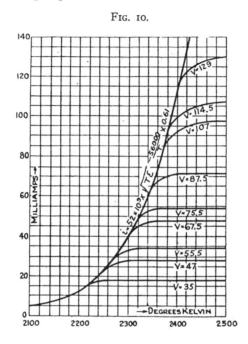

considerable negative potential with respect to the cathode. It will be seen therefore how readily one can expect to control the plate current by means of relatively small potentials, positive or negative, applied to this control electrode, the grid. But the presence of the grid between cathode and anode so complicates the electric field distribution that a theoretical analysis of the relation between the plate current and the plate voltage and the grid voltage (with respect to the filament) is too complicated to be of any practical use. Empirical formula for predetermining the characteristics of various types of audion have been evolved. These, unlike the greater mass of mathematical writings already flooding this new art, are actually proving of some real help to the tube engineer.

In a highly exhausted bulb the so-called " space charge " ordinarily acts to very quickly limit the thermionic current flowing from the hot cathode to the cold anode, but the present of positive ions partly neutralizes this space charge. Now if a small positive charge be applied to the grid the velocity of the electrons passing through it is increased, and consequently they produce more ions by bombardment. Moreover, the number of electrons passing the grid is increased, which in turn again increases the ionization. If too much gas is present, permitting too large a plate current to flow, the relaying action of the audion disappears, because the grid charges are then unable to control the large ionic currents. This condition is usually evidenced by the visible blue glow. In the region between these two limits the audion may possess an extraordinary sensitiveness, as is usual with any condition of instability.

In the early days when audions were exhausted, like the ordinary incandescent lamp, by oil pumps merely, it was ordinarily impracticable to exhaust to such high vacuua as to permit the use of more than 40 to 80 volts of B potential, without producing this excessive ionization. However, in 1912, when I first began to construct larger bulbs for large amplification of telephonic currents, it became apparent that the higher voltage necessary for producing the loud amplifications desired required higher and higher potentials, which obviously necessitated higher vacuua, and better methods of exhaustion. It was then for the first time that I caused audion bulbs to be exhausted by X-ray tube methods, enabling me.to apply several hundred volts of plate potential. So the perfecting of means for exhausting the bulb kept pace with the growing requirements for larger power to be handled. There was at no time in the evolution of the audion, from the original incandescent lamp vacuum to those high exhausts now necessary in the largest " power-tube," or oscillion, any definite demarkation in the degree of vacuum needed or obtained. The lampmakers' and glassblowers' skill kept pace with the radio engineers' requirements of larger bulbs and greater amounts of energy handled.

In the earlier types of audion detector a stopping condenser in the grid lead was usually, but not always, employed. There were

then sufficient ions in the bulb to ordinarily prevent the gradual
accumulation on the thus insulated grid of a large negative
charge, which would very rapidly completely cut off the plate-
filament current. However, when heavy static discharges struck
the receiving antenna it was frequently observed that the audion
would be ". paralyzed " for several seconds thereafter. We early
learned how to discharge this residual negative accumulation
on the grid, at first merely by putting the wetted fingers across the
grid and filament terminals. This primitive " grid leak " was
soon made in permanent form—for example, a wet string, and
later a high-resistance graphite pencil mark. As the degree of
vacuum of the bulb was increased the necessity for this grid leak
became more urgent, and since 1913 it has been generally ap-

FIG. 11.

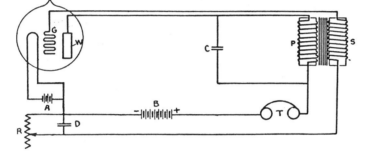

plied to all high vacuum tubes, when stopping condensers are
used; whether the bulb be a detector, amplifier, or generator of
alternating currents.

It was in the summer of 1912, when at work on the problem
of audion amplifiers in cascade arrangements for telephone repeaters, that I first discovered that if the input, or grid, circuit
was inductively coupled with the output, or plate inductance, the
audion became a generator of continuous alternating currents,
originally made evident by a shrill tone in the telephone receiver.
A typical regenerative circuit is illustrated in Fig. 11. The explanation of the operation is simple. An initial impulse in the
plate circuit, however produced, induces a similar one in the grid
circuit, which, if of proper polarity, will impress on the grid a
sudden change in potential which may in its turn produce an
impulse in the plate current in the opposite direction to the origi-

nal disturbance. This reaction then becomes self-sustaining, provided the resistance and hysteresis losses in the two circuits are not too great; and the amplitude of the oscillating current thus set up goes on increasing, taking energy supply from the B battery, until the losses in the circuits equal the increment of energy drawn from the battery. Whereafter an alternating current of perfectly constant amplitude and wave form is maintained. The frequency of this alternating current depends on the constants of the circuit, the inductance and capacity in the input or output circuits. But under certain conditions it depends to some extent also on the resistance in the grid leak, if this be used, and sometimes, but not usually, on the temperature of the filament and the B voltage.

A few months after this type of circuit was first used for the production of alternating currents of audible frequency I first demonstrated the fact that weak *high-frequency* currents could equally well be generated, simply by substituting radio-frequency coils for the original iron-cored coils, and small variable air condensers for the large telephone condensers of the original experiment. And quite naturally, also, since I was at the time engaged chiefly in work on undamped wave radio transmission, this generation of radio frequency waves was first demonstrated in receiving heterodyne, or more exactly autodyne, signals. The circuit used at this early date, April, 1913, which was almost identical with that in Fig. 11—Fig. 12, shows the usual antenna receiving circuit, the usual secondary circuit connected across the grid and filament of the audion, but with another coil similar to the secondary in series with the telephone receiver, which in this case was abridged by a small condenser.

In the fall of that year my assistant, Mr. Longwood, and I discovered, largely by accident, that if the secondary receiving circuit be connected across the grid and plate, instead of as customary between the grid and filament of an audion, the circuit became a persistent oscillator, very simple and effective as a receiver of undamped wave signals. On account of the great sensitiveness of this combination the name "ultraudion" was applied to it. Countless modifications and adaptations of these two general types of oscillating audion circuits have been developed by radio men here and abroad. For their simplicity, the ease with which all the advantages of the beautiful heterodyne prin-

ciple of Prof. Fessenden and Vreeland can be realized, the clarity
of note and range of pitch which the receiving operator can in-
stantly command—coupled with a degree of sensitiveness of a
different order from that of any other type of detector—these
advantages very quickly relegated to the scrap-heap the ticker and
tone-wheel; and at once placed the transmission by undamped
waves upon an altogether different level from that of the older
spark methods.

But the audion in an oscillating or an almost oscillating, or
unstable, condition is also of great utility in detecting damped
wave signals, or even radio telephone currents. If the two cir-

FIG. 12.

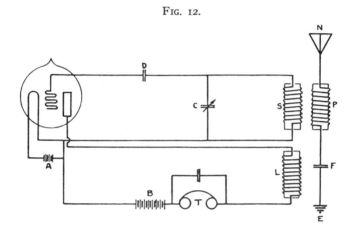

cuits, input and output, are so separated as to interact less ener-
getically the oscillations become weaker and finally just cease to
be generated. When in this condition a very feeble impulse, if
properly attuned, can set the system into vibration. The resulting response develops an energy almost unrelated to the cause.
Enormous magnifications are thus possible with a single audion,
and spark signals have thus been received over the greatest span
which it will ever be possible to reach on this earth—half way
around the globe.

In receiving undamped wave signals, when the local oscillating receiver circuit is slightly out of tune with the incoming
waves, the received currents on reaching the grid are amplified,
first by the ordinary processes of the audion, and then combine
with the local oscillations to produce " beat " notes, of audio-

frequency, which beat note currents are themselves amplified by the audion, before delivery to the telephone receiver. So sensitive is the pitch of this beat note to the slightest change of capacity or inductance in the circuit (when very high frequencies are employed) that in properly designed circuits a change of capacity of one-thousandth part of the electrostatic unit can be detected. A change of capacity which is caused by substituting coal gas for air in a condenser can thus be easily measured. Similarly can be demonstrated slight changes in resistance with temperature of conductors, the conductivity of flames, the permeability of liquids, etc. Very recently Prof. Blondel has utilized the audion in a balanced bridge method for measuring excessively slight differences in static potentials.

The uniform generation of electrical oscillations in a circuit by means of an audion is one of the most striking and fascinating of its applications. If these are of radio-frequency there is no sensible manifestation of their presence, but if of audio-frequency the telephone receiver or " loud speaker " reproducer may be made to give forth sounds from the highest pitch or volume to the softest and most soothing tones. Such wide range and variety of tone can be produced from suitably designed singing circuits that a few years ago I prophesied that at some future time a musical instrument, involving audions instead of strings or pipes, and batteries in place of air, would be created by the musicians' skill.

But lower frequencies, even to one oscillation per second, can be obtained from the audion. Pulsations suitable for submarine cable signalling, or for chronograph and time-pendulum work, can be had of remarkable constancy and reliability, free from all difficulties of speed regulation of motors, or of any moving parts. Or a combination of mechanical time-factors, and the electrical properties of the audion can be advantageously employed. For example, a tuning fork may be driven by electro-magnets, one connected in the grid circuit, the other in a plate circuit, as shown in Fig. 13.

The movement of the grid prong here induces an e. m. f. on the grid, which in turn controls the plate current through the other coil acting upon the other prong of the fork, thus sustaining its motion. If the two coils here shown are also closely coupled inductive reaction, or regeneration, is added to the mechanical.

and very powerful vibrations may be thus set up. Various modifications of this principle will suggest themselves to physicists, who desire sparkless generation of low frequencies of great constancy. Tuned relays, highly selective to definite frequencies, and where it is desirable to reduce the damping to zero or nearly zero, can thus be constructed. The above arrangement is due to Messrs. Eccles and Jordan.

A modification of this method of linking the audion with mechanical motion is the magnetic pendulum, actuated by the plate currents through electro-magnets and inducing in another coil properly tuned impulses which, if conveyed to the grid, control through almost senusoidal currents the successive pulls upon

FIG. 13.

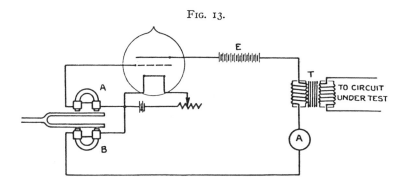

the pendulum (Fig. 14). When a second system identical with the first, but located at right angles thereto, is employed, the pendulum will be set into conical vibration, circular or elliptical as desired, and a revolving electric field will be produced, which can also be made to drive an armature or magnetized wheel at a certain definite speed.

There seems to be in fact no limit to the number of applications to which this three-electrode vacuum tube can be applied as a tool in the hands of the experimental physicist. Of especial value is the fact that it renders easily available devices having *negative electrical resistance,* as in the four-electrode device of Dr. Hull (styled the " dynatron ")—or its equivalent in some mechanical form. For one fundamental property of the audion is that an electrical influence in one circuit may, through the grid, be made to produce effects in another circuit without appreciable reaction. For the energy absorbed by the control electrode may be considered negligible—frequently less than that required in moving a galvanometer needle.

Then, and probably the most promising field of all, the arrangement of audions in cascade as amplifiers, of pulsating currents of any form or frequency—opens to the ear what the microscope has given to the eye—new regions of research in numerous and diversified fields, from physiology, for heart beats and breath sounds—to chemistry, where some even predict that we shall some day hear "the collision of individual atons with one another." During the war British army engineers used as many as nineteen audion bulbs in cascade circuit, amplifying preferably

FIG. 14.

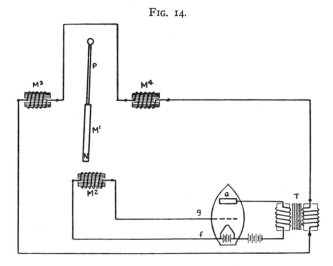

the radio instead of the audion frequency currents. With such a series it is possible to detect with certainty alternating currents of one-ten-thousandth-millionth of a volt on the input grid—involving magnifications of the order of twenty thousand times. It is an everyday occurrence now to receive radio messages from Norway or Honolulu, on a closed-loop antenna one metre in diameter, using three or more audion amplifiers in cascade between this antenna and the detector, and sometimes a similar multi-stage amplifier for audio frequencies, between the detector and the telephone receiver.

Principles which though long understood were impossible of application to radio signalling have been made realizable by the

audion amplifier, and the scope and value of the new art im-
measurably increased thereby. For example, the use of underground receiving antenna, the direction-finder, or radio-compass
loop, the elimination of static interference by either of the above,
or other methods—all such were compelled to await for their
successful application the introduction of the grid electrode. Starting with the small bulb used in 1912–13 as a telephone amplifier
and generator of minute electric oscillations for heterodyning
purposes, I began the construction of larger sizes to be used in
undamped wave transmission. At first spherical bulbs, three or
four inches in diameter, and taking 50 watts of plate input energy,
were considered large. Such rapid progress was made in improvement of design and construction of these so-called " power
tubes," notably by the engineers of the Western Electric Co.,
that by autumn of 1915 a bank of several hundred tubes, their
input and output electrodes connected in parallel, were installed
at the Arlington wireless station. By a pyramidal circuit arrangement, whereby one oscillion tube controlled a group in
parallel, these in turn controlling larger groups of oscillion tubes,
some twelve kilowatts of undamped wave energy was delivered
to the great antenna, all perfectly controlled or modulated by an
ordinary telephone microphone. By this arrangement the voice
was transmitted that year as far as Honolulu and Paris, thus ful-
filling predictions made in 1909 to a very skeptical world.

In these Arlington tests the entire system was one of three-
electrode tubes—for power generator, for current modulation
thereof at the transmitter, and for detector and amplifier at the
receiver. More recently Alexanderson, using his powerful high
frequency alternator at New Brunswick, has controlled 80 kilowatts of antenna energy by means of his magnetic amplifier.
This ingenious development of a Fessenden device was in turn
controlled by a bank of large audion amplifier tubes, nicknamed
" pliotrons," whereby the original microphone currents were sufficiently amplified to control the saturation currents necessary for
the magnetic controlling device.

There are to-day grave differences of opinion among radio
engineers as to what type of high-power radio transmitter will
prove the key to the future—the high-frequency alternator, the
Poulsen arc, or the oscillion. In my own opinion, the long-distance transmission art will shortly depart from true radiation

methods; and the a. c. generator, of comparatively *low* frequency, will be widely used for such subterranean, or submarine transmission, leaving for ship communication only the survival of

FIG. 15.

radio transmission, as it is known to-day. Such being the case, we will then have little use for *radio* transmitters of more than 20 to 50 kilowatts. For such transmitters I foresee the early use of a few large oscillion tubes, of say 5 kw. capacity each. Already we are making tubes capable of handling one and two kilowatts, using tungsten filaments and grids, and large anode plates of tungsten or molybdenum. The efficiency with which several such tubes can operate in parallel, the ease with which an amplified voice current, acting upon their grids in parallel, can control their combined output make such a system almost ideal as a radio telephone transmitter. A typical oscillion transmitter utilizing two half-kilowatt tubes is illustrated in Fig. 15. The schematic circuit diagram for such a transmitter is shown in Fig. 16.

In the construction of these large tubes a thousand details must be scrupulously observed—in addition to the calculated physical

FIG. 16.

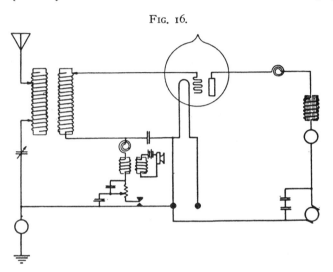

dimensions of the elements, the choice of materials, the method of seal, the preliminary treatment of the metals, their welding, the screening of the glass from bombardment, the various steps in the process of exhaustion—on careful observance of all these alone depends success in the manufacture of a high-power tube. A reasonably long life, of 500 to 1000 hours, is afforded by the tungsten filament, pure or alloyed with thorium; but this is by no means an ideal source of electrons. As such, tungsten, while preferable, is highly inefficient. By coating fine platinum ribbon with oxides of calicum, strontium, etc., or of the rare earths, similar to those in the Nernst glower, far higher emission efficiency is had, at lower temperatures, with resultant increase in life. But such oxide-coated filaments are fragile and very frequently damaged during exhaust. Moreover, many types of coating lose their power of electron emission after a time. This method seems at best an imperfect makeshift. What the audion art awaits is a ribbon filament of some new, well-conducting alloy, wire drawn or rolled, of non-crystalline structure, emitting floods of electrons at a heat even lower than visibility. Reward awaits the metallurgist who first produces such a filament. For to-day the audion is being produced in quantities which in pre-war days would have been considered fantastic exaggeration. During the last months of the war the world production of such bulbs had attained the incredible rate of 1,000,000 per annum. And now the demand in America alone, chiefly from radio amateurs and experimenters, is at the rate of some 5000 per month, and constantly growing. And most of these latter are used singly or in two-step amplifier arrangements. During the war, however, thousands of amplifier and transmitter instruments, each requiring 3 to 9 bulbs, were in use—in earth telegraphy, in submarine listening, in telegraphy by ultra-violet or infra-red rays, in gunspotting, airplane detection, etc., in addition to those required for ordinary radio telegraphy and telephony.

The necessary conditions for an audion to function as a generator of alternating currents have been the subject of exhaustive study by many investigators, notably by Hazeltine, Ballantine, and Mills in this country; Vallauri and Eccles abroad. There are to-day countless circuit arrangements whereby the audion may be caused to generate such currents; but in all of the practically useful ones, where considerable power is required, the inductive linking of the grid and plate circuits, analogous to that first used in 1912, is in one form or another employed. One of the simplest forms of such circuit is shown in Fig. 17. If there is no time lag in the electronic stream behind the pulsations of grid voltage, as is the case in a highly exhausted tube (up to frequencies of ten million per second), then the above arrangement becomes an alternating current source whose frequency depends upon the natural frequency in the LC circuit. The period of this oscillation is very nearly $2\pi\sqrt{LC}$ if the resistance, r, of the external plate circuit is small, the resistance, or reactance, p, of the plate-filament gap is great, and provided the mutual induction, m, between the inductances in the grid-filament and plate-filament circuits is just sufficient to maintain the oscillations.

If, then, $m \gtrless \frac{1}{k}\left(\frac{L}{p}+r.C\right)$ this oscillating condition is realized; and K in this formula can be defined as the "amplification factor."

One of the latest developments in the oscillion transmitter is the application of alternating current for the plate voltage supply. Sixty cycle current is taken from a lamp socket, stepped up to 500 or 10,000 volts (according to the size of the transmitter)—the two halves of the cycle rectified through two-electrode vacuum valves, this rectified current stored in a suitable condenser, smoothed out by an appropriate "filter" circuit, and finally delivered as high-voltage direct current to the plate-filament circuit of three-electrode oscillator tubes. The filaments of both recti-

FIG. 17.

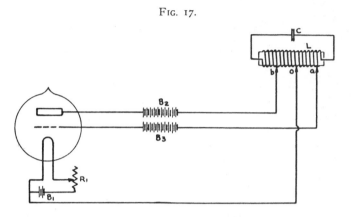

fiers and oscillators are lighted from the low-voltage windings on the one transformer. Such an arrangement does away with the motor generator converter, and even with 60 cycle supply gives surprisingly clear voice transmission. A small set of this type employing two rectifier and four small oscillating audions in multiple is shown in Fig. 18. With this small unit, consuming 50 watts and putting three-quarters of an ampere in an average antenna, one has recently telephoned fifty miles.

The developments by the engineering staff of the Western Electric Company of the audion amplifier as a telephone repeater, since my first demonstration to them of its possibilities in that field, are beyond all praise. The zeal and rare understanding of the elements of the problem with which this staff of trained men developed the amplifier and applied it to the long-sought trans-

VOL. 190, No. 1135—3

continental telephone line stand unique in the annals of brilliant achievement in electrical engineering.

The time was ripe. Had the audion amplifier been presented at a much earlier date it is unlikely it would have then met the warm welcome which twenty years of futile search for the telephone repeater had earned for it. It was the irony of inventive fate that this revolutionary telephone device was to come, not

FIG. 18.

from those whose efforts had for years spun in the old rut of the receiver-microphone "siameesed" together, but from an art younger than telephony, from a device conceived for a quite different application—a wireless telegraph detector.

"From small beginnings the transcontinental line has been evolved. One element after another came. First the telephone receiver of Bell; then the Berliner—Edison microphone; then adequate line construction; the Pupin coil to prevent voice distortion—and finally the one missing link, the Audion Amplifier. Try to imagine one of the electronic carriers of the voice currents in this amplifier, and contrast it with a carbon granule of a microphone transmitter of the early telephone relays. Compare a soap bubble with a load of coal, and you will have some relative idea of the distinction between the delicacy and elegance of the audion and that of the old microphonic relay." A more revolutionary step was never taken in the history of electrical engineering.

A repeater suitable for our present wire telephone system should supply energy amplification sufficient to restore the attenuation produced by twenty miles of standard cable. This actually means that the repeater must be capable of delivering 256 times as much energy as it receives; that is, possess a telephone efficiency of some 26,000 per cent., and this without appreciable distortion of the most intricate of voice current waves, involving all frequencies from 100 to 3000 per second. Any repeater or amplifier which produces distortion of the speech currents is to that extent unfitted for use in tandem operation, because the distortion is cumulative in the successive repeaters; and mechanical amplifiers generally, and even the best of that type, produce distortion.

A large amount of unnecessary secretiveness or mystery was for some time thrown around the type of telephone repeater which made possible transcontinental telephony.

A well-known telephone engineer has recently stated that the audion amplifiers used by the American Telephone and Telegraph Company are practically distortionless, and are commercially used in tandem operation in regular installations, and were so used in the first transcontinental line, which would have been impossible without the use of the tandem arrangement. By actual trial over cable circuits approximately one thousand miles in length it has been found that as many as thirty of these audion amplifiers can be connected in tandem and produce excellent speech at the receiving end of the line. This engineer is authority for the statement that computation shows the attentuation of a cable circuit of this length to be so enormous that if all the power received on the earth from the sun could be applied in the form of telephone waves to one end of the line, without destruction of the apparatus, the energy received at the other end would be insufficient to produce audible speech without the use of amplifiers; whereas with 30 amplifiers used in tandem the relatively minute energy of ordinary telephone speech currents at the transmitting end produced speech in the receiver at the opposite end which was both loud and clear, the amplification due to such a tandem arrangement of tubes being of the order of 10^{50}

The audion which has been evolved to meet these requirements, most rigorous of all its numerous applications, differs in many details from the detector or the oscillating audion. The presence here of gas ionization sufficient to cause appreciable dis-

FIG. 19.

tortion cannot be tolerated, neither must the grid be permitted to be positive at any phase of the cycle of impressed voltage. A hundred other minor requirements, small yet difficult of realization, have been patiently achieved by our telephone engineers, who now state that "the amount by which it (the audion amplifier) fails to meet all the requirements for a perfect repeater is so small as to be negligible except under the most rigorous conditions."

The illustration (Fig. 19) conveys a more vivid idea than any description of the thorough completeness with which the

American Telegraph and Telephone engineers have applied the audion repeater to the commercial long-distance telephone service. It illustrates a typical group of repeater racks, each rack carrying two complete repeaters. This view was taken at one of the main repeater stations on the Boston-Washington underground cable line, located at Princeton, N. J.

Fig. 20.

Two-stage audion amplifier.

Popular attention has been attracted to the success of the recently announced application to line wires of wireless methods of transmission, reception and tuning, whereby multiplex telegraphy and telephony have been made possible over wires already loaded down with their ordinary communication. The original ideas of such multiplex telephony date back to the early nineties, when John Stone Stone, Hutin and Leblanc disclosed methods all involving the same principle, that several alternating currents of superaudio-frequency, each from a separate source, could be directed over the same wire or pair of wires, each be modulated or controlled by its own microphone, or Morse key, and at the receiving station each frequency taken off by its own properly tuned circuit, and there retransformed into its own original telephone or telegraph current. But none of these early investigators utilized at that time the all-necessary integrating detector which

FIG. 21.

50-Watt amplifier.

was alone capable of retransforming the modulated high-frequency wave-trains back into their original audio-frequency currents. Here again the wire telephone requirements had to await the advent of a radio-detector.

General (then Captain) George O. Squier in 1910 carried out certain experiments which are destined to become classic as the new art of wired-wireless attains the important commercial proportions to which it is unquestionably destined. He, for the first time, used a constant, reliable source of undamped electric currents of high frequency for the transmitter, and an audion detector between each tuned receiving circuit and its telephone receiver. By this combination multiplex telephony became at once a realized fact.

But so long as a high-frequency alternator was required at each transmitter station the wired-wireless idea could not become commercialized. Its first cost, the size and weight of it with its motor, its delicacy of speed regulation, its limitation to relatively low frequencies, all made this impossible. So again an important development was compelled to await the advent of the oscillating audion.

Supplied from a common filament-lighting battery, a common " B " battery, or d. c. generator, any desired number of tiny alternating current generators, each driving its own easily tuned circuit, can now be assembled in a small central station. The grid of each oscillator is voice-controlled from its local telephone circuit, and as many high-frequency " carrier " wave-trains superimposed upon a single trunk line pair, as it may be feasible to use without interferences between the modulated frequencies of the several conversations.

At present carrier frequencies ranging from 5000 to 25,000 have been used commercially over a single pair of telephone wires, between Baltimore and Pittsburgh. A zone of frequencies of 2500 is allotted to each conversation, which permits of eight simultaneous telephone conversations over the line, in addition to the usual " physical circuit " conversations. The constant frequency generated by each individual oscillion lies in the middle of each allotted zone of wave-frequencies, but the modulation of this " carrier wave " by the voice currents results in a wide band of frequencies (analogous to a spectrum band) on each side of the particular carrier-wave frequency. This means that at the receiving station it is preferable to employ, instead of a circuit attuned to the single frequency of the carrier-wave, a " band-filter," or combination of several tuning elements (inductance and capacity). This band-filter, then, is equally receptive to any wave-frequency lying within the prescribed limits, say 1,250 cycles on each side of the carrier-frequency, but offers very high impedence to all frequencies above or below the limits of the band-frequencies. By eight such band-filter receiving circuits the eight conversations are segregated, each delivered to its own proper audion detector, and sent out on its own local telephone line.

But it is by no means necessary to limit wired-wireless to the use of such low frequencies as we have been considering. Certain tests were recently carried out in Canada which proved conclusively that frequencies as high as 500,000 per second can be used over telephone lines, including several miles of cable, without harmful attenuation. This demonstration widens very

FIG. 22.

Western electric audion amplifier.

greatly the range of frequencies available for wired-wireless, with hope for a corresponding increase in the number of conversations, or telegraph communications, which can be placed upon a single pair of wires, or group of pairs. Moreover, with such high frequencies (say from 100,000 to 300,000 per second) the necessity for complicated band-filter receiving circuits vanishes, with obvious attendant advantages.

Wired-wireless is the youngest of the large family of methods for electrical communication of intelligence. He is indeed a bold prophet who will to-day attempt to foretell the limits of its application. That the great saving in line costs, the vast multiplication of available channels of long distance communication which it makes possible will work profound changes in our present methods of business, cannot be questioned. Thus again it seems evident that the audion is destined to play a leading rôle in the work of knitting more closely the people of this land, and of all lands.

We have briefly recounted some of the main achievements which the three-electrode audion, or triode, has to its credit. Let us now consider some of the possibilities of its future. From its invention until 1912 it attracted an almost negligible interest in the scientific world. A year after the audion was first brought to the attention of the engineers of the American Telegraph and Telephone Company that corporation acquired exclusive license under all the audion patents for wire telephone purposes. Thereupon the research men of that organization initiated an elaborate line of investigation of the device, which about that time began to interest other scientists in America and abroad. Prior to 1914 not a dozen articles on the audion had appeared in scientific publications. To-day it is impossible to pick up a magazine directed to physics or electric communication without finding one or several papers dealing with some of what Dr. Eccles styes " the protean properties of the ubiquitous three-electrode tube."

Writing in the *Radio Review*, Dr. Eccles (who is affiliated with the British Marconi Co.) says: " The most important single instrument in modern wireless practice is the three-electrode thermionic vacuum valve, for it enters into every main division of the subject—it plays a dominant part in the generation of oscillations, the detections of signals, and in the amplification of feeble voltages and currents. Its arrival and devolpment have, besides, helped greatly towards the success of apparatus and methods that might otherwise have remained almost failures."

Dr. Eccles has outlined the present status and forecast of the future of the audion so clearly that I am constrained to quote further his words, as those of an unbiased observer: " During the war, hints reached the civilian that a revolution was taking place in wireless telegraphy, the principal agent in which was reported to be an instrument called a ' valve,' a ' lamp,' or a ' tube.' This instrument seemed to have arisen suddenly into a predominant position among all the apparatus of the wireless experimenter and operator, and appeared to be of use in every corner of his outfit. The complete name of the instrument is the three-electrode thermionic vacuum tube. It must be emphasized that it is the three-electrode valve, and not the valve with two electrodes, that has been responsible for the overthrowing of the old

FIG. 23.

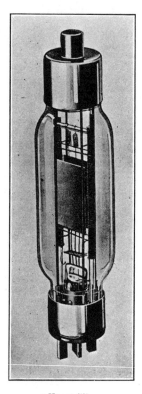

1 Kw oscillion.

methods and apparatus. That it has been a veritable revolution can be seen by comparing the common practice in wireless telegraphy of 1914 with that of 1919. In 1914 practically all the most powerful transmitting stations in the world generated waves by sparks and signals were receivel at nearly all stations by means of crystal or magnetic detectors. The spark method of generating waves involved the use of very large antennæ for spanning great distances; and at the receiving stations which wished to listen to stations more than even 100 miles away very large aerial structures were customary. But if we look at the state of affairs to-day we find most of the high-power stations for long-distance transmission are 'continuous wave' stations; that is, they produce uniform uninterrupted waves instead of a series of short gushes made by sparks; while at the receiving end new modes of detecting these continuous waves appropriate to, and taking advantage of, their uniformity in character have been introduced. This is where the three-electrode tube, in various adaptations, enters the arena. Taken together, the improvements at both ends of the span have made possible the use of smaller antennæ at transmitting stations, and have almost removed the necessity for any antenna at all at receiving stations. For example, under reasonable weather conditions, it is quite easy to listen to the messages coming from stations on the other side of the Atlantic by using a receiving circuit of which the receptive element is a small coil of wire, three to four feet square. Thus, so far as receiving goes, it is possible to intercept all the great stations on one-half of the globe by means of apparatus contained wholly in one room, or even in a cupboard. In accomplishing this the magnifications in use amount to several hundred-thousand-fold. All this is the work of a thing which looks like an ordinary electric-light bulb with a few extra pieces of metal in it—the three-electrode tube."

Years ago what physicist did not look at the simple, self-contained, noiseless incandescent lamp, consider it as an ideal source of electro-magnetic waves of a wide spectrum—of heat, visible, and ultra-violet radiation, and wonder why it should not be made to generate also waves of any length? To-day that incandescent lamp, with the addition of a metal plate and wire grid, has become such a generator. Undamped Hertzien radiations of a few centimetres' wave-length can be generated by audions specially designed to give minimum capacity between the three-electrodes and their lead-in wires. From these short waves, representing alternating current frequencies of some hundreds of millions, down to those of one or two per second, the electric-wave spectrum afforded by the oscillating audion is continuous. Consider this fact in connection with the almost infinite sensitivity of the device as a detector, and its unlimited power as a magnifier, or amplifier, and one realizes something of the value of the three-electrode vacuum tube to the physicist and the inventor. To the former, however, the keenest interest lies perhaps in the audion itself, because there is no known piece of electrical apparatus linked so directly with the most recent work on the structure

Fig. 24.

"VT-21" signal corps audion.

of matter. A prominent British physicist has recently remarked: " It is probable that there is no other sphere where research work has had such a combination of immediate practical value and intense theoretical interest."

Many an early experiment in telegraph transmission or reception by wire or wireless, long since abandoned as too limited in range, can to-day be revived to the great benefit of man. Calculations have shown that with a littoral cable stretched for 50 miles on each side of the Atlantic, and carrying some forty amperes of 20-cycle alternating current, telegraphic communication by conduction or leakage currents should be possible, using the audion

Fig. 25.

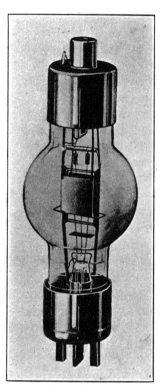

¼-Kw oscillion tube.

as detector and amplifier. I venture to prophesy that within a few years the tall towers and the atmospheric disturbances, which have for two decades been esteemed necessary evils in transoceanic wireless signalling, will be regarded with those sentiments which we now bestow upon the coherer and the spark.

But more than this. Signalling by conduction currents of relatively low frequency will soon be practiced through the earth as well as water; and we will find the antennæ of the future thrust upside down, as into abandoned oil-well borings, and making contact with deep semi-conducting strata, at points separated by a few miles; the two inverted antennæ of such a transmitter connected by an overhead power transmission line containing the alternating current generator and signalling device; and a similar arrangement for receiving. Then our wireless messages will go through the earth's crust, or possibly by a more direct path, and not around the earth's surface, to be tangled up as at present with a bewildering snarl of static ravellings. The audion amplifier stands ready to lead us back to the simpler methods of Morse and Lindsay, meritorious methods long ago abandoned because of the lack of an electric ear of indefinitely great sensitiveness.

The future of radio signalling at sea lies with the telephone rather than the telegraph. The simplicity, the reliability with which the medium of an undamped wave-carrier, ideally suited for voice transmission, can now be had will rapidly limit the crudity and laboriousness of the Morse code signalling between ships. Yet to-day scarcely the dawn of this new epoch has been seen. Vessel owners are to-day almost as skeptical regarding the practicability and utility of the radiophone as we pioneers found them towards the wireless telegraph sixteen years ago!

In the future during fogs at sea a short-wave radio telephone will be used to prevent collisions, distances being determined (as wall as direction) by conversation, whistled signal or bell, and a calibrated stop-watch. This service will be quite independent of the long-range wireless signalling. The new radio has also a wide field of usefulness in telephoning between islands, thousands of which will never be linked by cable. Other useful fields await in sparsely peopled countries, between mines, oil wells, forest patrols, from express trains, etc. The future of aviation will be found linked with radio telephone, for a score of different purposes. Telephony by audion transmitter, receiver, and amplifier not only carries the complexes of human speech without distortion, but delivers them where human speech itself is impossible otherwise—amid the deafening motor and propellor noises of the airplane, from one to five miles above the earth.

Little imagination is required to depict new developments in radio telephone communication, all of which have lain fallow heretofore awaiting a simple lamp by which one can speak instead of read.

12

Reprinted from pp. 83–92 of *IRE Proc.* **10**:83–109 (1922)

THE PIEZO-ELECTRIC RESONATOR*

BY

W. G. CADY

(WESLEYAN UNIVERSITY, MIDDLETOWN, CONNECTICUT)

In the course of experiments with piezo-electric crystals, extending over a number of years, certain radio frequency phenomena were brought to light, the practical application of which appeared worthy of development. The two applications that seem most promising at present are (1) as a frequency-standard, and (2) as a frequency-stabilizer, or means of generating electric oscillations of very constant frequency. It is with these that this paper is chiefly concerned. The fundamental phenomena will first be described, followed by the mathematical theory, and finally an account of the applications will be given.[1]

I. FUNDAMENTAL PHENOMENA

1. A plate or rod suitably prepared from a piezo-electric crystal, and provided with metallic coatings, can be brought into a state of vigorous longitudinal vibration when the coatings are connected to a source of alternating emf. of the right frequency. Under these conditions the plate reacts upon the electric circuit in a remarkable manner. Owing to the piezo-electric polarization produced by the vibrations, and to the absorption of energy in the plate, the apparent electrostatic capacity and resistance of the plate are not constant, but depend upon the frequency somewhat as does the motional impedance of a telephone receiver.[2] Over a certain very narrow range in frequency the capacity becomes negative. An analogy may also be drawn between the vibrating plate and a synchronous motor. The man-

---

*Received by the Editor, October 11, 1921. Presented before THE INSTITUTE OF RADIO ENGINEERS, New York, November 2, 1921.

[1] Preliminary reports on this work have appeared in "The Physical Review," 17, page 531, 1921, and 18, page 142, 1921. The writer wishes to acknowledge the aid that he has received thru a grant from the American Association for the Advancement of Science.

[2] For an explanation of the motional impedance of a telephone receiver, see PROCEEDINGS OF THE INSTITUTE OF RADIO ENGINEERS, volume 6, 1918, page 40.—Editor.

ner in which the reactions upon the circuit are utilized will be described below.   It is necessary, however, to consider the theory of the phenomenon first.

II.   PIEZO-ELECTRIC THEORY

2. Four decades have elapsed since the discovery of piezo-electricity by the Curie brothers, and the prediction of the converse effect by Lippmann, which the Curies promptly verified. During this time much has been accomplished, both theoretically and experimentally, in systematizing and extending our knowledge of the behavior of crystals under static mechanical or electric stress.   Only in very recent years, however, has consideration been given to rapidly varying stresses in piezo-electric crystals.

Nicolson[3] has had marked success in the use of suitably treated Rochelle salt crystals at telephonic frequencies, both as transmitters (direct piezo-electric effect) and as receivers (converse effect).   The writer has also experimented with crystals at audio frequencies, but has devoted his attention chiefly to radio frequency vibrations in the neighborhood of the natural frequency of the crystal plates or rods.

We now summarize briefly those features of Voigt's theory of which we shall make use hereafter.[4]

When a piezo-electric crystal is mechanically strained, there results a dielectric polarization, the magnitude of which is proportional to the strain, and the direction and magnitude of which depend upon the direction of the strain and upon the class to which the crystal belongs.   Except in the case of the class of crystals of lowest symmetry (triclinic), not all of the six components of strain are effective in producing a polarization.   The higher the degree of symmetry, the smaller does this number become.   Of the 32 classes, ten are devoid of piezo-electric properties.

The only two crystals the piezo-electric applications of which have hitherto been considered important are quartz and Rochelle salt; the latter, because it is far more strongly piezo-electric than any other crystal thus far examined; and quartz, because of its excellent mechanical qualities, which make it for most purposes decidedly preferable to Rochelle salt, in spite of its

[3]Nicolson, "Proceedings of the American Institute of Electrical Engineers," 38, page 1315, 1919; "Electrical World," June 12, page 1358, 1920.

[4]For a more complete statement, see Voigt, "Lehrbuch der Kristallphysik," Leipzig, 1910; Graetz, "Handbuch der Elektricität und des Magnetismus," Leipzig, 1914, volume 1, page 342: or Winkelmann, "Handbuch der Physik," 1905, volume 4, part 1, page 774.

being only moderately piezo-electric. The present paper has to do only with quartz, tho obviously the theory applies to any piezo-electric crystal.

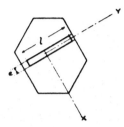

FIGURE 1—Section of a Quartz Crystal perpendicular to the Optical Axis

3. PIEZO-ELECTRIC PROPERTIES OF QUARTZ—Quartz belongs to the trigonal trapezohedral class of crystals. Figure 1 shows a cross-section of a quartz crystal, of which the Z-axis (optical axis) is perpendicular to the paper. The Y-axis is normal to two opposite prismatic faces. Owing to the threefold symmetry of quartz, the Y-axis may be drawn in any one of three directions 120° apart. The three X-axes (electric axes) are perpendicular to the Z- and Y-axes. For piezo-electric experiments, a plate is usually cut from the crystal with its length l, breadth b, and thickness e parallel respectively to the Y-, Z-, and X-axes. The two faces perpendicular to the X-axis are provided with conductive coatings, which may or may not be in actual contact with the quartz.

DIRECT EFFECT—If the plate is compressed in a direction parallel to the X-axis (*longitudinal effect*), the resulting polarization induces equal and opposite charges on the coatings, and the charges change sign with the pressure. Similarly, in the *transverse effect*, an endwise compression of the plate, parallel to the Y-axis, causes the coatings to become charged. A *compresions* of the plate parallel with the X-axis causes a polarization in the same direction as an *extension* parallel with the Y-axis.

CONVERSE EFFECT—In terms of the *converse effect*, if the plate is polarized by an external electric field in the same direction in which it would become polarized by compression along the X-axis, it tends to contract along the X-axis and to expand along the Y-axis.

From what has been said, two important conclusions should

be borne in mind: first, that, in quartz, just as the *direct* effect may be produced by compression along either one of two directions (longitudinal and transverse effects), so both of these effects manifest themselves in connection with the *converse* effect; and second, that in both the direct and converse effects, a given strain is always associated with an electric polarization *in the same direction* and *of the same algebraic sign*.

SYMBOLS

$l, b, e,$	length, breadth, and thickness of quartz plate or rod.
$\varepsilon, \delta,$	piezo-electric constant and modulus respectively. From section 11 on, a special meaning is attached to δ.
$M, N, g,$	equivalent mass, resistance, and stiffness of resonator.
$x,$	displacement of end of resonator.
$F,$	equivalent mechanical force on resonator.
$E,$	voltage impressed on circuit.
$V,$	potential difference across resonator.
$D,$	piezo-electric polarization in resonator.
$I, i,$	currents in coil and resonator branches, Figure 3.
$C_1,$	normal capacity of resonator, vibrations damped.
$C_2,$	capacity of tuning condenser.
$C_1', C_1'',$	equivalent series and parallel capacity of resonator.
$R_1', R_1'',$	equivalent series and parallel resistance of resonator.
$C_a,$	"apparent" capacity of resonator.
$C_t, R_t,$	equivalent series capacity and resistance of entire circuit, Figure 3.
$R_{12},$	equivalent series resistance of resonator and C_2, together. When printed without subscripts, x, F, E, V, D, I, and i denote instantaneous values. x_o and so on, denote maximum values.
$f,$	frequency.
$\omega,$	angular velocity $= 2\pi f$. ω_o and f_o denote resonance values.

4. In the case of quartz, the general polarization-strain equations reduce to the following form:

$$P_1 = \varepsilon_{11} x_x + \varepsilon_{12} y_y + \varepsilon_{14} y_z \tag{1}$$

$$P_2 = \varepsilon_{25} z_x + \varepsilon_{26} x_y. \tag{2}$$

P_1 and P_2 are X and Y components, respectively, of polarization (electric moment per unit volume), and the ε's are the *piezo-electric constants*. x_x and x_y are, in Voigt's notation, the components of extension (elongation or contraction per unit length), and y_z and so on, the components of shearing strain.

If, instead of the components of strain, we have given the components of *stress*, (1) and (2) become

$$-P_1 = \partial_{11} X_x + \partial_{12} Y_y + \partial_{14} Y_z \qquad (3)$$

$$-P_2 = \partial_{25} Z_x + \partial_{26} X_y. \qquad (4)$$

The ∂'s are the *piezo-electric moduli*, which are related to the piezo-electric constants ϵ by equations involving also the elastic constants.

As is evident from equations (2) and (4), the polarization P_2 is produced only by shears, which may be neglected in the present paper, as may also the third term in (1) and (3). Of the two remaining terms on the right-hand side of (1) and (3) the first expresses the longitudinal effect, the second the transverse effect.

We shall need also the following expressions for the *converse effect*, in which the stresses along the X- and Y-axis are given in terms of the X-component E_1 of impressed electric intensity:

$$-X_x = \epsilon_{11} E_1 \qquad (5)$$

$$-Y_y = \epsilon_{12} E_1. \qquad (5a)$$

The other stress-components are of no concern here. The equation (5) expresses the longitudinal effect, and (5a) the transverse. In the applications described in the present paper, only the *transverse effect* is utilized.

One more fundamental equation must be added, namely the strain-equation for the transverse converse effect, which is analo- · gous to (5a):

$$y_y = \partial_{12} E_1 \qquad (6)$$

According to Voigt's theory, in the case of the class of crystals to which quartz belongs, $\epsilon_{26} = \epsilon_{12} = -\epsilon_{11}$, $\epsilon_{25} = -\epsilon_{14}$, $\partial_{12} = -\partial_{11}$, $\partial_{25} = -\partial_{14}$, and $\partial_{26} = -2\partial_{11}$. Hence in all only two different numerical values of ∂ and ϵ have to be known, and of these only one occurs in the present investigation. The following values of ϵ_{11} and ∂_{11} were determined by Riecke and Voigt:[5]

$$\epsilon_{11} = -4.77 \times 10^4, \quad \partial_{11} = -6.45 \times 10^{-8}.$$

The ϵ's and ∂'s as indeed all electric and magnetic quantities in this paper, unless otherwise stated, are in c. g. s. electrostatic units. As is evident from (1) and (2), ϵ has the dimensions of an electrostatic polarization, while from (3) and (4) it may be seen that ∂ has the dimensions of the reciprocal of an electric intensity. Hence

$$\epsilon = [k^{\frac{1}{2}} M^{\frac{1}{2}} L^{-\frac{1}{2}} T^{-1}], \quad \partial = [k^{\frac{1}{2}} M^{-\frac{1}{2}} L^{\frac{1}{2}} T].$$

Other observers have obtained slightly different values for

[5]Voigt, previous citation, pages 869-870.

ϵ_{11} and δ_{11}. Fortunately, in the practical applications under consideration, the absolute values need not be accurately known.

III. Theory of Longitudinal Vibrations in Rods

5. The theory of electric reactions of vibrating piezo-electric plates is a structure built upon two main piers. First, there is the fundamental piezo-electric theory which has just been set forth; and second, the theory of longitudinal mechanical vibrations in rods, which will now be briefly summarized. The "plates" which the writer uses are, as far as mechanical considerations permit, in the form of thin rods. The advantage of this procedure, in addition to economy of material, is that the fundamental vibration together with harmonics of considerable purity may be secured, free from the disturbing effects of other modes of vibration. The theroretical treatment is also greatly simplified.

In a paper which is to appear in "The Physical Review," the general theory of forced longitudinal vibrations in rods is developed. The characterizing feature is the insertion in the equations of a symbol representing the *viscosity* of the material composing the rod; for that property of the rod whereby it absorbs energy and damps its own vibrations is as important here as is the resistance in an oscillating electric circuit. It is possible to measure the actual value of the viscosity by a purely electrical method, at any desired frequency; this, as well as the effect upon the resultant viscosity of air friction and of restraints imposed by the method of mounting, need not concern us here. It is only necessary to remark that a successful piezo-electric resonator must be prepared and mounted as to reduce the damping to a minimum.

6. When an alternating emf. is applied to the metallic coatings of a rod of this sort, an alternating mechanical stress is set up in the rod in accordance with equation (5a), which is uniform throut the mass of the rod. In this statement we neglect the "edge effect" of the condenser formed by the quartz and its coatings. Considering the thinness of the quartz and its high dielectric constant—about 4.5—this procedure is justifiable as a first approximation. In the paper referred to above, it is shown that the vibrations are the same as if the rod had impressed upon its ends two alternating forces, numerically equal to the actual internal stress, of like amplitude but opposite phases, and it is on this basis that the theory of forced vibrations is developed.

The general equation of motion is

$$\frac{\partial^2 \xi}{\partial t^2} = P \frac{\partial^2 \xi}{\partial u^2} + Q \frac{\partial^3 \xi}{\partial u^2 \partial t}. \tag{7}$$

ξ is the displacement, at the time t, of that cross-section of the rod whose undisturbed co-ordinate is u. P is defined by the equation $P = G/\rho$, where G is Young's modulus and ρ the density; P is therefore the square of the wave-velocity in absence of damping.[6] For brevity, we call Q the "viscosity," and treat it as a constant of the material, implying thereby that it is independent of the frequency. Its possible dependence upon frequency can be tested experimentally. The dimensions of Q are $[L^2 T^{-1}]$.

7. In the paper referred to, equation (7) is solved, but its application to actual cases of forced vibration is somewhat cumbersome. It is, however, shown that, for the fundamental vibration in the neighborhood of resonance, the rod may be replaced by a *fictitious "equivalent mass" M possessing one degree of freedom*. The equation of motion then has the familiar form

$$M \frac{d^2 x}{d t^2} + N \frac{d x}{d t} + g x = F = F_o \cos \omega t. \tag{8}$$

Here M is half the actual mass of the rod, or $M = \frac{1}{2} \rho b l e$. In place of Young's modulus G in (7) we use the "equivalent stiffness," $g = M \omega_o^2$, which is related to G by the equation $g = \pi^2 b e G/2l$. This follows from the equation $\omega_o = 2\pi f_o$, and $\sqrt{G/\rho} = 2 l f_o$, $2l$ being the fundamental wave-length.[7]

x is the mechanical displacement at time t of the end of the rod, so that the actual elongation (or contraction) of the entire rod at any instant is $2x$.

M and g correspond to L and $1/C$ in an electric circuit hav-

---

[6] In crystalline media, the elastic constants depend, of course, upon the direction with respect to the axis of the crystal. Slight differences are found between individual crystals. Moreover, in the case of our rods, the elastic modulus is modified by lateral effects, unless the rod is extremely narrow, and by any discrepancy between the axis of the rod and the true Y-axis of the crystal. The effective value of G with the rods employed by the writer ranges from 8×10^{11} to 10×10^{11}. The value for quartz as given by Voigt is 8.51×10^{11}.

[7] Strictly, ω^o is the angular velocity when the amplitude of the *velocity* of the equivalent mass M is a maximum under forced vibrations; it is also the free angular velocity in absence of damping. The maximum amplitude of equivalent *displacement* x (equation (10)) comes (under forced vibrations) at the angular velocity $\sqrt{\dfrac{g}{M} - \dfrac{N^2}{2 M^2}}$, while the angular velocity of free damped vibrations is $\sqrt{\dfrac{g}{M} - \dfrac{N^2}{4 M^2}}$. The distinction between these three values may under ordinary circumstances be ignored.

ing concentrated, as contrasted with distributed, constants. N is the equivalent resistance, and bears to the viscosity Q the relation $N = \pi^2 \rho \, b \, e \, Q/2l$. For the proof of this the paper on longitudinal vibrations must be consulted. F is the equivalent impressed force. If the actual stress acts thruout the entire length of the rod, it may be proven that F is twice the actual force at any cross-section, or $F = 2beX$, where X is the instantaneous stress. The expression for X in terms of the piezo-electric constant is given below, section 11.

We now write the steady-state solution of equation (8), which is of prime importance for the graphical method described in section 12:

$$x = x_o \sin (\omega t - \theta), \tag{9}$$

in which the maximum displacement is

$$x_o = \frac{F_o}{\omega \sqrt{N^2 + \left(\omega M - \dfrac{g}{\omega}\right)^2}}, \tag{10}$$

and

$$\tan \theta = \frac{\omega M - \dfrac{g}{\omega}}{N} = \frac{\pi (\omega - \omega_o)}{\omega_o \Delta} \tag{11}$$

approximately, since $g/\omega = \omega_o M$ very nearly, and the logarithmic decrement per period, Δ, is, as in the electrical analogy, $N/2fM$.

The *power expended in maintaining vibrations*, as in the case of the electrical analogy, is easily proved to be

$$p = \frac{1}{2} \cdot \frac{F_o{}^2}{N} \ (ergs \ per \ sec.) \tag{12}$$

The *maximum stress when in resonance* may easily become so great as to break the quartz rod. On the assumption that the distribution of stress is sinusoidal, being zero at the ends, and' for the fundamental, a maximum at the center, we find that the maximum stress at the center is $\pi x_o G/l$, where x_o is half the maximum elongation of the rod of length l, and G is Young's modulus.

IV.   The Resonance Circle

8. In applying the foregoing theory to investigations with piezo-electric resonators, it is advantageous to employ a graphical method, based on the properties of what may be called, for brevity, the resonance circle. In principle, this curve is similar to the "motional impedance" circle which has been used by

Kennelly and his collaborators in their studies of the telephone receiver.[3]

The equation of the curve in question is obtained by eliminating $\omega M - g/\omega$ between equations (10) and (11):

$$x_0 = \frac{F_0}{\omega N} \cos \theta. \tag{13}$$

If ω were constant, this would be the polar equation of a circle passing thru the origin. In reality, as θ varies from $-90°$ thru zero to $+90°$, ω varies from zero to infinity. Nevertheless, when N is very small, as is the case with quartz, not only is the "diameter" of the "circle" in Figure 2 large, but that portion of the curve corresponding to the neighborhood of resonance comprises nearly the entire curve. For all other values of ω, θ is nearly equal either to $-90°$ or to $+90°$, so that with quartz, to the precision attainable by ordinary graphical methods, the curve cannot be distinguished from a perfect circle. The distortion of the curve owing to varying ω in Figure 2 is very greatly exaggerated in order to illustrate the principle.

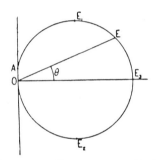

FIGURE 2—The Resonance Circle

$O E$ represents one value of the modulus x_0, with the corresponding argument θ. It has been found most convenient to draw the maximum modulus $O E_3$ horizontally to the right from the origin, and to lay off positive values of θ *below* the horizontal axis, so that *increasing frequency* is represented by a *clockwise movement* of the point E around the curve. Strictly speaking, the maximum modulus, $x_0 = F_0/\omega N$, should be inclined slightly upwards, corresponding to a small negative value of θ. Here

---

[3]"Proc. Am. Acad. Arts and Sci.," 48, page 113, 1912; 51, page 421, 1915; "Proc. Am. Phil. Soc.," 54, page 96, 1915; 55, page 415, 1916. The circle diagram is also used by Hahnemann and Hecht, "Phys. Zeitschr.," 20, page 104, 1919, and 21, page 264, 1920, and by Wegel, "Journal of the American Institute of Electrical Engineers," 40, page 791, 1921.

again the damping in the case of quartz is so slight that the maximum is practically the line $O\,E_3$.

$O\,A$ represents the "amplitude" when $\omega = 0$, that is, $O\,A$ is the equilibrium elongation under static stress at zero frequency, and must therefore have the value $X_o\,l/2\,G$. In the case of a typical quartz plate (Quartz Resonator N 2, to which further reference will be made), $3.07 \times 0.41 \times 0.14$ cm. ($1.21 \times 0.16 \times 0.055$ inch), the fundamental frequency of which is 89,870, the equilibrium elongation at either end under a potential difference of one electrostatic unit (300 volts) is 6.5×10^{-7} cm., while the maximum amplitude of vibration at either end is (by calculation) 0.0025 cm. (0.001 inch). Thus we see that $O\,E$ is about 4,000 times as large as $O\,A$: in other words, the growth of amplitude at resonance is 4,000-fold.

For our purposes, the advantage of the resonance circle as outlined above is two-fold: the moduli $O\,E$, being proportional to elongations of the plate, are thereby also proportional to the piezo-electric polarization; and since the argument θ is a phase angle, the resonance circle can be incorporated into an ordinary alternating current vector diagram in studying the reaction of the plate upon the circuit. We now come to a consideration of the latter.

[Editor's Note: Material has been omitted at this point.]

13

Reprinted from *Phys. Rev.* **75**(8):1208–1225 (1949)

Physical Principles Involved in Transistor Action

J. Bardeen and W. H. Brattain

Bell Telephone Laboratories, Murray Hill, New Jersey

(Received December 27, 1948)

The transistor in the form described herein consists of two point-contact electrodes, called emitter and collector, placed in close proximity on the upper face of a small block of germanium. The base electrode, the third element of the triode, is a large area, low resistance contact on the lower face. Each point contact has characteristics similar to those of the high back-voltage rectifier. When suitable d.c. bias potentials are applied, the device may be used to amplify a.c. signals. A signal introduced between the emitter and base appears in amplified form between collector and base. The emitter is biased in the positive direction, which is that of easy flow. A larger negative or reverse voltage is applied to the collector. Transistor action depends on the fact that electrons in semiconductors can carry current in two different ways: by excess or conduction electrons and by defect "electrons" or holes. The germanium used is n-type, i.e., the carriers are conduction electrons. Current from the emitter is composed in large part of holes, i.e., of carriers of opposite sign to those normally in excess in the body of the block. The holes are attracted by the field of the collector current, so that a large part of the emitter current, introduced at low impedance, flows into the collector circuit and through a high impedance load. There is a voltage gain and a power gain of an input signal. There may be current amplification as well.

The influence of the emitter current, I_e, on collector current, I_c, is expressed in terms of a current multiplication factor, α, which gives the rate of change of I_c with respect to I_e at constant collector voltage. Values of α in typical units range from about 1 to 3. It is shown in a general way how α depends on bias voltages, frequency, temperature, and electrode spacing. There is an influence of collector current on emitter current in the nature of a positive feedback which under some operating conditions may lead to instability.

The way the concentrations and mobilities of electrons and holes in germanium depend on impurities and on temperature is described briefly. The theory of germanium point contact rectifiers is discussed in terms of the Mott-Schottky theory. The barrier layer is such as to raise the levels of the filled band to a position close to the Fermi level at the surface, giving an inversion layer of p-type or defect conductivity. There is considerable evidence that the barrier layer is intrinsic and occurs at the free surface, independent of a metal contact. Potential probe tests on some surfaces indicate considerable surface conductivity which is attributed to the p-type layer. All surfaces tested show an excess conductivity in the vicinity of the point contact which increases with forward current and is attributed to a flow of holes into the body of the germanium, the space charge of the holes being compensated by electrons. It is shown why such a flow is to be expected for the type of barrier layer which exists in germanium, and that this flow accounts for the large currents observed in the forward direction. In the transistor, holes may flow from the emitter to the collector either in the surface layer or through the body of the germanium. Estimates are made of the field produced by the collector current, of the transit time for holes, of the space charge produced by holes flowing into the collector, and of the feedback resistance which gives the influence of collector current on emitter current. These calculations confirm the general picture given of transistor action.

I. INTRODUCTION

THE transistor, a semiconductor triode which in its present form uses a small block of germanium as the basic element, has been described briefly in the Letters to the Editor columns of the Physical Review.[1] Accompanying this letter were two further communications on related subjects.[2, 3] Since these initial publications a number of talks describing the characteristics of the device and the theory of its operation have been given by the authors and by other members of the Bell Telephone Laboratories staff.[4] Several articles have appeared in the technical literature.[5] We plan to give

here an outline of the history of the development, to give some further data on the characteristics and to discuss the physical principles involved. Included is a review of the nature of electrical conduction in germanium and of the theory of the germanium point-contact rectifier.

A schematic diagram of one form of transistor is shown in Fig. 1. Two point contacts, similar to those used in point-contact rectifiers, are placed in close proximity ($\sim$0.005–0.025 cm) on the upper surface of a small block of germanium. One of these, biased in the forward direction, is called the emitter. The second, biased in the reverse direction, is called the collector. A large area, low resistance contact on the lower surface, called the base electrode, is the third element of the triode. A physical embodiment of the device, as designed in large part by W. G. Pfann, is shown in Fig. 2. The transistor can be used for many functions now performed by vacuum tubes.

During the war, a large amount of research on

[1] J. Bardeen and W. H. Brattain, Phys. Rev. **74**, 230 (1948).

[2] W. H. Brattain and J. Bardeen, Phys. Rev. **74**, 231 (1948).

[3] W. Shockley and G. L. Pearson, Phys. Rev. **74**, 232 (1948).

[4] This paper was presented in part at the Chicago meeting of the American Physical Society, Nov. 26, 27, 1948. W. Shockley and the authors presented a paper on "The Electronic Theory of the Transistor" at the Berkeley meeting of the National Academy of Sciences, Nov. 15–17, 1948. A talk was given by one of the authors (W.H.B.) at the National Electronics Conference at Chicago, Nov. 4, 1948. A number of talks have been given at local meetings by J. A. Becker and other members of the Bell Telephone Laboratories Staff, as well as by the authors.

[5] Properties and characteristics of the transistor are given by J. A. Becker and J. N. Shive in Elec. Eng. **68**, 215 (1949).

A coaxial form of transistor is described by W. E. Kock and R. L. Wallace, Jr. in Elec. Eng. **68**, 222 (1949). See also "The Transistor, A Crystal Triode," D. G. F. and F. H. R., Electronics, September (1948) and a series of articles by S. Young White in Audio Eng., August through December (1948).

the properties of germanium and silicon was carried out by a number of university, government, and industrial laboratories in connection with the development of point-contact rectifiers for radar. This work is summarized in the book of Torrey and Whitmer.[6] The properties of germanium as a semiconductor and as a rectifier have been investigated by a group working under the direction of K. Lark-Horovitz at Purdue University. Work at the Bell Laboratories[7] was initiated by R. S. Ohl before the war in connection with the development of silicon rectifiers for use as detectors at microwave frequencies. Research and development on both germanium and silicon rectifiers during and since the war has been done in large part by a group under J. H. Scaff. The background of information obtained in these various investigations has been invaluable.

The general research program leading to the transistor was initiated and directed by W. Shockley. Work on germanium and silicon was emphasized because they are simpler to understand than most other semiconductors. One of the investigations undertaken was the study of the modulation of conductance of a thin film of semiconductor by an electric field applied by an electrode insulated from the film.[3] If, for example, the film is made one plate of a parallel plate condenser, a charge is induced on the surface. If the individual charges which make up the induced charge are mobile, the conductance of the film will depend on the voltage applied to the condenser. The first experiments performed to measure this effect indicated that most of the induced charge was not mobile. This result, taken along with other unexplained phenomena such as the small contact potential difference between n- and p-type silicon[8] and the independence of the rectifying properties of the point-contact rectifier on the work function of the metal point led one of the authors to an explanation in terms of surface states.[9] This work led to the concept that space-charge barrier layers may be present at the free surfaces of semiconductors such as germanium and silicon, independent of a metal contact. Two experiments immediately suggested were to measure the dependence of contact potential on impurity concentration[10] and to measure the change of contact potential on illuminating the surface with light.[11] Both of these experiments were successful and confirmed the theory. It was while studying the latter effect with a silicon surface immersed in a

liquid that it was found that the density of surface charges and the field in the space-charge region could be varied by applying a potential across an electrolyte in contact with the silicon surface.[12] While studying the effect of field applied by an electrolyte on the current voltage characteristic of a high back-voltage germanium rectifier, the authors were led to the concept that a portion of the current was being carried by holes flowing near the surface. Upon replacing the electrolyte with a metal contact transistor action was discovered.

The germanium used in the transistor is an n-type or excess semiconductor with a resistivity of the order of 10 ohm cm and is the same as the material used in high back-voltage germanium rectifiers.[13] All of the material we have used was prepared by J. C. Scaff and H. C. Theuerer of the metallurgical group of the Laboratories.

While different metals may be used for the contact points, most work has been done with phosphor bronze points. The spring contacts are made with wire from 0.002 to 0.005″ in diameter. The ends are cut in the form of a wedge so that the two contacts can be placed close together. The actual contact area is probably no more than about 10^{-6} cm².

The treatment of the germanium surface is similar to that used in making high back-voltage rectifiers.[14] The surface is ground flat and then etched. In some cases special additional treatments such as anodizing the surface or oxidation at 500°C have been used. The oxide films formed in these processes wash off easily and contact is made to the germanium surface.

The circuit of Fig. 1 shows how the transistor may be used to amplify a small a.c. signal. The emitter is biased in the forward (positive) direction so that a small d.c. current, of the order of 1 ma, flows into the germanium block. The collector is biased in the reverse (negative) direction with a

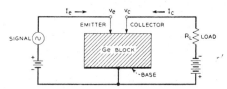

FIG. 1. Schematic of transistor showing circuit for amplification of an a.c. signal and conventional direction for currents. Note bias currents I_e and V_e are normally positive, I_c and V_c negative.

[6] H. C. Torrey and C. A. Whitmer, *Crystal Rectifiers* (McGraw-Hill Book Company, Inc., New York, 1948).
[7] J. H. Scaff and R. S. Ohl, Bell Sys. Tech. J. **26**, 1 (1947).
[8] Walter E. Meyerhof, Phys. Rev. **71**, 727 (1947).
[9] John Bardeen, Phys. Rev. **71**, 717 (1947).
[10] W. H. Brattain and W. Shockley, Phys. Rev. **72**, 345(L) (1947).
[11] Walter H. Brattain, Phys. Rev. **72**, 345(L) (1947).

[12] R. B. Gibney, formerly of Bell Telephone Laboratories, now at Los Alamos Scientific Laboratory, worked on chemical problems for the semiconductor group, and the authors are grateful to him for a number of valuable ideas and for considerable assistance.
[13] J. H. Scaff and H. C. Theuerer *Preparation of High Back Voltage Germanium Rectifiers* NDRC 14–155, Oct. 24, 1945. See reference 6, Chap. 12.
[14] The surface treatment is described in reference 6, p. 369.

FIG. 2. Microphotograph of a cutaway model of a transistor.

higher voltage so that a d.c. current of a few milli-amperes flows out through the collector point and through the load circuit. It is found that the current in the collector circuit is sensitive to and may be controlled by changes of current from the emitter. In fact, when the emitter current is varied by changing the emitter voltage, keeping the collector voltage constant, the change in collector current may be larger than the change in emitter current. As the emitter is biased in the direction of easy flow, a small a.c. voltage, and thus a small power input, is sufficient to vary the emitter current. The collector is biased in the direction of high resistance and may be matched to a high resistance load. The a.c. voltage and power in the load circuit are much larger than those in the input. An over-all power gain of a factor of 100 (or 20 db) can be obtained in favorable cases.

Terminal characteristics of an experimental transistor[15] are illustrated in Fig. 3, which shows how the current-voltage characteristic of the collector is changed by the current flowing from the emitter. Transistor characteristics, and the way they change with separation between the points, with temperature, and with frequency, are discussed in Section II.

The explanation of the action of the transistor depends on the nature of the current flowing from the emitter. It is well known that in semiconductors there are two ways by which the electrons can carry electricity which differ in the signs of the effective mobile charges.[16] The negative carriers are excess

[15] The transistor whose characteristics are given in Fig. 3 is one of an experimented pilot production which is under the general direction of J. A. Morton.
[16] See, for example, A. H. Wilson, *Semi-Conductors and Metals* (Cambridge University Press, London, 1939) or F. Seitz, *The Modern Theory of Solids* (McGraw-Hill Book Company, Inc., New York, 1940), Sec. 68.

electrons which are free to move and are denoted by the term conduction electrons or simply electrons. They have energies in the conduction band of the crystal. The positive carriers are missing or defect "electrons" and are denoted by the term "holes." They represent unoccupied energy states in the uppermost normally filled band of the crystal. The conductivity is called n- or p-type depending on whether the mobile charges normally in excess in the material under equilibrium conditions are electrons (negative carriers) or holes (positive carriers). The germanium used in the transistor is n-type with about 5×10^{14} conduction electrons per cc; or about one electron per 10^8 atoms. Transistor action depends on the fact that the current from the emitter is composed in large part of *holes*; that is, of carriers of opposite sign to those normally in excess in the body of the semiconductor.

The collector is biased in the reverse, or negative direction. Current flowing in the germanium toward the collector point provides an electric field which is in such a direction as to attract the holes flowing from the emitter. When the emitter and collector are placed in close proximity, a large part of the hole current from the emitter will flow to the collector and into the collector circuit. The nature of the collector contact is such as to provide a high resistance barrier to the flow of electrons from the metal to the semiconductor, but there is little impediment to the flow of holes into the contact. This theory explains how the change in collector current might be as large as but not how it can be larger than the change in emitter current. The fact that the collector current may actually change more than the emitter current is believed to result from an alteration of the space charge in the barrier layer at the collector by the hole current flowing into the junction. The increase in density of space charge and in field strength make it easier for electrons to flow out from the collector, so that there is an increase in electron current. It is better to think of the hole current from the emitter as modifying the current-voltage characteristic of the collector, rather than as simply adding to the current flowing to the collector.

In Section III we discuss the nature of the conductivity of germanium, and in Section IV the theory of the current-voltage characteristic of a germanium-point contact. In the latter section we attempt to show why the emitter current is composed of carriers of opposite sign to those normally in excess in the body of germanium. Section V is concerned with some aspects of the theory of transistor action. A complete quantitative theory is not yet available.

There is evidence that the rectifying barrier in germanium is internal and occurs at the free surface, independent of the metal contact.[9,17] The barrier contains what Schottky and Spenke[18] call an inversion region; that is, a change of conductivity type. The outermost part of the barrier next to the surface is p-type. The p-type region is very thin, of the order of 10^{-5} cm in thickness. An important question is whether there is a sufficient density of holes in this region to provide appreciable lateral conductivity along the surface. Some evidence bearing on this point is described below.

Transistor action was first discovered on a germanium surface which was subjected to an anodic oxidation treatment in a glycol borate solution after it had been ground and etched in the usual way for diodes. Much of the early work was done on surfaces which were oxidized by heating in air. In both cases the oxide is washed off and plays no direct role. Some of these surfaces were tested for surface conductivity by potential probe tests. Surface conductivities, on a unit area basis, of the order of 0.0005 to 0.002 mhos were found.[2] The value of 0.0005 represents about the lower limit of detection possible by the method used. It is inferred that the observed surface conductivity is that of the p-type layer, although there has been no direct proof of this. In later work it was found that the oxidation treatment is not essential for transistor action. Good transistors can be made with surfaces prepared in the usual way for high back-voltage rectifiers provided that the collector point is electrically formed. Such surfaces exhibit no measurable surface conductivity.

One question that may be asked is whether the holes flow from the emitter to the collector mainly in the surface layer or whether they flow through the body of the germanium. The early experiments suggested flow along the surface. W. Shockley proposed a modified arrangement in which in effect the emitter and collector are on opposite sides of a thin slab, so that the holes flow directly across through the semiconductor. Independently, J. N. Shive made, by grinding and etching, a piece of germanium in the form of a thin flat wedge.[19] Point contacts were placed directly opposite each other on the two opposite faces where the thickness of the wedge was about 0.01 cm. A third large area contact was made to the base of the wedge. When the two points were connected as emitter and collector, and the collector was electrically formed, transistor action was obtained which was comparable to that found with the original arrangement. There is no doubt that in this case the holes are flowing directly through the n-type germanium from the emitter to the collector. With two points close together on a

plane surface holes may flow either through the surface layer or through the body of the semiconductor.

Still later, at the suggestion of W. Shockley, J. R. Haynes[20] further established that holes flow into the body of the germanium. A block of germanium was made in the form of a thin slab and large area electrodes were placed at the two ends. Emitter and collector electrodes were placed at variable separations on one face of the slab. The field acting between these electrodes could be varied by passing currents along the length of the slab. The collector was biased in the reverse direction so that a small d.c. current was drawn into the collector. A signal introduced at the emitter in the form of a pulse was detected at a slightly later time in the collector circuit. From the way the time interval, of the order of a few microseconds, depends on the field, the mobility and sign of the carriers were determined. It was found that the carriers are positively charged, and that the mobility is the same as that of holes in bulk germanium (1000 cm²/volt sec.).

These experiments clarify the nature of the excess conductivity observed in the forward direction in high back-voltage germanium rectifiers which has been investigated by R. Bray, K. Lark-Horovitz, and R. N. Smith[21] and by Bray.[22] These authors attributed the excess conductivity to the strong electric field which exists in the vicinity of the point contact. Bray has made direct experimental tests to observe the relation between conductivity and field strength. We believe that the excess conductivity arises from holes injected into the germanium at the contact. Holes are introduced because of the nature of the barrier layer rather than as a direct result of the electric field. This has been demonstrated by an experiment of E. J. Ryder and W. Shockley.[23] A thin slab of germanium was cut in the form of a pie-shaped wedge and electrodes placed at the narrow and wide boundaries of the wedge. When a current is passed between the electrodes, the field strength is large at the narrow end of the wedge and small near the opposite electrode. An excess conductivity was observed when the narrow end was made positive; none when the wide end was positive. The magnitude of the current flow was the same in both cases. Holes injected at the narrow end lower the resistivity in the region which contributes most to the over-all resistance. When the current is in the opposite direction, any holes injected enter in a region of low field and do not have sufficient lifetime to be drawn down to the narrow end and so do not alter the resistance very

[17] The nature of the barrier is discussed in Section IV.
[18] W. Schottky and E. Spenke, Wiss. Veroff. Siemens Werken, **18**, 225 (1939).
[19] John N. Shive, Phys. Rev. **75**, 689 (1949).

[20] J. R. Haynes and W. Shockley, Phys. Rev. **75**, 691 (1949).
[21] R. Bray, K. Lark-Horovitz, and R. N. Smith, Phys. Rev. **72**, 530 (1947).
[22] R. Bray, Phys. Rev. **74**, 1218 (1948).
[23] E. J. Ryder and W. Shockley, Phys. Rev. **75**, 310 (1949).

much. With some surface treatments, the excess conductivity resulting from hole injection may be enhanced by a surface conductivity as discussed above.

The experimental procedure used during the present investigation is of interest. Current voltage characteristics of a given point contact were displayed on a d.c. oscilloscope.[24] The change or modulation of this characteristic produced by a signal impressed on a neighboring electrode or point contact could be easily observed. Since the input impedance of the scope was 10 megohms, and the gain of the amplifiers such that the lower limit of sensitivity was of the order of a millivolt, the oscilloscope was also used as a very high impedance voltmeter for probe measurements. Means were included for matching the potential to be measured with an adjustable d.c. potential the value of which could be read on a meter. A micromanipulator designed by W. L. Bond was used to adjust the positions of the contact points.

II. SOME TRANSISTOR CHARACTERISTICS

The static characteristics of the transistor are completely specified by four variables which may be taken as the emitter and collector currents, I_e and I_c, and the corresponding voltages, V_e and V_c. As shown in the schematic diagram of Fig. 1, the conventional directions for current flow are taken as positive into the germanium and the terminal voltages are relative to the base electrode. Thus I_e and V_e are normally positive, I_c and V_c negative.

There is a functional relation between the four variables such that if two are specified the other two are determined. Any pair may be taken as the independent variables. As the transistor is essentially a current-operated device, it is more in accord with the physics involved to choose the currents rather than the voltages. All fields in the semiconductor outside of the space charge regions immediately surrounding the point contacts are determined by the currents, and it is the current flowing from the emitter which controls the current-voltage characteristic of the collector. The voltages are single-valued functions of the currents, but, because of inherent feedback, the currents may be double-valued functions of the voltages. In reference 1, the characteristics of an experimental transistor were shown by giving the constant voltage contours on a plot in which the independent variables I_e and I_c are plotted along the coordinate axes.

In the following we give further characteristics, and show in a general way how they depend on the spacing between the points, on the temperature, and on the frequency. The data were taken mainly on experimental set-ups on a laboratory bench, and are not to be taken as necessarily typical of the characteristics of finished units. They do indicate in a general way the type of results which can be obtained. Characteristics of units made in pilot production have been given elsewhere.[5]

The data plotted in reference 1 were taken on a transistor made with phosphor bronze points on a

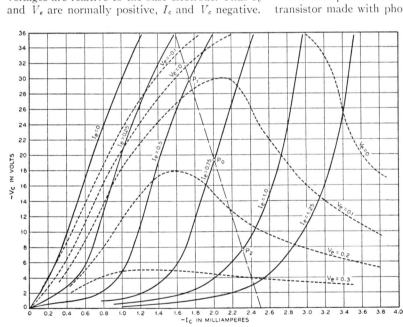

FIG. 3. Characteristics of an experimental transistor (see reference 15). The conventional directions for current and voltage are as in Fig. 1.

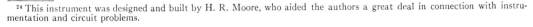

[24] This instrument was designed and built by H. R. Moore, who aided the authors a great deal in connection with instrumentation and circuit problems.

133

surface which was oxidized and on which potential probe tests gave evidence for considerable surface conductivity. The collector resistance is small in units prepared in this way. In Fig. 3 are shown the characteristics of a unit[15] in which the surface was prepared in a different manner. The surface was ground and etched in the usual way,[14] but was not subjected to the oxidation treatment. Phosphor bronze contact points made from 5-mil wire were used. The collector was electrically formed by passing large currents in the reverse direction. This reduced the resistance of the collector in the reverse direction, improving the transistor action. However, it remained considerably higher than that of the collector on the oxidized surface.

While there are many ways of plotting the data, we have chosen to give the collector voltage, V_c, as a function of the collector current, I_c, with the emitter current, I_e, taken as a parameter. This plot shows in a direct manner the influence of the emitter current on the current-voltage characteristic of the collector. The curve corresponding to $I_e = 0$ is just the normal reverse characteristic of the collector as a rectifier. The other curves show how the characteristic shifts to the right, corresponding to larger collector currents, with increase in emitter current. It may be noted that the change in collector current for fixed collector voltage is larger than the change in emitter current. The current amplification factor, α, defined by

$$\alpha = -(\partial I_c/\partial I_e)_{V_c=\text{const.}}, \qquad (\text{II}.1)$$

is between 2 and 3 throughout most of the plot.

The dotted lines on Fig. 3 correspond to constant values of the emitter voltage, V_e. By interpolating between the contours, all four variables corresponding to a given operating point may be obtained. The V_e contours reach a maximum for I_e about 0.7 ma and have a negative slope beyond. To the left of the maximum, V_e increases with I_e as one follows along a line corresponding to $V_c=\text{const.}$ To the right, V_e decreases as I_e increases, corresponding to a negative input admittance. For given values of V_e and V_c, there are two possible operating points. Thus for $V_e = 0.1$ and $V_c = -20$ one may have $I_e = 0.3$ ma, $I_c = -1.1$ ma or $I_e = 1.0$, $I_c = -2.7$.

The negative resistance and instability result from the effect of the collector current on the emitter current.[1] The collector current lowers the potential of the surface in the vicinity of the emitter and increases the effective bias on the emitter by an equivalent amount. This potential drop is $R_F I_c$, where R_F is a feed-back resistance which may depend on the currents flowing. The effective bias on the emitter is then $V_e - R_F I_c$, and we may write

$$I_e = f(V_e - R_F I_c), \qquad (\text{II}.2)$$

where the function gives the forward characteristic

of the emitter point. In some cases R_F is approximately constant over the operating range; in other cases R_F decreases with increasing I_e as the conductivity of the germanium in the vicinity of the points increases with forward current. Increase of I_e by a change of V_e increases the magnitude of I_c, which by the feedback still further increases I_e. Instability may result. Some consequences will be discussed further in connection with the a.c. characteristics.

Also shown in Fig. 3 is a load line corresponding to a battery voltage of -100 in the output circuit and a load, R_L, of 40,000 ohms, the equation of the line being

$$V_c = -100 - 40 \times 10^3 I_c. \qquad (\text{II}.3)$$

The load is an approximate match to the collector resistance, as given by the slope of the solid lines. If operated between the points P_1 and P_2, the output voltage is 8.0 volts r.m.s. and the output current is 0.20 ma. The corresponding values at the input are 0.07 and 0.18, so that the over-all power gain is

$$\text{Gain} \sim 8 \times 0.20/(0.07 \times 0.18) \sim 125, \quad (\text{II}.4)$$

which is about 21 db. This is the available gain for a generator with an impedance of 400 ohms, which is an approximate match for the input impedance.

We turn next to the equations for the a.c. characteristics. For small deviations from an operating point, we may write

$$\Delta V_e = R_{11}\Delta I_e + R_{12}\Delta I_c, \qquad (\text{II}.5)$$

$$\Delta V_c = R_{21}\Delta I_e + R_{22}\Delta I_c, \qquad (\text{II}.6)$$

in which we have taken the currents as the independent variables and the directions of currents and voltages as in Fig. 1. The differentials represent small changes from the operating point, and may be small a.c. signals. The coefficients are defined by:

$$R_{11} = (\partial V_e/\partial I_e)_{I_c=\text{const.}}, \qquad (\text{II}.7)$$

$$R_{12} = (\partial V_e/\partial I_c)_{I_e=\text{const.}}, \qquad (\text{II}.8)$$

$$R_{21} = (\partial V_c/\partial I_e)_{I_c=\text{const.}}, \qquad (\text{II}.9)$$

$$R_{22} = (\partial V_c/\partial I_c)_{I_e=\text{const.}}, \qquad (\text{II}.10)$$

These coefficients are all positive and have the dimensions of resistances. They are functions of the d.c. bias currents, I_e and I_c which define the operating point. For $I_e = 0.75$ ma and $I_c = -2$ ma the coefficients of the unit of Fig. 3 have the following approximate values:

$$\begin{aligned} R_{11} &= 800 \text{ ohms,} \\ R_{12} &= 300, \\ R_{21} &= 100,000, \\ R_{22} &= 40,000. \end{aligned} \qquad (\text{II}.11)$$

Equation (II.5) gives the emitter characteristic. The coefficient R_{11} is the input resistance for a fixed collector current (open circuit for a.c.). To a close approximation, R_{11} is independent of I_c, and is just the forward resistance of the emitter point when a current I_e is flowing. The coefficient R_{12} is the feedback or base resistance, and is equal to R_F as defined by Eq. (II.2) in case R_F is a constant. Both R_{11} and R_{12} are of the order of a few hundred ohms, R_{12} usually being smaller than R_{11}.

Equation (II.6) depends mainly on the collector and on the flow of holes from the emitter to the collector. The ratio R_{21}/R_{22} is just the current amplification factor α as defined by Eq. (II.1). Thus we may write

$$\Delta V_c = R_{22}(\alpha \Delta I_e + \Delta I_c). \qquad (II.12)$$

The coefficient R_{22} is the collector resistance for fixed emitter current (open circuit for a.c.), and is the order of 10,000–50,000 ohms. Except in the range of large I_e and small I_c, the value of R_{22} is relatively independent of I_e. The factor α generally is small when I_c is small compared with I_e, and increases with I_c, approaching a constant value the order of 1 to 4 when I_c is several times I_e.

The a.c. power gain with the circuit of Fig. 1 depends on the operating point (the d.c. bias currents) and on the load impedance. The positive feedback represented by R_{12} increases the available gain, and it is possible to get very large power gains by operating near a point of instability. In giving the gain under such conditions, the impedance of the input generator should be specified. Alternatively, one can give the gain which would exist with no feedback. The maximum available gain neglecting feedback, obtained when the load R_L is equal to the collector resistance R_{22} and the impedance of the generator is equal to the emitter resistance, R_{11}, is:

$$\text{Gain} = \alpha^2 R_{22}/4R_{11}, \qquad (II.13)$$

which is the ratio of the collector to the emitter resistance multiplied by $\frac{1}{4}$ the square of the current amplification factor. This gives the a.c. power delivered to the load divided by the a.c. power fed into the transistor. Substituting the values listed above (Eqs. (II.11)) for the unit whose characteristics are shown in Fig. 3 gives a gain of about 80 times (or 19 db) for the operating point P_0. This is to be compared with the gain of 21 db estimated above for operation between P_1 and P_2. The difference of 2 db represents the increase in gain by feedback, which was omitted in Eq. (II.13).

Equations (II.5) and (II.6) may be solved to express the currents as functions of the voltages, giving

$$\Delta I_e = Y_{11}\Delta V_e + Y_{12}\Delta V_c, \qquad (II.14)$$

$$\Delta I_c = Y_{21}\Delta V_e + Y_{22}\Delta V_c, \qquad (II.15)$$

where

$$Y_{11} = R_{22}/D, \qquad Y_{12} = -R_{12}/D, \qquad (II.16)$$
$$Y_{21} = -R_{21}/D, \qquad Y_{22} = R_{11}/D,$$

and D is the determinant of the coefficients

$$D = R_{11}R_{22} - R_{12}R_{21}. \qquad (II.17)$$

The admittances, Y_{11} and Y_{22} are negative if D is negative, and the transistor is then unstable if the terminals are short-circuited for a.c. currents. Stability can be attained if there is sufficient impedance in the input and output circuits exterior to the transistor. Feedback and instability are increased by adding resistance in series with the base electrode. Further discussion of this subject would carry us too far into circuit theory and applications. From the standpoint of transistor design, it is desirable to keep the feed-back resistance, R_{12}, as small as possible.

Variation with Spacing

One of the important parameters affecting the operation of the transistor is the spacing between the point electrodes. Measurements to investigate this effect have been made on a number of germanium surfaces. Tests were made with use of a micromanipulator to adjust the positions of the points. The germanium was generally in the form of a slab from 0.05 to 0.20 cm thick, the lower surface of which was rhodium plated to form a low resistant contact, and the upper plane surface ground and etched, or otherwise treated to give a surface suitable for transistor action. The collector point was usually kept fixed, since it is more critical, and the emitter point moved. Measurements were made with formed collector points. Most of the data have been obtained on surfaces oxidized as described below.

As expected, the emitter current has less and less influence on the collector as the separation, s,[25] is increased. This is shown by a decrease in R_{21}, or α, with s. The effect of the collector current on the emitter, represented by the feed-back resistance R_{12}, also decreases with increase in s. The other coefficients, R_{11} and R_{22}, are but little influenced by spacing. Figures 4, 5, and 6 illustrate the variation of R_{12} and α with the separation. Shown are results for two different collector points A and B on different parts of the same germanium surface.[26] In making the measurements, the bias currents were kept fixed as the spacing was varied. For collector A, $I_e = 1.0$ ma and $I_c = 3.8$ ma; for collector B, $I_e = 1.0$ ma and $I_c = 4.0$ ma. The values of R_{11} and R_{22} were about 300 and 10,000, respectively, in both cases.

---

[25] Measured between centers of the contact areas.
[26] The surface had been oxidized, and potential probe measurements (reference 2) gave evidence for considerable surface conductivity.

Figure 5 shows that α decreases approximately exponentially with s for separations from 0.005 cm to 0.030 cm, the rate of decrease being about the same in all cases. Extrapolating down to $s=0$ indicates that a further increase of only about 25 percent in α could be obtained by decreasing the spacing below 0.005 cm.

Figure 6 shows that the decrease of α with distance is dependent on the germanium sample used. Curve 1 is similar to the results in Fig. 5. Curve 2 is for a germanium slice with the same surface treatment but from a different melt.

Figure 4 shows the corresponding results for R_{12}. There is an approximate inverse relationship between R_{12} and s.

Another way to illustrate the decreased influence of the emitter on the collector with increase in spacing is to plot the collector characteristic for fixed emitter current at different spacings. Figure 7 is such a plot for a different surface which was ground flat, etched, and then oxidized at 500°C in moist air for 1 hour. The resultant oxide film was washed off.[27] The emitter current, I_e, was kept constant at 1.0 ma.

Data taken on the same surface have been plotted in other ways. As the spacing increases, more emitter current is required to produce the same change in collector current. The fraction of the emitter current which is effective at the collector decreases with spacing. It is of interest to keep V_c and I_c fixed by varying I_e as s is changed and to plot the values of I_e so obtained as a function of s. Such a plot is shown in Fig. 8. The collector voltage, V_c, is fixed at -15 volts. Curves are shown for $I_c = -3$, -4, -6, and -8 ma. We may define a geometrical factor, g, as the ratio of I_e extrapolated to zero spacing to the value of g at the separation s:

$$g(s) = (I_r(0)/I_e(s))_{V_c,\ I_c=\text{const}.} \quad (11.18)$$

It is to be expected that $g(s)$ will depend on I_c, as it is the collector current which provides the field which draws the holes into the collector. For the same reason, it is expected that $g(s)$ will be relatively independent of V_c. This was indeed found to be true in this particular case and values $V_c = -5$, -10, and -15 were used in Fig. 9 which gives a plot of g versus s for several values of I_c. The dotted lines give the extrapolation to $s=0$. As expected, g increases with I_c for a fixed s. The different curves can be brought into approximate agreement by taking $s/I_c^{\frac{1}{3}}$ as the independent variable, and this is done in Fig. 10. As will be discussed in Section V, such a relation is to be expected if g depends on the transit time for the holes.

---

[27] Potential probe measurements on the same surface, given in reference 2, gave evidence of surface conductivity.

Variation with Temperature

Only a limited amount of data has been obtained on the variation of transistor characteristics with temperature.[5] It is known that the reverse characteristic of the germanium diode varies rapidly with temperature, particularly in the case of units with high reverse resistance. In the transistor, the collector is electrically formed in such a way as to have relatively low reverse resistance, and its characteristic is much less dependent on temperature. Both R_{22} and R_{11} decrease with increase in T, R_{22} usually decreasing more rapidly than R_{11}. The feedback resistance, R_{12}, is relatively independent of temperature. The current multiplication factor, α, increases with temperature, but the change is not extremely rapid. Figure 11 gives a plot of α versus T for two experimental units. The d.c. bias currents are kept fixed as the temperature is varied. The over-all change in α from $-50°C$ to $+50°C$ is only about 50 percent. The increase in α with T results in an increase in power gain with temperature. This may be nullified by a decrease in the ratio R_{22}/R_{11}, so that the over-all gain at fixed bias current may have a negative temperature coefficient.

Variation with Frequency

Equations (II.5) and (II.6) may be used to describe the a.c. characteristics at high frequencies if the coefficients are replaced by general impedances. Thus if we use the small letters i_e, v_e, i_c, v_c, to denote the amplitude and phase of small a.c. signals about a given operating point, we may write

$$v_e = Z_{11}i_e + Z_{12}i_c, \quad (2.19)$$

$$v_c = Z_{21}i_e + Z_{22}i_c. \quad (2.20)$$

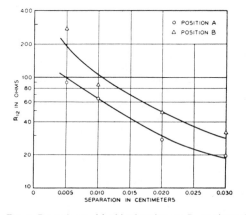

FIG. 4. Dependence of feed-back resistance R_{12} on electrode separation for two different parts A and B, of the same germanium surface. The surface had been oxidized by heating in air.

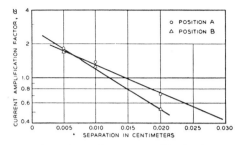

FIG. 5. Dependence of current amplification factor α on electrode separation for formed and unformed collector points. Positions A and B as in Fig. 4.

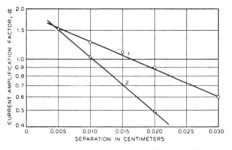

FIG. 6. Dependence of current amplification factor α on electrode separation for germanium surfaces from two different melts, 1 and 2.

Measurements of A. J. Rack and others[28] show that the over-all power gain drops off between 1 and 10 mc/sec. and few units have positive gain above 10 mc/sec. The measurements showed further that the frequency variation is confined almost entirely to Z_{21} or α. The other coefficients, Z_{11}, Z_{12}, and Z_{22} are real and independent of frequency, at least up to 10 mc/sec. Figure 12 gives a plot of α versus frequency for an experimental unit. Associated with the drop in amplitude is a phase shift which varies approximately linearly with the frequency. A phase shift in Z_{21} of 90° occurs at a frequency of about 4 mc/sec., corresponding to a delay of about 5×10^{-8} second. Estimates of transit time for the holes to flow from the emitter to the collector, to be made in Section V, are of the same order. These results suggest that the frequency limitation is associated with transit time rather than electrode capacities. Because of the difference in transit times for holes following different paths, there is a drop in amplitude rather than simply a phase shift.

III. ELECTRICAL CONDUCTIVITY OF GERMANIUM

Germanium, like carbon and silicon, is an element of the fourth group of the periodic table, with the same crystal structure as diamond. Each germanium atom has four near neighbors in a tetrahedral configuration with which it forms covalent bonds. The specific gravity is about 5.35 and the melting point 958°C.

The conductivity at room temperature may be either n- or p-type, depending on the nature and concentration of impurities. Scaff, Theuerer, and Schumacher[29] have shown that group III elements with one less valence electron, give p-type conductivity, group V elements, with one more valence electron, give n-type conductivity. This applies to both germanium and silicon. There is evidence that

both acceptor (p-type) and donor (n-type) impurities are substitutional.[30]

A schematic energy level diagram[31] which shows the allowed energy levels for the valence electrons in a semiconductor like germanium is given in Fig. 13. There is a continuous band of levels, the filled band, normally occupied by the electrons in the valence bonds, an energy gap, E_G, in which there are no levels of the ideal crystal, and then another continuous band of levels, the conduction band, normally unoccupied. There are just sufficient levels in the filled band to accommodate the four valence electrons per atom. The acceptor impurity levels, which lie just above the filled band, and the donor levels, just below the conduction band, correspond to electrons localized about the impurity atoms. Donors are normally neutral, but become positively charged by excitation of an electron to the conduction band, an energy E_D being acquired. Acceptors, normally neutral, are negatively ionized by excitation of an electron from the filled band, an energy E_A being required. Both E_D and E_A are so small in germanium that practically all donors and acceptors are ionized at room temperature. If only donors are present, the concentration of conduction electrons is equal to the concentration of donors, and the conductivity is n-type. If only acceptors are present, the concentration of missing electrons, or holes, is equal to that of the acceptors, and the conductivity is p-type.

It is possible to have both donor and acceptor type impurities present in the same crystal. In this case, electrons will be transferred from the donor levels to the lower lying acceptor levels. The conductivity type then depends on which is in excess, and the concentration of carriers is equal to the difference between the concentrations of donors and acceptors. It is probable that impurities of both types are present in high back-voltage germanium. The relative numbers in solid solution can be

[28] Unpublished data.
[29] J. H. Scaff, H. C. Theuerer, and E. E. Schumacher, "P-type and N-type Silicon and the Formation of the Photovoltaic Barrier in Silicon" (in publication).

[30] G. L. Pearson and J. Bardeen, Phys. Rev. 75, 865 (1949).
[31] See, for example, reference 6, Chap. 3.

changed by heat treatment, thus changing the conductivity and even the conductivity type.[13]

The conductivity depends on the concentrations and mobilities of the carriers: Let μ_e and μ_h be the mobilities, expressed in cm²/volt sec., and n_e and n_h the concentrations (number/cm³) of the electrons and holes, respectively. If both types of carriers are present, the conductivity, in mhos/cm, is

$$\sigma = n_e e \mu_e + n_h e \mu_h, \qquad (III.1)$$

where e is the electronic charge in coulombs (1.6×10^{-19}).

Except for relatively high concentrations ($\sim 10^{17}$/cm³ or larger), or at low temperatures, the mobilities in germanium are determined mainly by lattice scattering and so should be approximately the same in different samples. Approximate values, estimated from Hall and resistivity data obtained at Purdue University[32] and at the Bell Telephone Laboratories[33] are:

$$\mu_h = 5 \times 10^6 T^{-\frac{3}{2}}, \qquad (III.2)$$

$$\mu_e = 6.5 \times 10^6 T^{-\frac{3}{2}} (\text{cm}^2/\text{volt sec.}), \qquad (III.3)$$

in which T is the absolute temperature. There is a considerable spread among the different measurements, possibly arising from inhomogeneity of the samples. The temperature variation is as indicated by theory. These equations give $\mu_h \sim 1000$ and $\mu_e \sim 1300$ cm²/volt sec. at room temperature. The resistivity of the germanium used varies from about 1 to 30 ohm cm, corresponding to values of n_e between 1.5×10^{14} and 4×10^{15}/cm³.

At high temperatures, electrons may be thermally excited from the filled band to the conduction band, an energy E_G being required. Both the excited electron and the hole left behind contribute to the conductivity. The conductivities of all samples approach the same limiting values, regardless of impurity concentration, given by an equation of the form

$$\sigma = \sigma_\infty \exp(-E_G/2kT), \qquad (III.4)$$

where k is Boltzmann's constant. For germanium, σ_∞ is about 3.3×10^4 mhos/cm and E_G about 0.75 ev.

[32] Lark-Horovitz, Middleton, Miller, and Walerstein, Phys. Rev. 69, 258 (1946).

[33] Hall and resistivity data at the Bell Laboratories were obtained by G. L. Pearson on samples furnished by J. H. Scaff and H. C. Theuerer. *Added in proof:* Recent hall measurements of G. L. Pearson on single crystals of *n*- and *p*-type germanium give values of 2700 and 1600 cm²/volt sec. for electrons and holes, respectively, at room temperature. The latter value has been confirmed by J. R. Haynes by measurements of the drift velocity of holes injected into *n*-type germanium. These values are higher, particularly for electrons, than earlier measurements on polycrystalline samples. Use of the new values will modify some of the numerical estimates made herein, but the orders of magnitude, which are all that are significant, will not be affected. W. Ringer and H. Welker, Zeits. f. Naturforschung 1, 20 (1948), give a value of 2000 cm²/volt sec. for high resistivity *n*-type germanium.

The exponential factor comes from the variation of concentration with temperature. Statistical theory[34] indicates that n_e and n_h depend on temperature as

$$n_e = C_e T^{\frac{3}{2}} \exp(-\varphi_e/kT), \qquad (III.5a)$$

$$n_h = C_h T^{\frac{3}{2}} \exp(-\varphi_h/kT), \qquad (III.5b)$$

where φ_e is the energy difference between the bottom of the conduction band and the Fermi level and φ_h is the difference between the Fermi level and the top of the filled band. The position of the Fermi level depends on the impurity concentration and on temperature. The theory gives

$$C_e \sim C_h \sim 2(2\pi m k/h^2)^{\frac{3}{2}} \sim 5 \times 10^{15}, \qquad (III.6)$$

where m is an effective mass for the electrons (or holes) and h is Plank's constant. The numerical value is obtained by using the ordinary electron mass for m.

The product $n_e n_h$ is independent of the position of the Fermi level, and thus of impurity concentration, and depends only on the temperature. From Eqs. (III.5a) and (III.5b)

$$n_e n_h = C_e C_h T^3 \exp(-E_G/kT). \qquad (III.7)$$

In the intrinsic range, we may set $n_e = n_h = n$, and find, using (III.1), (III.2), and (III.3), an expression of the form (III.4) for σ with

$$\sigma_\infty = 11.5 \times 10^6 e (C_e C_h)^{\frac{1}{2}}. \qquad (III.8)$$

Using the theoretical value (III.6) for $(C_e C_h)^{\frac{1}{2}}$, we find

$$\sigma_\infty = 0.9 \times 10^4 \text{ mhos/cm},$$

as compared with the empirical value of 3.3×10^4, a difference of a factor of 3.6. A similar discrepancy for silicon appears to be related to a variation of E_G with temperature. With an empirical value of

$$C_e C_h = 25 \times 10^{30} \times 3.6^2 \sim 3 \times 10^{32}, \qquad (III.9)$$

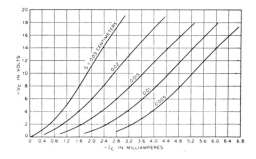

FIG. 7. Collector characteristic V_c vs. I_c for fixed I_e but variable distance of separation.

[34] See R. H. Fowler, *Statistical Mechanics* (Cambridge University Press, London, 1936), second edition.

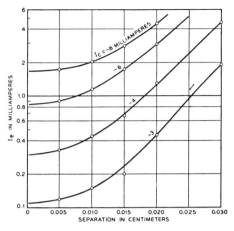

FIG. 8. Emitter current I_e vs. separation for fixed I_c and V_e.

Eq. (III.7) gives

$$n_e n_h \sim 10^{27}/\text{cm}^6 \qquad (\text{III.10})$$

when evaluated for room temperature. Thus for $n_e \sim 10^{15}/\text{cm}^3$, n_h is the order of 10^{12}. The equilibrium concentration of holes is small.

Below the intrinsic temperature range, n_e is approximately constant and n_h varies as

$$n_h = (C_e C_h T^3/n_e) \exp(-E_G/kT). \qquad (\text{III.11})$$

IV. THEORY OF THE DIODE CHARACTERISTIC

Characteristics of metal point-germanium contacts include high forward currents, as large as 5 to 10 ma at 1 volt, small reverse currents, corresponding to resistances as high as one megohm or more at reverse voltages up to 30 volts, and the ability to withstand large voltages in the reverse direction without breakdown. A considerable variation of rectifier characteristics is found with changes in preparation and impurity content of the germanium, surface treatment, electrical power or forming treatment of the contacts, and other factors.

A typical d.c. characteristic of a germanium rectifier[35] is illustrated in Fig. 14. The forward voltages are indicated on an expanded scale. The forward current at 1-volt bias is about 3.5 ma and the differential resistance is about 200 ohms. The reverse current at 30 volts is about 0.02 ma and the differential resistance about 5×10^5 ohms. The ratio of the forward to the reverse current at 1-volt bias is about 500. At a reverse voltage of about 160 the differential resistance drops to zero, and with further increase in current the voltage across the unit drops. The nature of this negative resistance portion of the curve is not completely understood, but it is believed to be associated with thermal

[35] From unpublished data of K. M. Olsen.

effects. Successive points along the curve correspond to increasingly higher temperatures of the contact. The peak value of the reverse voltage varies among different units. Values of more than 100 volts are not difficult to obtain.

Theories of rectification as developed by Mott,[36] Schottky,[37] and others[38] have not been successful in explaining the high back-voltage characteristic in a quantitative way. In the following we give an an outline of the theory and its application to germanium. It is believed that the high forward currents can now be explained in terms of a flow of holes. The type of barrier which gives a flow of carriers of conductivity type opposite to that of the base material is discussed. It is possible that a hole current also plays an important role in the reverse direction.

The Space-Charge Layer

According to the Mott-Schottky theory, rectification results from a potential barrier at the

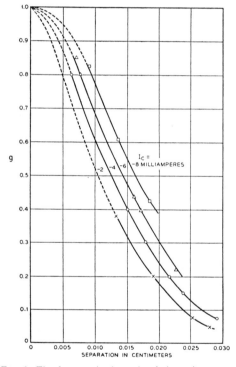

FIG. 9. The factor g is the ratio of the emitter current extrapolated to $s = 0$ to that at electrode separation s required to give the same collector current, I_c and voltage, V_c. Plot shows variation of g with s for different I_c. The factor is independent of V_c over the range plotted.

[36] N. F. Mott, Proc. Roy. Soc. 171A, 27 (1939).
[37] W. Schottky, Zeits. f. Physik 113, 367 (1939); Physik. Zeits. 41, 570 (1940); Zeits. f. Physik. 118, 539 (1942). Also see reference 18.
[38] See reference 6, Chap. 4.

contact which impedes the flow of electrons between the metal and the semiconductor. A schematic energy level diagram of the barrier region, drawn roughly to scale for germanium, is given in Fig. 15. There is a rise in the electrostatic potential energy of an electron at the surface relative to the interior which results from a space-charge layer in the semiconductor next to the metal contact. The space charge arises from positively ionized donors, that is from the same impurity centers which give the conduction electrons in the body of the semiconductor. In the interior, the space charge of the donors is neutralized by the space charge of the conduction electrons, which are present in equal numbers. Electrons are drained out of the space-charge layer near the surface, leaving the immobile donor ions.

The space-charge layer may be a result of the metal-semiconductor contact, in which case the positive charge in the layer is compensated by an induced charge of opposite sign on the metal surface. Alternatively, the charge in the layer may be compensated by a surface charge density of electrons trapped in surface states on the semiconductor.[9] It is believed, for reasons to be discussed below, that the latter situation applies to high back-voltage germanium, and that a space-charge layer exists at the free surface, independent of the metal contact. The height of the conduction band above the Fermi level at the surface, φ_s, is then determined by the distribution in energy of the surface states.

That the space-charge layer which gives the rectifying barrier in germanium arises from surface states, is indicated by the following:

(1) Characteristics of germanium-point contacts do not depend on the work function of the metal, as would be expected if the space-charge layer were determined by the metal contact.

(2) There is little difference in contact potential between different samples of germanium with varying impurity concentration. Benzer[39] found less than 0.1-volt difference between samples ranging from n-type with 2.6×10^{18} carriers/cm³ to p-type with 6.4×10^{18} carriers/cm³. This is much less than the difference of the order of the energy gap, 0.75 volt, which would exist if there were no surface effects.

(3) Benzer[40] has observed the characteristics of contacts formed from two crystals of germanium. He finds that in both directions the characteristic is similar to the reverse characteristic of one of the crystals in contact with a highly conducting metal-like germanium crystal.

(4) One of the authors[11] has observed a change in contact potential with light similar to that expected for a barrier layer at the free surface.

Prior to Benzer's experiments, Meyerhof[8] had shown that the contact potential difference measured between different metals and silicon showed

[39] S. Benzer, *Progress Report*, Contract No. W-36-039-SC-32020, Purdue University (Sept. 1–Nov. 30, 1946).
[40] S. Benzer, Phys. Rev. **71**, 141 (1947).

little correlation with rectification, and that the contact potential difference between n- and p-type silicon surfaces was small. There is thus evidence that the barrier layers in both germanium and silicon are internal and occur at the free surface.[41]

In the development of the mathematical theory of the space-charge layer at a rectifier contact, Schottky and Spenke[18] point out the possibility of a change in conductivity type between the surface and the interior if the potential rise is sufficiently large. The conductivity is p-type if the Fermi level is closest to the filled band, n-type if it is closest to the conduction band. In the illustration (Fig. 15), the potential rise is so large that the filled band is raised up to a position close to the Fermi level at the surface. This situation is believed to apply to germanium. There is then a thin layer near the surface whose conductivity is p-type, superimposed on the n-type conductivity in the interior. Schottky

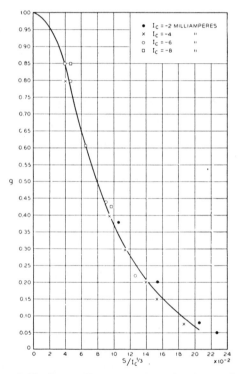

FIG. 10. The factor g (Fig. 9) plotted as a function of $s/I_c^{\frac{1}{3}}$, with s in cm and I_c in amp.

[41] Further evidence that the barrier is internal comes from some unpublished experiments of J. R. Haynes with the transistor. Using a fixed collector point, and keeping a fixed distance between emitter and collector, he varied the material used for the emitter point. He used semiconductors as well as metals for the emitter point. While the impedance of the emitter point varied, it was found that equivalent emitter currents give changes in current at the collector of the same order for all materials used. It is believed that in all cases a large part of the forward current consists of holes.

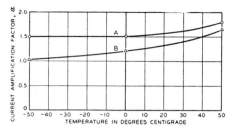

FIG. 11. Current amplification factor α vs. temperature for two experimental units A and B.

and Spenke call the layer of opposite conductivity type an inversion region.

Referring to Eqs. (III.5a) and (III.5b) for the concentrations, it can be seen that since C_e and C_h are of the same order of magnitude, the conductivity type depends on whether φ_e is larger or smaller than φ_h. The conductivity is n-type when

$$\varphi_e < \tfrac{1}{2}E_G, \quad \varphi_h > \tfrac{1}{2}E_G, \tag{IV.1}$$

and is p-type when the reverse situation applies. The maximum resistivity occurs at the position where the conductivity type changes and

$$\varphi_e \sim \varphi_h \sim \tfrac{1}{2}E_G. \tag{IV.2}$$

The change from n- to p-type will occur if

$$\varphi_s > \tfrac{1}{2}E_C, \tag{IV.3}$$

or if the over-all potential rise, φ_b, is greater than

$$\tfrac{1}{2}E_G - \varphi_{e0}, \tag{IV.4}$$

where φ_{e0} is the value of φ_e in the interior. Since for high back-voltage germanium, $E_G \sim 0.75$ ev and $\varphi_{e0} \sim 0.25$ ev, a rise of more than 0.12 ev is sufficient for a change of conductivity type to occur. A rise of 0.50 ev will bring the filled band close to the Fermi level at the surface.

Schottky[37] relates the thickness of the space-

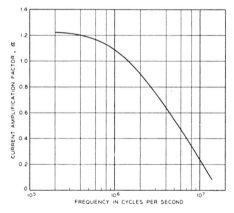

FIG. 12. Current amplification factor α vs. frequency.

charge layer with a potential rise as follows. Let ρ be the average change density, assumed constant for simplicity, in the space-charge layer. In the interior ρ is compensated by the space charge of the conduction electrons. Thus, if n_0 is the normal concentration of electrons,[42]

$$\rho = en_0. \tag{IV.5}$$

Integration of the space-charge equations gives a parabolic variation of potential with distance, and the potential rise, φ_b, is given in terms of the thickness of the space-charge layer, l, by the equation

$$\varphi_b = 2\pi e\rho l^2/\kappa = 2\pi e^2 n_0 l^2/\kappa. \tag{IV.6}$$

For

$$\varphi_b = \varphi_s - \varphi_{e0} \sim 0.5 \text{ ev} \sim 8 \times 10^{-13} \text{ erg},$$

and

$$n_0 \sim 10^{15}/\text{cm}^3,$$

the barrier thickness, l, is about 10^{-4} cm. The dielectric constant, κ, is about 18 in germanium.

When a voltage V_a is applied to a rectifying contact, there will be a drop V_b across the space-charge layer itself and an additional drop, IR_s, in the body of the germanium which results from the spreading resistance, R_s, so that

$$V_a = V_b + IR_s. \tag{IV.7}$$

The potential energy drop, $-eV_b$, is superimposed on the drop φ_b which exists under equilibrium conditions. For this case Eq. (IV.6) becomes

$$\varphi_b - eV_b = 2\pi e^2 n_0 l^2/\kappa. \tag{IV.8}$$

The potential V_b is positive in the forward direction, negative in the reverse. A reverse voltage increases the thickness of the layer, a forward voltage decreases the thickness of the layer. The barrier disappears when $eV_b = \varphi_b$, and the current is then limited entirely by the spreading resistance in the body of the semiconductor.

The electrostatic field at the contact is

$$F = 4\pi en_0 l/\kappa = (8\pi n_0(\varphi_b - eV_b)/\kappa)^{\frac{1}{2}}. \tag{IV.9}$$

For $n_0 \sim 10^{15}$, $l \sim 10^{-4}$, and $\kappa \sim 18$, the field F is about 30 e.s.u. or 10,000 volts/cm. The field increases the current flow in much the way the current from a thermionic emitter is enhanced by an external field.

Previous theories of rectification have been based on the flow of only one type of carrier, i.e., electrons in an n-type or holes in a p-type semiconductor. If the barrier layer has an inversion region, it is necessary to consider the flow of both types of carriers. Some of the hitherto puzzling features of the germanium diode characteristic can be explained by the hole current. While a complete

---

[42] The space charge of the holes in the inversion region of the barrier layer is neglected for simplicity.

theoretical treatment has not been carried out, we will give an outline of the factors involved and then give separate discussions for the reverse and forward directions.

The current of holes may be expected to be important if the concentration of holes at the semiconductor boundary of the space-charge layer is as large as the concentration of electrons at the metal-semiconductor interface. In equilibrium, with no current flow, the former is just the hole concentration in the interior, n_{h0}, which is given by

$$n_{h0} = C_h T^{\frac{3}{2}} \exp(-\varphi_{h0}/kT), \qquad (IV.10)$$

where φ_{h0} is the energy difference between the Fermi level in the interior and the top of the filled band. The concentration of electrons at the interface is given by:

$$n_{em} = C_e T^{\frac{3}{2}} \exp(-\varphi_s/kT). \qquad (IV.11)$$

Since C_h and C_e are of the same order, n_{h0} will be larger than n_{em} if φ_s is larger than φ_{h0}. This latter condition is met if the hole concentration at the metal interface is larger than the electron concentration in the interior. The concentrations will, of course, be modified when a current is flowing, but the criterion just given is nevertheless a useful guide. The criterion applies to an inversion barrier layer regardless of whether it is formed by the metal contact or is of the surface states type. In the latter case, as discussed in the Introduction, a lateral flow of holes along the surface layer into the contact may contribute to the current.

Two general theories have been developed for the current in a rectifying junction which apply in different limiting cases. The diffusion theory applies if the current is limited by the resistance of the space-charge layer. This will be the case if the mean free path is small compared with the thickness of the layer, or, more exactly, small compared with the distance required for the potential energy to drop kT below the value at the contact. The diode theory applies if the current is limited by the thermionic emission current over the barrier. In germanium, the mean free path (10^{-5} cm) is of the same order as the barrier thickness. Analysis shows, however, that scattering in the barrier is unimportant and that it is the diode theory which should be used.[43]

Reverse Current

Different parts of the d.c. current-voltage characteristics require separate discussion. We deal first with the reverse direction. The applied voltages are assumed large compared with kT/e (0.025 volt at room temperature), but small compared with the peak reverse voltage, so that thermal effects are

[43] Reference 6, Chapter 4.

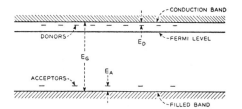

FIG. 13. Schematic energy level diagram for germanium showing filled and conduction bands and donor and acceptor levels.

unimportant. Electrons flow from the metal point contact to the germanium, and holes flow in the opposite direction.

Benzer[44] has made a study of the variation of the reverse characteristic with temperature. He divides the current into three components whose relative magnitudes vary among different crystals and which vary in different ways with temperature. These are:

(1) A saturation current which rises very rapidly with applied voltage, approaching a constant value at a fraction of a volt.
(2) A component which increases linearly with the voltage.
(3) A component which increases more rapidly than linearly with the voltage.

The first two increase rapidly with increasing temperature, while the third component is more or less independent of ambient temperature. It is the saturation current, and perhaps also the linear component, which are to be identified with the theoretical diode current.

The third component is the largest in units with low reverse resistance. It is probable that in these units the barrier is not uniform. The largest part of the current, composed of electrons, flows through patches in which the height of the barrier is small.

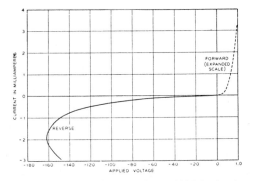

FIG. 14. Current-voltage characteristic of high back-voltage germanium rectifier. Note that the voltage scale in the forward direction has been expanded by a factor of 20.

[44] S. Benzer, *Temperature Dependence of High Voltage Germanium Rectifier D.C. Characteristics*, NDRC 14–579, Purdue University, October 31, 1945. See reference 6, p. 376.

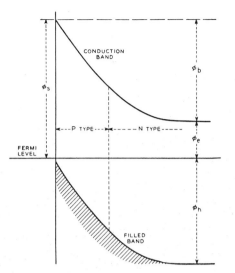

FIG. 15. Schematic energy level diagram of barrier layer at germanium surface showing inversion layer of p-type conductivity.

The electrically formed collector in the transistor may have a barrier of this sort.

Benzer finds that the saturation current predominates in units with high reverse resistance, and that this component varies with temperature as

$$I_s = -I_0 e^{\epsilon/kT}, \qquad (IV.12)$$

with ϵ nearly 0.7 ev. The negative sign indicates a reverse current. According to the diode theory,[43] one would expect it to vary as

$$I_s = -BT^2 e^{\epsilon/kT}. \qquad (IV.13)$$

Since ϵ is large, the observed current can be fitted just about as well with the factor T^2 as without. The value of ϵ obtained using (IV. 13) is about 0.6 ev. The saturation current[43] at room temperature varies from 10^{-7} to 10^{-6} amp., which corresponds to values of B in the range of 0.01 to 0.1 amp./deg.[2].

The theoretical value of B is 120 times the contact area, A_c. Taking $A_c \sim 10^{-6}$ cm^2 as a typical value for the area of a point contact gives $B \sim 10^{-4}$ amp./deg.^{2} which is only about 1/100 to 1/1000 of the observed. It is difficult to reconcile the magnitude of the observed current with the large temperature coefficient, and it is possible that an important part of the total flow is a current of holes into the contact. Such a current particularly is to be expected on surfaces which exhibit an appreciable surface conductivity.

Neglecting surface effects for the moment, an estimate of the saturation hole current might be obtained as follows. The number of holes entering the space-charge region per second is[45]

$$n_{hb} v_a A_c / 4,$$

where n_{hb} is the hole concentration at the semiconductor boundary of the space-charge layer and v_a is an average thermal velocity ($\sim 10^7$ cm/sec.). The hole current, I_h, is obtained by multiplying by the electronic charge, giving

$$I_h = -n_{hb} e v_a A_c / 4, \qquad (IV.14)$$

If we set n_{hb} equal to the equilibrium value for the interior, say 10^{12}/cm^3, we get a current $I_h \sim 4 \times 10^{-7}$ amp., which is of the observed order of magnitude of the saturation current at room temperature. With this interpretation, the temperature variation of I_s is attributed to that of n_h, which, according to Eq. (III.11) varies as $\exp(-E_G/kT)$. The observed value of ϵ is indeed almost equal to the energy gap.

The difficulty with this picture is to see how n_{hb} can be as large as n_{h0} when a current is flowing. Holes must move toward the contact area primarily by diffusion, and the hole current will be limited by a diffusion gradient. The saturation current depends on how rapidly holes are generated, and reasonable estimates based on the mean lifetime, τ, yield currents which are several orders of magnitude too small. A diffusion velocity, v_D, of the order

$$v_D \sim (D/\tau)^{\frac{1}{2}}, \qquad (IV.15)$$

replaces $v_a/4$ is Eq. (IV.14). Setting $D \sim 25$ cm^2/sec. and $\tau \sim 10^{-6}$ sec. gives $v_D \sim 5 \times 10^3$, which would give a current much smaller than the observed. What is needed, then, is some other mechanism which will help maintain the equilibrium concentration near the barrier. Surface effects may be important in this regard.

Forward Current

The forward characteristic is much less dependent on such factors as surface treatment than the reverse. In the range from 0 to 0.4 volt in the forward direction, the current can be fitted quite closely by a semi-empirical expression[46] of the form:

$$I = I_0 (e^{\beta V_b} - 1), \qquad (IV.16)$$

where V_b is the drop across the barrier resulting from the applied voltage, as defined by Eq. (IV.7). Equation (IV.16) is of the general form to be expected from theory, but the measured value of β is generally less than the theoretical value e/kT (40 volts^{-1} at room temperature). Observed values of β may be as low as 10, and in other units are nearly as high as the theoretical value of 40. The factor I_0 also varies among different units and is of

---

[45] See, for example, E. H. Kennard, *Kinetic Theory of Gases* (McGraw-Hill Book Company, Inc., New York, 1938), p. 63.
[46] Reference 6, p. 377.

the order 10^{-7} to 10^{-6} ampere. While both experiment and theory indicate that the forward current at large forward voltages is largely composed of holes, the composition of the current at very small forward voltages is uncertain. Small areas of low φ_s, unimportant at large forward voltages, may give most of the current at very small voltages. Currents flowing in these areas will consist largely of electrons.

Above about 0.5 volt in the forward direction, most of the drop occurs across the spreading resistance, R_s, rather than across the barrier. The theoretical expression for R_s for a circular contact of diameter d on the surface of a block of uniform resistivity ρ is:

$$R_s = \rho/2d. \qquad (IV.17)$$

Taking as typical values for a point contact on high back-voltage germanium, $\rho = 10$ ohm cm and $d = 0.0025$ cm, we obtain $R_s = 2000$ ohms, which is the order of ten times the observed.

As discussed in the Introduction, Bray and others[21, 22] have attempted to account for this discrepancy by assuming that the resistivity decreases with increasing field, and Bray has made tests to observe such an effect. The authors have investigated the nature of the forward current by making potential probe measurements in the vicinity of a point contact.[2] These measurements indicate that there may be two components involved in the excess conductivity. Some surfaces, prepared by oxidation at high temperatures, give evidence for excess conductivity in the vicinity of the point in the reverse as well as in the forward direction. This ohmic component has been attributed to a thin p-type layer on the surface. All surfaces investigated exhibit an excess conductivity in the forward direction which increases with increasing forward current. This second component is attributed to an increase in the concentration of carriers, holes and electrons, in the vicinity of the point with increase in forward current. Holes flow from the point into the germanium and their space charge is compensated by electrons.

The ohmic component is small, if it exists at all, on surfaces treated in the normal way for high back-voltage rectifiers (i.e., ground and etched). The nature of the second component on such surfaces has been shown by more recent work of Shockley, Haynes,[20] and Ryder[23] who have investigated the flow of holes under the influence of electric fields. These measurements prove that the forward current consists at least in large part of holes flowing into the germanium from the contact.

It is of interest to consider the way the concentrations of holes and electrons vary in the vicinity of the point. An exact calculation, including the effect of recombination, leads to a non-linear differential equation which must be solved by numerical methods. A simple solution can be obtained, however, if it is assumed that all of the forward current consist of holes and if recombination is neglected.

The electron current then vanishes everywhere, and the electric field is such as to produce a conduction current of electrons which just cancels the current from diffusion, giving

$$n_e F = -(kT/e) \, \text{grad} \, n_e. \qquad (IV.18)$$

This equation may be integrated to give the relation between the electrostatic potential, V, and n_e,

$$V = (kT/e) \log(n_e/n_{e0}). \qquad (IV.19)$$

The constant of integration has been chosen so that $V = 0$ when n_e is equal to the normal electron concentration n_{e0}. The equation may be solved for n_e to give:

$$n_e = n_{e0} \exp(eV/kT). \qquad (IV.20)$$

If trapping is neglected, electrical neutrality requires that

$$n_e = n_h + n_{e0}. \qquad (IV.21)$$

Using this relation, and taking n_{e0} a constant, we can express the field F in terms of n_h

$$F = -(kT/e(n_h + n_{e0})) \, \text{grad} \, n_h. \qquad (IV.22)$$

The hole current density, i_h, is the sum of a conduction current resulting from the field F and a diffusion current:

$$i_h = n_h e \mu_h F - kT \mu_h \, \text{grad} \, n_h. \qquad (IV.23)$$

Using Eq. (IV.22), we may write this in the form

$$i_h = -kT \mu_h((2n_h + n_{e0})/(n_h + n_{e0})) \, \text{grad} \, n_h. \qquad (IV.24)$$

The current density can be written

$$i_h = -\text{grad} \, \psi, \qquad (IV.25)$$

where

$$\psi = kT \mu_h(2n_h - n_{e0} \log((n_h + n_{e0})/n_{e0})). \qquad (IV.26)$$

Since i_h satisfies a conservation equation,

$$\text{div} \, i_h = 0, \qquad (IV.27)$$

ψ satisfies Laplace's equation.

If surface effects are neglected and it is assumed that holes flow radially in all directions from the point contact, ψ may be expressed simply in terms of the total hole current, I_h, flowing from the contact:

$$\psi = -I_h/2\pi r. \qquad (IV.28)$$

Using (IV.26), we may obtain the variation of n_h with r. We are interested in the limiting case in which n_h is large compared with the normal electron concentration, n_{e0}. The logarithmic term in (IV.26) can then be neglected, and we have

$$n_h = I_h/4\pi r \mu_h kT. \qquad (IV.29)$$

For example, if $I_h = 10^{-3}$ amp., $\mu_h = 10^3$ cm²/volt sec., and $kT/e = 0.025$ volt, we get, approximately,

$$n_h = 2 \times 10^{13}/r. \tag{IV.30}$$

For $r \sim 0.0005$ cm, the approximate radius of a point contact,

$$n_h \sim 4 \times 10^{16}/\text{cm}^3, \tag{IV.31}$$

which is about 40 times the normal electron concentration in high back-voltage germanium. Thus the assumption that n_h is large compared with n_{e0} is valid, and remains valid up to a distance of the order of 0.005 cm, the approximate distance the points are separated in the transistor.

To the same approximation, the field is

$$F = kT/er, \tag{IV.32}$$

independent of the magnitude of I_h.

The voltage drop outside of the space-charge region can be obtained by setting n_e in (IV.19) equal to the value at the semiconductor boundary of the space-charge layer. This result holds generally, and does not depend on the particular geometry we have assumed. It depends only on the assumption that the electron current i_e is everywhere zero. Actually i_h will decrease and i_e increase by recombination, and there will be an additional spreading resistance for the electron current.

If it is assumed that the concentration of holes at the metal-semiconductor interface is independent of applied voltage and that the resistive drop in the barrier layer itself is negligible, that part of the applied voltage which appears across the barrier layer itself is:

$$V_b = (kT/e) \log(n_{hb}/n_{h0}), \tag{IV.33}$$

where n_{hb} is the hole concentration at the semiconductor boundary of the space-charge layer and n_{h0} is the normal concentration. For $n_{hb} \sim 5 \times 10^{16}$ and $n_{h0} \sim 10^{12}$, V_b is about 0.30 volt.

The increased conductivity caused by hole emission accounts not only for the large forward currents, but also for the relatively small dependence of spreading resistance on contact area. At a small distance from the contact, the concentrations and voltages are independent of contact area. The voltage drop within this small distance is a small part of the total and does not vary rapidly with current.

We have assumed that the electron current, I_e, at the contact is negligible compared with the hole current, I_h. An estimate of the electron current can be obtained as follows. From the diode theory,

$$I_e = (en_{eb}v_a A_c/4) \exp(-(\varphi_b - eV_b)/kT), \tag{IV.34}$$

since the electron concentration at the semiconductor boundary of the space-charge layer is n_{eb} and the height of the barrier with the voltage applied is $\varphi_b - eV_b$. For simplicity we assume that both n_{eb}

and n_{hb} are large compared with n_{e0} so that we may replace n_{eb} by n_{hb} without appreciable error. The latter can be obtained from the value of ψ at the contact:

$$\psi = I_h/4a. \tag{IV.35}$$

Expressing ψ in terms of n_{hb}, we find

$$n_{hb} = I_h/8kT\mu_h a. \tag{IV.36}$$

Using (IV.33) for V_b, and (III.5b) for n_{h0} we find after some reduction,

$$I_e = I_h^2/I_{crit}, \tag{IV.37}$$

where

$$I_{crit} = \frac{256 C_h (kT\mu_h)^2 T^{\frac{3}{2}}}{\pi e v_a} \exp(-\varphi_{hm}/kT). \tag{IV.38}$$

The energy difference φ_{hm} is the difference between the Fermi level and the filled band at the metal-semiconductor interface. Evaluated for germanium at room temperature, (IV.38) gives

$$I_{crit} = 0.07 \exp(-\varphi_{hm}/kT) \text{ amp.} \tag{IV.39}$$

which is a fairly large current if φ_{hm} is not too large compared with kT. If I_h is small compared with I_{crit}, the electron current will be negligible.

V. THEORETICAL CONSIDERATIONS ON TRANSISTOR ACTION

In this section we discuss some of the problems connected with transistor action, such as:

(1) fields produced by the collector current,
(2) transit times for the holes to flow from emitter to collector,
(3) current multiplication in collector,
(4) feed-back resistance.

We do no more than estimate orders of magnitude. An exact calculation, taking into account the change of conductivity introduced by the emitter current, loss of holes by recombination, and effect of surface conductivity is difficult and is not attempted.

To estimate the field produced by the collector, we assume that the collector current is composed mainly of conduction electrons, and that the electrons flow radially away from the collector. This assumption should be most nearly valid when the collector current is large compared with the emitter current. The field at a distance r from the collector is,

$$F = \rho I_c/2\pi r^2. \tag{V.1}$$

For example, if, $\rho = 10$ ohm cm, $I_c = 0.001$ amp., and $r = 0.005$ cm, F is about 100 volts/cm.

The drift velocity of a hole in the field F is $\mu_h F$. The transit time is

$$T = \int \frac{dr}{\mu_h F} = \frac{2\pi}{\mu_h \rho I_c} \int_0^s r^2 dr, \tag{V.2}$$

where s is the separation between the emitter and collector. Integration gives,

$$T = 2\pi s^3 / 3\mu_{h0} I_c. \qquad (V.3)$$

For $s = 0.005$ cm, $\mu_h = 1000$ cm²/volt sec., $\rho = 10$ ohm cm, and $I_c = 0.001$ amp. T is about 0.25×10^{-7} sec. This is of the order of magnitude of the transit times estimated from the phase shift in α or Z_{21}.

The hole current, I_h, is attenuated by recombination in going from the emitter to the collector. If τ is the average lifetime of a hole, I_h will be decreased by a factor, $e^{-T/\tau}$. In Section II it was found that the geometrical factor, g, which gives the influence of separation on the interaction between emitter and collector depends on the variable $s/I_c^{\frac{1}{3}}$. This suggests that the transit time is the most important factor in determining g. An estimate[47] of τ, obtained from the data of Fig. 10, is 2×10^{-7} sec.

Because of the effect of holes in increasing the conductivity of the germanium in the vicinity of the emitter and collector, it can be expected that the field, the lifetime, and the geometrical factor will depend on the emitter current. The effective value of ρ to be used in Eqs. (V.1) and (V.2) will decrease with increase in emitter current. This effect is apparently not serious with the surface used in obtaining the data for Figs. 8 to 10.

Next to be considered is the effect of the space charge of the holes on the barrier layer of the collector. An estimate of the hole concentration can be obtained as follows. The field in the barrier layer is of the order of 10^4 volts/cm. Multiplying by the mobility gives a drift velocity, v_d of 10^7 cm/sec., which is approximately thermal velocity.[48] The hole current is

$$I_h = n_h e v_d A_c, \qquad (V.4)$$

where A_c is the area of the collector contact, and n_h the concentration of holes in the barrier. Solving for the latter, we get

$$n_h = I_h / e v_d A_c. \qquad (V.5)$$

For $I_h = 0.001$ amp., $v_d = 10^7$ cm/sec. and $A_c = 10^{-6}$ cm², n_h is about 0.6×10^{15}, which is of the same order as the concentration of donors. Thus the hole current can be expected to alter the space charge in the barrier by a significant amount, and correspondingly alter the flow of electrons from the collector. It is believed that current multiplication

(values of $\alpha > 1$) can be accounted for along these lines.

As discussed in Section II, there is an influence of collector current on emitter current of the nature of a positive feedback. The collector current lowers the potential of the surface in the vicinity of the emitter by an amount

$$V = \rho I_c / 2\pi s. \qquad (V.6)$$

The feed-back resistance R_F as used in Eq. (II.2) is

$$R_F = \rho / 2\pi s. \qquad (V.7)$$

For $\rho = 10$ ohm cm and $s = 0.005$ cm, the value of R_F is about 300 ohms, which is of the observed order of magnitude. It may be expected that R_F will decrease as ρ decreases with increase in emitter current.

The calculations made in this section confirm the general picture which has been given of the way the transistor operates.

VI. CONCLUSIONS

Our discussion has been confined to the transistor in which two point contacts are placed in close proximity on one face of a germanium block. It is apparent that the principles can be applied to other geometrical designs and to other semiconductors. Some preliminary work has shown that transistor action can be obtained with silicon and undoubtedly other semiconductors can be used.

Since the initial discovery, many groups in the Bell Laboratories have contributed to the progress that has been made. This work includes investigation of the physical phenomena involved and the properties of the materials used, transistor design, and measurements of characteristics and circuit applications. A number of transistors have been made for experimental use in a pilot production. Obviously no attempt has been made to describe all of this work, some of which has been reported on in other publications.[5, 19, 20, 23]

In a device as new as the transistor, various problems remain to be solved. A reduction in noise and an increase in the frequency limit are desirable. While much progress has been made toward making units with reproducible characteristics, further improvement in this regard is also desirable.

It is apparent from reading this article that we have received a large amount of aid and assistance from other members of the Laboratories staff, for which we are grateful. We particularly wish to acknowledge our debt to Ralph Bown, Director of Research, who has given us a great deal of encouragement and aid from the inception of the work and to William Shockley, who has made numerous suggestions which have aided in clarifying the phenomena involved.

---

[47] Obtained by plotting $\log g$ *versus* S^3/I_c. This plot is not a straight line, but has an upward curvature corresponding to an increase in τ with separation. The value given is a rough average, corresponding to S^3/I_c the order of 10^{-3} cm³/amp.
[48] One may expect that the mobility will depend on field strength when the drift velocity is as large as or is larger than thermal velocity. Since ours is a borderline case, the calculation using the low field mobility should be correct at least as to order of magnitude.

Part II

ENERGY CONTROL WITH FEEDBACK

Editor's Comments
on Papers 14 Through 18

The common domestic thermostat has already been mentioned in the Introduction. It is probably the simplest form of energy control mechanism known to most people. A small amount of energy in the system under control is fed back as information to the source to increase or decrease appropriately the output, thus providing automatic instead of manual control. Search for the origin of the simple bimetallic form of this thermostat has not been too rewarding, but it is believed that announcement of its existence was first communicated to a meeting of the Royal Society of London in 1830 by the Scottish physician and author of popular works on science, Andrew Ure (1778–1857). A very brief account of his presentation was printed in the *Philosophical Magazine* for July–December, 1831, and is reproduced here as Paper 14.

The most famous early example of a feedback energy control device is without doubt the steam engine governor of James Watt (1736–1819) described briefly in the Introduction. This invention was probably made around 1788, though Watt at that time did not publish an account of it in any technical journal. References to it can be found in his correspondence with Matthew Boulton around that period. An early description by Watt appears in the Appendix of the work "The Articles Steam and Steam Engines, Written for the Encyclopedia Britannica by the late John Robison, with Notes and Additions by James Watt" (Edinburgh, 1818, p. 154f). We reproduce this brief extract as Paper 15. We also include a reproduction of the earliest known drawing of the governor (1788), reproduced by permission from Plate LXXX in the book *James Watt and the Steam Engine*, by H.W. Dickinson and R. Jenkins (Oxford-Clarendon Press, 1927). Further description of the mode of operation of the Watt governor can be found in practically all books on automatic control, e.g., *The Elements of Feedback Control*, by A..M. Hardie (Oxford University Press, London, 1964), p. 10f. Reference is also made to pages 220 and the following from the Dickinson and Jenkins book.

The first general analytical treatment of feedback energy control was provided by J. C. Maxwell (1831–1879) in his article "On Governors" (1868), which we reproduce as Paper 16. Maxwell placed special emphasis on the use of attachments on governors to prevent too large swings in the energy transfer rate and so assure maximum stability of operation of the system being controlled. He analyzes the behavior of such attachments through the study of the roots of the solutions of the differential equations for their motion. Most of the devices considered by Maxwell operate on the basis of the centrifugal force action like that used in the Watt governor. All authorities on automatic control consider this a fundamental article that strongly influenced subsequent work.

The use of the word "servomechanism" to describe the devices through which automatic feedback energy control is carried out in all its manifold applications has been stressed in the Introduction. The theory of automatic control may in a certain sense be termed the theory of servomechanisms. Perhaps the first to provide a complete analytical description of the most significant types of such mechanisms was H. L. Hazen (b. 1901) of the Massachusetts Institute of Technology. We reproduce in full his article "Theory of Servomechanisms" (1934) as Paper 17. Hazen calls attention to the many possible uses of servomechanisms and includes a useful bibliography. He analyzes in detail the operation of the relay-type servomechanism as well as the definite-correction type. He also considers the two principal defects in the action of such mechanisms, namely oscillation and lag. The reader should not be misled by the apparent lack of emphasis on the basic function of these devices as controllers of energy. This idea is basic throughout. Harold Hazen has had a distinguished career as an electrical engineer and as an administrator in graduate education.

We conclude our introductory commentary on feedback energy control in nonliving systems with consideration of a paper by A. Porter (b. 1910) "Basic Principles of Automatic Control Systems" (1948). We reproduce this as Paper 18. Professor Porter's paper has been helpful in the standardization of terminology, which has threatened to become unwieldy in the face of the ever-increasing applications of control devices. Naturally there has been a very extensive development of the whole subject of control during the last quarter of a century, but Porter's paper still remains fundamental.

14

Reprinted from *Philos. Mag.* **10**, ser. 2:295 (1831)

ON THE THERMOSTAT OR HEAT GOVERNOR, A SELF-ACTING PHYSICAL APPARATUS FOR REGULATING TEMPERATURE

Andrew Ure

A paper was read, "On the Thermostat or Heat Governor, a self-acting physical Apparatus for regulating Temperature;" constructed by Andrew Ure, M.D., F.R.S.

The principle of the instrument here described is the unequal expansion of different metals by heat. A bar of zinc, alloyed with four or five per cent. of copper, and one of tin, about an inch in breadth, one quarter of an inch thick, and two feet long, is firmly and closely riveted along its face to the face of a similar bar of steel of about one third in thickness. The product of the rigidity and strength should be nearly the same, so that the texture of each may pretty equally resist the strains of flexure. Twelve such compound bars are united in pairs by a hinge joint at each of their ends; having the zinc or alloy bars fronting one another. At ordinary temperatures these bars will be parallel, and nearly in contact; but when heated, they bend outwards, receding from each other at their middle parts, like two bows tied together at their ends. When a more considerable expansion is wanted, a series of such bars is laid one over the other. The movement thus resulting is applied by the author in various ways to regulate the opening of dampers, letting in either cold air or cold water, or closing the draught of a fireplace, as the case may be. He proposes its employment to regulate the safety valves of steam boilers, as working with more certainty than the common expedients.

15

Reprinted from pp. 154–155 of *The Articles, Steam and Steam Engines, Written for the Encyclopaedia Britannica, by the late John Robison*, D. Brewster, ed., John Murray, London, Edinburgh, 1818.

[Editor's Note: In the original, material precedes this excerpt.]

A second article omitted to be described, is the method of regulating the speed of the rotative engines, a matter essential to their application to cotton-spinning, and many other manufactories.

Throttle-valve.

It is performed by admitting the steam into the cylinder more or less freely, by means of what is called a *Throttle-valve*, which is commonly a circular plate of metal A, having a spindle B fixed across its diameter.

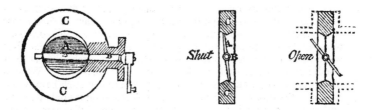

This plate is accurately fitted to an aperture in a metal ring CC, of some thickness, through the edgeway of which the spindle is fitted steam-tight, and the ring is fixed between the two flanches of the joint of the steam-pipe which is next to the cylinder. One end of the spindle, which has a square upon it, comes through the ring, and has a spanner fixed upon it, by which it can be turned in either direction.

When the valve is parallel to the outsides of the ring, it shuts the opening nearly perfectly; but when its plane lies at an angle to the ring, it admits more or less steam according to the degree it has opened; consequently the piston is acted upon with more or less force. For many purposes engines are thus regulated by hand at the pleasure of the attendant; but where a regular velocity is required, other means must be applied to open and shut it, without any attention on the part of those who have the care of it. For this purpose Mr Watt had various methods, but at last fixed upon what he calls the *Governor,* (shewn at W. Plate IV. Governor. and VII.) consisting of a perpendicular axis, turned by the engine: To a joint near the top of this axis are suspended two iron rods carrying heavy balls of metal at their lower ends, in the nature of pendulums. When this axis is put in motion by the engine, the balls recede from the perpendicular by the centrifugal force, and by means of a combination of levers fixed to their upper end, raise the end of a lever which acts upon the spanner of the throttle-valve, and shuts it more or less according to the speed of the engine, so that as the velocity augments, the valve is shut, until the speed of the engine and the opening of the valve come to a maximum and balance each other.

The application of the centrifugal *principle* was not a new invention, but had been applied by others to the regulation of water and wind-mills, and other things; but Mr Watt improved the mechanism by which it acted upon the machines, and adapted it to his engines.

[*Editor's Note:* Material has been omitted at this point.]

Reprinted with permission from H. W. Dickinson and R. Jenkins, *James Watt and the Steam Engine*, Oxford-Clarendon Press, 1927.

16

Reprinted from *R. Soc. (London) Proc.* **16**:270–283 (Mar. 5, 1868)

ON GOVERNORS

J. Clerk Maxwell

A Governor is a part of a machine by means of which the velocity of the machine is kept nearly uniform, notwithstanding variations in the driving-power or the resistance.

Most governors depend on the centrifugal force of a piece connected with a shaft of the machine. When the velocity increases, this force increases, and either increases the pressure of the piece against a surface or moves the piece, and so acts on a break or a valve.

In one class of regulators of machinery, which we may call *moderators* *, the resistance is increased by a quantity depending on the velocity. Thus in some pieces of clockwork the moderator consists of a conical pendulum revolving within a circular case. When the velocity increases, the ball of the pendulum presses against the inside of the case, and the friction checks the increase of velocity.

In Watt's governor for steam-engines the arms open outwards, and so contract the aperture of the steam-valve.

In a water-break invented by Professor J. Thomson, when the velocity is increased, water is centrifugally pumped up, and overflows with a great velocity, and the work is spent in lifting and communicating this velocity to the water.

In all these contrivances an increase of driving-power produces an increase of velocity, though a much smaller increase than would be produced without the moderator.

But if the part acted on by centrifugal force, instead of acting directly on the machine, sets in motion a contrivance which continually increases the resistance as long as the velocity is above its normal value, and reverses its action when the velocity is below that value, the governor will bring the velocity to the same normal value whatever variation (within the working limits of the machine) be made in the driving-power or the resistance.

I propose at present, without entering into any details of mechanism, to direct the attention of engineers and mathematicians to the dynamical theory of such governors.

It will be seen that the motion of a machine with its governor consists in general of a uniform motion, combined with a disturbance which may be expressed as the sum of several component motions. These components may be of four different kinds :—

1. The disturbance may continually increase.
2. It may continually diminish.
3. It may be an oscillation of continually increasing amplitude.
4. It may be an oscillation of continually decreasing amplitude.

The first and third cases are evidently inconsistent with the stability of the motion; and the second and fourth alone are admissible in a good governor. This condition is mathematically equivalent to the condition that all the possible roots, and all the possible parts of the impossible roots, of a certain equation shall be negative.

I have not been able completely to determine these conditions for equa-

* See Mr. C. W. Siemens "On Uniform Rotation," Phil. Trans. 1866, p. 657.

tions of a higher degree than the third; but I hope that the subject will obtain the attention of mathematicians.

The actual motions corresponding to these impossible roots are not generally taken notice of by the inventors of such machines, who naturally confine their attention to the way in which it is *designed* to act; and this is generally expressed by the real root of the equation. If, by altering the adjustments of the machine, its governing power is continually increased, there is generally a limit at which the disturbance, instead of subsiding more rapidly, becomes an oscillating and jerking motion, increasing in violence till it reaches the limit of action of the governor. This takes place when the possible part of one of the impossible roots becomes positive. The mathematical investigation of the motion may be rendered practically useful by pointing out the remedy for these disturbances.

This has been actually done in the case of a governor constructed by Mr. Fleeming Jenkin, with adjustments, by which the regulating power of the governor could be altered. By altering these adjustments the regulation could be made more and more rapid, till at last a dancing motion of the governor, accompanied with a jerking motion of the main shaft, showed that an alteration had taken place among the impossible roots of the equation.

I shall consider three kinds of governors, corresponding to the three kinds of moderators already referred to.

In the first kind, the centrifugal piece has a constant distance from the axis of motion, but its pressure on a surface on which it rubs varies when the velocity varies. In the *moderator* this friction is itself the retarding force. In the *governor* this surface is made moveable about the axis, and the friction tends to move it; and this motion is made to act on a break to retard the machine. A constant force acts on the moveable wheel in the opposite direction to that of the friction, which takes off the break when the friction is less than a given quantity.

Mr. Jenkin's governor is on this principle. It has the advantage that the centrifugal piece does not change its position, and that its pressure is always the same function of the velocity. It has the disadvantage that the normal velocity depends in some degree on the coefficient of sliding friction between two surfaces which cannot be kept always in the same condition.

In the second kind of governor, the centrifugal piece is free to move further from the axis, but is restrained by a force the intensity of which varies with the position of the centrifugal piece in such a way that, if the velocity of rotation has the normal value, the centrifugal piece will be in equilibrium in every position. If the velocity is greater or less than the normal velocity, the centrifugal piece will fly out or fall in without any limit except the limits of motion of the piece. But a break is arranged so that it is made more or less powerful according to the distance of the centrifugal piece from the axis, and thus the oscillations of the centrifugal piece are restrained within narrow limits.

Governors have been constructed on this principle by Sir W. Thomson and by M. Foucault.   In the first, the force restraining the centrifugal piece is that of a spring acting between a point of the centrifugal piece and a fixed point at a considerable distance, and the break is a friction-break worked by the reaction of the spring on the fixed point.

In M. Foucault's arrangement, the force acting on the centrifugal piece is the weight of the balls acting downward, and an upward force produced by weights acting on a combination of levers and tending to raise the balls. The resultant vertical force on the balls is proportional to their depth below the centre of motion, which ensures a constant normal velocity. The break is :—in the first place, the variable friction between the combination of levers and the ring on the shaft on which the force is made to act ; and, in the second place, a centrifugal air-fan through which more or less air is allowed to pass, according to the position of the levers.   Both these causes tend to regulate the velocity according to the same law.

The governors designed by the Astronomer Royal on Mr. Siemens's principle for the chronograph and equatorial of Greenwich Observatory depend on nearly similar conditions.   The centrifugal piece is here a long conical pendulum, not far removed from the vertical, and it is prevented from deviating much from a fixed angle by the driving-force being rendered nearly constant by means of a differential system.   The break of the pendulum consists of a fan which dips into a liquid more or less, according to the angle of the pendulum with the vertical.   The break of the principal shaft is worked by the differential apparatus ; and the smoothness of motion of the principal shaft is ensured by connecting it with a fly-wheel.

In the third kind of governor a liquid is pumped up and thrown out over the sides of a revolving cup.   In the governor on this principle, described by Mr. C. W. Siemens, the cup is connected with its axis by a screw and a spring, in such a way that if the axis gets ahead of the cup the cup is lowered and more liquid is pumped up.   If this adjustment can be made perfect, the normal velocity of the cup will remain the same through a considerable range of driving-power.

It appears from the investigations that the oscillations in the motion must be checked by some force resisting the motion of oscillation.   This may be done in some cases by connecting the oscillating body with a body hanging in a viscous liquid, so that the oscillations cause the body to rise and fall in the liquid.

To check the variations of motion in a revolving shaft, a vessel filled with viscous liquid may be attached to the shaft.   It will have no effect on uniform rotation, but will check periodic alterations of speed.

Similar effects are produced by the viscosity of the lubricating matter in the sliding parts of the machine, and by other unavoidable resistances ; so that it is not always necessary to introduce special contrivances to check oscillations.

I shall call all such resistances, if approximately proportional to the velocity, by the name of " viscosity," whatever be their true origin.

In several contrivances a differential system of wheelwork is introduced between the machine and the governor, so that the driving-power acting on the governor is nearly constant.

I have pointed out that, under certain conditions, the sudden disturbances of the machine do not act through the differential system on the governor, or *vice versâ*. When these conditions are fulfilled, the equations of motion are not only simple, but the motion itself is not liable to disturbances depending on the mutual action of the machine and the governor.

Distinction between Moderators and Governors.

In regulators of the first kind, let P be the driving-power and R the resistance, both estimated as if applied to a given axis of the machine. Let V be the normal velocity, estimated for the same axis, and $\frac{dx}{dt}$ the actual velocity, and let M be the moment of inertia of the whole machine reduced to the given axis.

Let the governor be so arranged as to increase the resistance or diminish the driving-power by a quantity $F\left(\frac{dx}{dt} - V\right)$, then the equation of motion will be

$$\frac{d}{dt}\left(M\frac{dx}{dt}\right) = P - R - F\left(\frac{dx}{dt} - V\right). \quad \cdots \quad (1)$$

When the machine has obtained its final rate the first term vanishes, and

$$\frac{dx}{dt} = V + \frac{P - R}{F}. \quad \cdots \cdots \cdots \cdots \quad (2)$$

Hence, if P is increased or R diminished, the velocity will be permanently increased. Regulators of this kind, as Mr. Siemens [*] has observed, should be called moderators rather than governors.

In the second kind of regulator, the force $F\left(\frac{dx}{dt} - V\right)$, instead of being applied directly to the machine, is applied to an independent moving piece, B, which continually increases the resistance, or diminishes the driving-power, by a quantity depending on the whole motion of B.

If y represents the whole motion of B, the equation of motion of B is

$$\frac{d}{dt}\left(B\frac{dy}{dt}\right) = F\left(\frac{dx}{dt} - V\right), \quad \cdots \cdots \cdots \quad (3)$$

and that of M

$$\frac{d}{dt}\left(M\frac{dx}{dt}\right) = P - R - F\left(\frac{dx}{dt} - V\right) + Gy, \quad \cdots \cdots \quad (4)$$

where G is the resistance applied by B when B moves through one unit of space.

[*] "On Uniform Rotation," Phil. Trans. 1866, p. 657.

We can integrate the first of these equations at once, and we find

$$B \frac{dy}{dt} = F(x - Vt) ; \qquad \dots \dots \dots \quad (5)$$

so that if the governor B has come to rest $x = Vt$, and not only is the velocity of the machine equal to the normal velocity, but the position of the machine is the same as if no disturbance of the driving-power or resistance had taken place.

Jenkin's Governor.—In a governor of this kind, invented by Mr. Fleeming Jenkin, and used in electrical experiments, a centrifugal piece revolves on the principal axis, and is kept always at a constant angle by an appendage which slides on the edge of a loose wheel, B, which works on the same axis. The pressure on the edge of this wheel would be proportional to the square of the velocity; but a constant portion of this pressure is taken off by a spring which acts on the centrifugal piece. The force acting on B to turn it round is therefore

$$F^1 \left(\frac{dx}{dt}\right)^2 - C^1 ;$$

and if we remember that the velocity varies within very narrow limits, we may write the expression

$$F \left(\frac{dx}{dt} - V_1\right)'$$

where F is a new constant, and V_1 is the lowest limit of velocity within which the governor will act.

Since this force necessarily acts on B in the positive direction, and since it is necessary that the break should be taken off as well as put on, a weight W is applied to B, tending to turn it in the negative direction; and, for a reason to be afterwards explained, this weight is made to hang in a viscous liquid, so as to bring it to rest quickly.

The equation of motion of B may then be written

$$B \frac{d^2 y}{dt^2} = F \left(\frac{dx}{dt} - V_1\right) - Y \frac{dy}{dt} - W, \qquad \dots \dots \quad (6)$$

where Y is a coefficient depending on the viscosity of the liquid and on other resistances varying with the velocity, and W is the constant weight.

Integrating this equation with respect to *t*, we find

$$B \frac{dy}{dt} = F(x - V_1 t) - Yy - Wt. \qquad \dots \dots \dots \quad (7)$$

If B has come to rest, we have

$$x = \left(V_1 + \frac{W}{F}\right) t + \frac{Y}{F} y, \qquad \dots \dots \dots \quad (8)$$

or the position of the machine is affected by that of the governor, but the final velocity is constant, and

$$V_1 + \frac{W}{F} = V, \qquad \dots \dots \dots \dots \quad (9)$$

where V is the normal velocity.

The equation of motion of the machine itself is

$$M\frac{d^2x}{dt^2}=P-R-F\left(\frac{dx}{dt}-V_1\right)-Gy. \quad . \quad . \quad . \quad . \quad (10)$$

This must be combined with equation (7) to determine the motion of the whole apparatus. The solution is of the form

$$x=A_1e^{n_1t}+A_2e^{n_2t}+A_3e^{n_3t}+Vt, \quad . \quad . \quad . \quad . \quad . \quad (11)$$

where n_1, n_2, n_3 are the roots of the cubic equation

$$MBn^3+(MY+FB)n^2+FYn+FG=0. \quad . \quad . \quad . \quad (12)$$

If n be a pair of roots of this equation of the form $a\pm\sqrt{-1}\,b$, then the part of x corresponding to these roots will be of the form

$$e^{at}\cos(bt+\beta).$$

If a is a negative quantity, this will indicate an oscillation the amplitude of which continually decreases. If a is zero, the amplitude will remain constant, and if a is positive, the amplitude will continually increase.

One root of the equation (12) is evidently a real negative quantity. The condition that the real part of the other roots should be negative is

$$\left(\frac{F}{M}+\frac{Y}{B}\right)\frac{Y}{B}-\frac{G}{B}=\text{a positive quantity.}$$

This is the condition of stability of the motion. If it is not fulfilled there will be a dancing motion of the governor, which will increase till it is as great as the limits of motion of the governor. To ensure this stability, the value of Y must be made sufficiently great, as compared with G, by placing the weight W in a viscous liquid if the viscosity of the lubricating materials at the axle is not sufficient.

To determine the value of F, put the break out of gear, and fix the moveable wheel; then, if V and V' be the velocities when the driving-power is P and P',

$$F=\frac{P-P'}{V-V'}.$$

To determine G, let the governor act, and let y and y' be the positions of the break when the driving-power is P and P', then

$$G=\frac{P-P'}{y-y'}.$$

General Theory of Chronometric Centrifugal Pieces.

Sir W. Thomson's and M. Foucault's Governors.—Let A be the moment of inertia of a revolving apparatus, and θ the angle of revolution. The equation of motion is

$$\frac{d}{dt}\left(A\frac{d\theta}{dt}\right)=L, \quad . \quad . \quad . \quad . \quad . \quad (1)$$

where L is the moment of the applied force round the axis.

Now, let A be a function of another variable ϕ (the divergence of the centrifugal piece), and let the kinetic energy of the whole be

$$\frac{1}{2} A \overline{\frac{d\theta}{dt}}^2 + \frac{1}{2} B \overline{\frac{d\phi}{dt}}^2,$$

where B may also be a function of ϕ, if the centrifugal piece is complex.

If we also assume that P, the potential energy of the apparatus, is a function of ϕ, then the force tending to *diminish* ϕ, arising from the action of gravity, springs, &c., will be $\dfrac{dP}{d\phi}$.

The whole energy, kinetic and potential, is

$$E = \frac{1}{2} A \overline{\frac{d\theta}{dt}}^2 + \frac{1}{2} B \overline{\frac{d\phi}{dt}}^2 + P. = \int L d\theta. \quad\quad\cdots\cdots (2)$$

Differentiating with respect to t, we find

$$\frac{d\phi}{dt}\left(\frac{1}{2}\frac{dA}{d\phi}\overline{\frac{d\theta}{dt}}^2 + \frac{1}{2}\frac{dB}{d\phi}\overline{\frac{d\phi}{dt}}^2 + \frac{dP}{d\phi}\right) + A\frac{d\theta}{dt}\frac{d^2\theta}{dt^2} + B\frac{d\phi}{dt}\frac{d^2\phi}{dt^2} \left.\begin{array}{c} \\ \\ \end{array}\right\}$$
$$= L\frac{d\theta}{dt} = \frac{d\theta}{dt}\left(\frac{dA}{d\phi}\frac{d\theta}{dt}\frac{d\phi}{dt} + A\frac{d^2\theta}{dt^2}\right), \left.\begin{array}{c} \\ \end{array}\right\} \quad\cdots (3)$$

whence we have, by eliminating L,

$$\frac{d}{dt}\left(B\frac{d\phi}{dt}\right) = \frac{1}{2}\frac{dA}{d\phi}\overline{\frac{d\theta}{dt}}^2 + \frac{1}{2}\frac{dB}{d\phi}\overline{\frac{d\phi}{dt}}^2 - \frac{dP}{d\phi}. \quad\cdots\cdots (4)$$

The first two terms on the right-hand side indicate a force tending to *increase* ϕ, depending on the squares of the velocities of the main shaft and of the centrifugal piece. The force indicated by these terms may be called the centrifugal force.

If the apparatus is so arranged that

$$P = \tfrac{1}{2} A\omega^2 + \text{const.}, \quad\cdots\cdots\cdots (5)$$

where ω is a constant velocity, the equation becomes

$$\frac{d}{dt}\left(B\frac{d\phi}{dt}\right) = \frac{1}{2}\frac{dA}{d\phi}\left(\overline{\frac{d\theta}{dt}}^2 - \omega^2\right) + \frac{1}{2}\frac{dB}{d\phi}\overline{\frac{d\phi}{dt}}^2. \quad\cdots\cdots (6)$$

In this case the value of ϕ cannot remain constant unless the angular velocity is equal to ω.

A shaft with a centrifugal piece arranged on this principle has only one velocity of rotation without disturbance. If there be a small disturbance, the equations for the disturbances θ and ϕ may be written

$$A\frac{d^2\theta}{dt^2} + \frac{dA}{d\phi}\omega\frac{d\phi}{dt} = L, \quad\cdots\cdots\cdots (7)$$

$$B\frac{d^2\phi}{dt^2} - \frac{dA}{d\phi}\omega\frac{d\theta}{dt} = 0. \quad\cdots\cdots\cdots (8)$$

The period of such small disturbances is $\dfrac{dA}{d\phi}$ $(AB)^{-\frac{1}{2}}$ revolutions of the shaft. They will neither increase nor diminish if there are no other terms in the equations.

To convert this apparatus into a governor, let us assume viscosities X and Y in the motions of the main shaft and the centrifugal piece, and a resistance $G\phi$ applied to the main shaft. Putting $\dfrac{dA}{d\phi}\omega = K$, the equations become

$$A\frac{d^2\theta}{dt^2} + X\frac{d\theta}{dt} + K\frac{d\phi}{dt} + G\phi = L, \quad \cdots \cdots \cdots \quad (9)$$

$$B\frac{d^2\phi}{dt^2} + Y\frac{d\phi}{dt} - K\frac{d\theta}{dt} = 0. \quad \cdots \cdots \cdots \quad (10)$$

The condition of stability of the motion indicated by these equations is that all the possible roots, or parts of roots, of the cubic equation

$$ABn^3 + (AY + BX)n^2 + (XY + K^2)n + GK = 0 \quad \cdots \cdots \quad (11)$$

shall be negative ; and this condition is

$$\left(\frac{X}{A} + \frac{Y}{B}\right)(XY + K^2) > GK. \quad \cdots \cdots \cdots \quad (12)$$

Combination of Governors.—If the break of Thomson's governor is applied to a moveable wheel, as in Jenkin's governor, and if this wheel works a steam-valve, or a more powerful break, we have to consider the motion of three pieces. Without entering into the calculation of the general equations of motion of these pieces, we may confine ourselves to the case of small disturbances, and write the equations

$$\left.\begin{aligned}
A\frac{d^2\theta}{dt^2} + X\frac{d\theta}{dt} + K\frac{d\phi}{dt} + T\phi + J\psi &= P - R, \\
B\frac{d^2\phi}{dt^2} + Y\frac{d\phi}{dt} - K\frac{d\theta}{dt} &= 0, \\
C\frac{d^2\psi}{dt^2} + Z\frac{d\psi}{dt} - T\phi &= 0,
\end{aligned}\right\} \quad \cdots \quad (13)$$

where θ, ϕ, ψ are the angles of disturbance of the main shaft, the centrifugal arm, and the moveable wheel respectively, A, B, C their moments of inertia, X, Y, Z the viscosity of their connexions, K is what was formerly denoted by $\dfrac{dA}{d\phi}\omega$, and T and J are the powers of Thomson's and Jenkin's breaks respectively.

The resulting equation in n is of the form

$$\begin{vmatrix} An^2 + Xn & Kn + T & J \\ -K & Bn + Y & 0 \\ 0 & -T & Cn^2 + Zn \end{vmatrix} = 0, \quad \cdots \quad (14)$$

or

$$\left.\begin{aligned}
&n^5 + n^4\left(\frac{X}{A} + \frac{Y}{B} + \frac{Z}{C}\right) + n^3\left[\frac{XYZ}{ABC}\left(\frac{A}{X} + \frac{B}{Y} + \frac{C}{Z}\right) + \frac{K^2}{AB}\right] \\
&+ n^2\left(\frac{XYZ + KTC + K^2Z}{ABC}\right) + n\frac{KTZ}{ABC} + \frac{KTJ}{ABC} = 0.
\end{aligned}\right\} \quad \cdots \quad (15)$$

I have not succeeded in determining completely the conditions of stability of the motion from this equation; but I have found two necessary conditions, which are in fact the conditions of stability of the two governors taken separately. If we write the equation

$$n^5 + pn^4 + qn^3 + rn^2 + sn + t, \quad . \quad . \quad . \quad . \quad . \quad . \quad (16)$$

then, in order that the possible parts of all the roots shall be negative, it is necessary that

$$pq > r \text{ and } ps > t. \quad . \quad . \quad . \quad . \quad . \quad . \quad (17)$$

I am not able to show that these conditions are sufficient. This compound governor has been constructed and used.

On the Motion of a Liquid in a Tube revolving about a Vertical Axis.

Mr. C. W. Siemens's Liquid Governor.—Let ρ be the density of the fluid, k the section of the tube at a point whose distance from the origin measured along the tube is s, r, θ, z the coordinates of this point referred to axes fixed with respect to the tube, Q the volume of liquid which passes through any section in unit of time. Also let the following integrals, taken over the whole tube, be

$$\int \rho k r^2 ds = A, \quad \int \rho r^2 d\theta = B, \quad \int \rho \frac{1}{a} ds = C, \quad . \quad . \quad . \quad . \quad (1)$$

the lower end of the tube being in the axis of motion.

Let ϕ be the angle of position of the tube about the vertical axis, then the moment of momentum of the liquid in the tube is

$$H = A \frac{d\phi}{dt} + BQ. \quad . \quad . \quad . \quad . \quad . \quad . \quad (2)$$

The moment of momentum of the liquid thrown out of the tube in unit of time is

$$\frac{dH'}{dt} = \rho r^2 Q \frac{d\phi}{dt} + \rho \frac{r}{k} Q^2 \cos a, \quad . \quad . \quad . \quad . \quad . \quad (3)$$

where r is the radius at the orifice, k its section, and a the angle between the direction of the tube there and the direction of motion.

The energy of motion of the fluid in the tube is

$$W = \frac{1}{2} A \overline{\frac{d\phi}{dt}}^2 + BQ \frac{d\phi}{dt} + \frac{1}{2} CQ^2. \quad . \quad . \quad . \quad . \quad (4)$$

The energy of the fluid which escapes in unit of time is

$$\frac{dW'}{dt} = \rho g Q (h + z) + \frac{1}{2} \rho r^2 Q \overline{\frac{d\phi}{dt}}^2 + \rho \frac{r}{k} \cos a Q^2 \frac{d\phi}{dt} + \frac{1}{2} \frac{\rho}{k^2} Q^3. \quad . \quad . \quad (5)$$

The work done by the prime mover in turning the shaft in unit of time is

$$L \frac{d\phi}{dt} = \frac{d\phi}{dt} \left(\frac{dH}{dt} + \frac{dH'}{dt} \right). \quad . \quad . \quad . \quad . \quad . \quad (6)$$

The work spent on the liquid in unit of time is

$$\frac{dW}{dt} + \frac{dW'}{dt}.$$

Equating this to the work done, we obtain the equations of motion

$$A \frac{d^2\phi}{dt^2} + B \frac{dQ}{dt} + \rho r^2 Q \frac{d\phi}{dt} + \rho \frac{r}{k} \cos \alpha Q^2 = L, \quad \ldots \ldots \quad (7)$$

$$B \frac{d^2\phi}{dt^2} + C \frac{dQ}{dt} + \frac{1}{2} \frac{\rho}{k^2} Q^2 + \rho g(h+z) - \frac{1}{2}\rho r^2 \overline{\frac{d\phi}{dt}}^2 = 0. \quad \ldots \ldots \quad (8)$$

These equations apply to a tube of given section throughout. If the fluid is in open channels, the values of A and C will depend on the depth to which the channels are filled at each point, and that of k will depend on the depth at the overflow.

In the governor described by Mr. C. W. Siemens in the paper already referred to, the discharge is practically limited by the depth of the fluid at the brim of the cup.

The resultant force at the brim is $f = \sqrt{g^2 + \omega^4 r^2}$.

If the brim is perfectly horizontal, the overflow will be proportional to $x^{\frac{3}{2}}$ (where x is the depth at the brim), and the mean square of the velocity relative to the brim will be proportional to x, or to $Q^{\frac{2}{3}}$.

If the breadth of overflow at the surface is proportional to x^n, where x is the height above the lowest point of overflow, then Q will vary as $x^{n+\frac{3}{2}}$, and the mean square of the velocity of overflow relative to the cup as x or as

$$\frac{1}{Q^{n+\frac{3}{2}}}.$$

If $n = -\frac{1}{2}$, then the overflow and the mean square of the velocity are both proportional to x.

From the second equation we find for the mean square of velocity

$$\frac{Q^2}{k^2} = -\frac{2}{\rho}\left(B \frac{d^2\phi}{dt^2} + C \frac{dQ}{dt}\right) + r^2 \overline{\frac{d\phi}{dt}}^2 - 2g(h+r). \quad \ldots \quad (9)$$

If the velocity of rotation and of overflow is constant, this becomes

$$\frac{Q^2}{k^2} = r^2 \overline{\frac{d\phi}{dt}}^2 - 2g(h+r). \quad \ldots \ldots \ldots \quad (10)$$

From the first equation, supposing, as in Mr. Siemens's construction, that $\cos \alpha = 0$ and $B = 0$, we find

$$L = \rho r^2 Q \frac{d\phi}{dt}. \quad \ldots \ldots \ldots \ldots \quad (11)$$

In Mr. Siemens's governor there is an arrangement by which a fixed relation is established between L and z,

$$L = -Sz, \quad \ldots \ldots \ldots \ldots \quad (12)$$

whence

$$\frac{Q^2}{k^2} = r^2 \overline{\frac{d\phi}{dt}}^2 - 2gh + \frac{2g\rho}{S} r^2 Q \frac{d\phi}{dt}. \quad \ldots \ldots \quad (13)$$

If the conditions of overflow can be so arranged that the mean square of the velocity, represented by $\frac{Q^2}{k^2}$, is proportional to Q, and if the strength of the spring which determines S is also arranged so that

$$\frac{Q^2}{k^2}=\frac{2g\rho}{S}r^2\omega Q, \quad . \quad . \quad . \quad . \quad . \quad . \quad . \quad (14)$$

the equation will become, if $2gh=\omega^2r^2$,

$$0=r^2\left(\frac{d\phi}{dt}\right)^2-\omega^2\right)+\frac{2g\rho}{S}r^2Q\left(\frac{d\phi}{dt}-\omega\right), \quad . \quad . \quad . \quad . \quad (15)$$

which shows that the velocity of rotation and of overflow cannot be constant unless the velocity of rotation is ω.

The condition about the overflow is probably difficult to obtain accurately in practice; but very good results have been obtained within a considerable range of driving-power by a proper adjustment of the spring. If the rim is uniform, there will be a *maximum* velocity for a certain driving-power. This seems to be verified by the results given at p. 667 of Mr. Siemens's paper.

If the flow of the fluid were limited by a hole, there would be a *minimum* velocity instead of a maximum.

The differential equation which determines the nature of small disturbances is in general of the fourth order, but may be reduced to the third by a proper choice of the value of the mean overflow.

Theory of Differential Gearing.

In some contrivances the main shaft is connected with the governor by a wheel or system of wheels which are capable of rotation round an axis, which is itself also capable of rotation about the axis of the main shaft. These two axes may be at right angles, as in the ordinary system of differential bevel wheels; or they may be parallel, as in several contrivances adapted to clockwork.

Let ξ and η represent the angular position about each of these axes respectively, θ that of the main shaft, and ϕ that of the governor; then θ and ϕ are linear functions of ξ and η, and the motion of any point of the system can be expressed in terms either of ξ and η or of θ and ϕ.

Let the velocity of a particle whose mass is m resolved in the direction of x be

$$\frac{dx}{dt}=p_1\frac{d\xi}{dt}+q_1\frac{d\eta}{dt}, \quad . \quad . \quad . \quad . \quad . \quad (1)$$

with similar expressions for the other coordinate directions, putting suffixes 2 and 3 to denote the values of p and q for these directions. Then Lagrange's equation of motion becomes

$$\Xi\delta\xi+H\delta\eta-\Sigma m\left(\frac{d^2x}{dt^2}\delta x+\frac{d^2y}{dt^2}\delta y+\frac{d^2z}{dt^2}\delta z\right)=0, \quad . \quad . \quad (2)$$

where Ξ and H are the forces tending to increase ξ and η respectively, no force being supposed to be applied at any other point.

Now putting

$$\delta x=p_1\,\delta\xi+q_1\,\delta\eta, \quad . \quad . \quad . \quad . \quad . \quad (3)$$

and

$$\frac{d^2x}{dt^2}=p_1\frac{d^2\xi}{dt^2}+q_1\frac{d^2\eta}{dt^2}, \quad . \quad . \quad . \quad . \quad (4)$$

the equation becomes

$$\left(\Xi-\Sigma mp^2\frac{d^2\xi}{dt^2}-\Sigma mpq^2\frac{d^2\eta}{dt^2}\right)\delta\xi+\left(\text{H}-\Sigma mpq\frac{d^2\xi}{dt^2}-\Sigma mq^2\frac{d^2\eta}{dt^2}\right)\delta\eta=0;\quad(5)$$

and since $\delta\xi$ and $\delta\eta$ are independent, the coefficient of each must be zero.

If we now put

$$\Sigma(mp^2)=\text{L},\quad \Sigma(mpq)=\text{M},\quad \Sigma(mq^2)=\text{N},\quad \ldots\quad(6)$$

where

$$p^2=p_1^2+p_2^2+p_3^2,\ \ pq=p_1q_1+p_2q_2+p_3q_3,\ \text{and}\ q^2=q_1^2+q_2^2+q_3^2,$$

the equations of motion will be

$$\Xi=\text{L}\frac{d^2\xi}{dt^2}+\text{M}\frac{d^2\eta}{dt^2},\quad \ldots\ldots\ldots\quad(7)$$

$$\text{H}=\text{M}\frac{d^2\xi}{dt^2}+\text{N}\frac{d^2\eta}{dt^2}.\quad \ldots\ldots\ldots\quad(8)$$

If the apparatus is so arranged that $\text{M}=0$, then the two motions will be independent of each other; and the motions indicated by ξ and η will be about conjugate axes—that is, about axes such that the rotation round one of them does not tend to produce a force about the other.

Now let Θ be the driving-power of the shaft on the differential system, and Φ that of the differential system on the governor; then the equation of motion becomes

$$\Theta\delta\theta+\Phi\delta\phi+\left(\Xi-\text{L}\frac{d^2\xi}{dt^2}-\text{M}\frac{d^2\eta}{dt^2}\right)\delta\xi+\left(\text{H}-\text{M}\frac{d^2\xi}{dt^2}-\text{N}\frac{d^2\eta}{dt^2}\right)\delta\eta=0;\quad(9)$$

and if

$$\left.\begin{aligned}\delta\xi&=\text{P}\delta\theta+\text{Q}\delta\phi,\\ \delta\eta&=\text{R}\delta\theta+\text{S}\delta\phi,\end{aligned}\right\}\quad\ldots\ldots\ldots\quad(10)$$

and if we put

$$\left.\begin{aligned}\text{L}'&=\text{LP}^2+2\text{MPR}&&+\text{NR}^2,\\ \text{M}'&=\text{LPQ}+\text{M(PS}+\text{QR)}+\text{NRS},\\ \text{N}'&=\text{LQ}^2+2\text{MQS}&&+\text{NS}^2,\end{aligned}\right\}\quad\ldots\quad(11)$$

the equations of motion in θ and ϕ will be

$$\left.\begin{aligned}\Theta+\text{P}\,\Xi+\text{QH}&=\text{L}'\frac{d^2\theta}{dt^2}+\text{M}'\frac{d^2\phi}{dt^2},\\ \Phi+\text{R}\,\Xi+\text{SH}&=\text{M}'\frac{d^2\theta}{dt^2}+\text{N}'\frac{d^2\phi}{dt^2}.\end{aligned}\right\}\quad\ldots\quad(12)$$

If $\text{M}'=0$, then the motions in θ and ϕ will be independent of each other. If M is also 0, then we have the relation

$$\text{LPQ}+\text{NRS}=0;\quad\ldots\ldots\ldots\quad(13)$$

and if this is fulfilled, the disturbances of the motion in θ will have no effect on the motion in ϕ. The teeth of the differential system in gear with the main shaft and the governor respectively will then correspond to the centres of percussion and rotation of a simple body, and this relation will be mutual.

In such differential systems a constant force, H, sufficient to keep the governor in a proper state of efficiency, is applied to the axis η, and the motion of this axis is made to work a valve or a break on the main shaft of the machine. Ξ in this case is merely the friction about the axis of ξ. If the moments of inertia of the different parts of the system are so arranged that $M'=0$, then the disturbance produced by a blow or a jerk on the machine will act instantaneously on the valve, but will not communicate any impulse to the governor.

17

Reprinted from *Franklin Inst. J.* **218**(3):279–331 (1934)

THEORY OF SERVO–MECHANISMS.

BY

H. L. HAZEN, Sc.D.,

Massachusetts Institute of Technology.

INTRODUCTION.

In this age characterized by huge resources of mechanical and electrical power, these agencies have in many fields almost completely replaced human muscular power. In a similar way the functions of human operators are being taken over by mechanisms that automatically control the performance of machines and processes. Automatic control is often more reliable and accurate, as well as cheaper, than human control. Consequently the study of the performance of automatic control devices is of particular interest at the present time.

Automatic control devices are of two principal kinds. In the first kind the control is actuated by some quantity such as time which is more or less independent of the result of the control operation. Time-operated traffic-signal control is an example of this first kind. With this type of control the flow of traffic resulting from the action of the signal lights in no way affects the cycle of operation of the lights. In the second kind, the control is actuated by a quantity that is affected by the control operation, and for this reason it may be called a "closed-cycle" control. An example of this kind is the thermostat in which the temperature-sensitive element controls the amount of heat supplied to some object such as a process furnace or a house, while the temperature of the object actuates the thermostat. The first kind of control may be superimposed upon the second kind as in the case of a house

(Note—The Franklin Institute is not responsible for the statements and opinions advanced by contributors to the JOURNAL.)

thermostat whose setting is changed periodically by a clock. The second kind is also illustrated by certain traffic light controls in which the cycle of light operation is automatically adjusted by the flow of traffic. Another example of the second kind of control is automatic ship steering. Here the rudder operation is actuated by the angle between compass and hull, while this angle is in turn affected by the rudder operation.

The distinction between these two kinds of control is made because of important differences in the nature of the control mechanism required in the two cases. In the first kind the source of the control usually has sufficient power to operate the control mechanism directly. Thus synchronous motors are available which can cheaply and accurately effect time control for practically any device. In the second kind, however, the source of control is some form of measuring instrument which very seldom can deliver any appreciable force or power and give usefully accurate indications. The operation of the actual control element nearly always requires much greater power than that available from the measuring instrument. Consequently some intermediate device for amplifying the power of the measuring instrument, while preserving its indications, is essential. When the output element of such a device is so actuated as to make the difference between the output and input indications tend to zero the device is called a "follow up" mechanism or servo-mechanism. In what follows a "servo-mechanism" is frequently called merely a "servo."

In the thermostat or controlling pyrometer,[1, 2, 3] the measuring instrument takes various forms. One of the simplest is the bimetal element or other differential expansion device in which the indication is a mechanical displacement. Other forms are the thermocouple and the temperature-sensitive resistance in which the indication appears as an electric potential difference or current. In nearly all cases, the energy associated with these indications is very small while the actual control is effected by a valve, switch, rheostat, damper, or other device requiring considerable energy for its operation. A servo-mechanism or other amplifier bridges this gap in energy level.

[1] For numbered references see bibliography.

Considering the ship-steering example, this disparity exists between the energy magnitude associated with the measuring instrument, a compass, and that associated with driving the rudder.   In small craft with a direct wheel-to-rudder drive, the helmsman serves as a human servo-mechanism.   In larger craft, the steering engine is a servo-mechanism between wheel and rudder, while the helmsman is still a human servo-mechanism between compass and wheel.   With automatic steering the helmsman is replaced by a servo-mechanism which in turn controls the steering engine.   Alternatively, the entire compass-to-rudder drive may be considered as a single servo-mechanism.[4, 5, 6, 7, 8]

Thus it is seen that a servo-mechanism is very likely to form an essential part of any closed-cycle control system. Servo-mechanisms are also used merely to produce the indications of measuring instruments at a relatively high power level. Many recording instruments contain illustrations of this use.[1, 2, 3, 22, 23]

Applications of servo-mechanisms, in addition to those already mentioned, include the speed control of steam turbines and water wheels by governors;[9] the control of these governors by master clocks or by power-indicating instruments;[9, 10, 11, 12] the stabilization of ships by gyroscopes;[13, 14, 15, 16, 17] the operation of gyro-compass repeaters;[6, 7] the automatic stabilization and guiding of aircraft;[18, 19, 20, 21] and in fact the automatic recording or control of almost any measurable or measurable and controlable physical quantity.[1, 2, 3, 22, 23, 24, 25, 26, 27, 28]

It is apparent that servo-mechanisms form a vital link in the application of automatic control and that a study of their performance is of importance.   The purpose of this paper is to present an analysis of the operating characteristics of certain important types of servo-mechanisms.

Although the subject of servo-mechanisms has been treated in a qualitative or semi-quantitative way in some of the references given above, to the writer's knowledge no systematic quantitative treatment of even the simple common types has previously been given.*   It seems worth while, therefore, in

---

* Minorsky's first paper, ref. 4, gives an excellent analysis of the rudder-hull dynamic system in the ship-steering problem which bears some resemblance to the problem of the continuous-control type of servo here treated.   Specific references to his results are given later in the paper.

view of the rapidly expanding field of application of servo-mechanisms, to present the beginnings of such a treatment, outlining quantitatively at least a few of the important properties of the familiar types. Preceding the quantitative treatment, a general discussion of servos and of properties of three important types will be given.*

GENERAL CHARACTERISTICS OF SERVO-MECHANISMS.

Before going into further detail regarding the performance of servos it will be well to define more explicitly what is meant by the term servo-mechanism. As stated before, a servo-mechanism is a power-amplifying device. However, its action differs in one essential particular from that of a simple vacuum-tube or mechanical power amplifier.† Such an amplifier preserves approximately a given functional relation, usually linear, between input and output quantities, due to the properties of the amplifier itself. Thus in a good vacuum-tube amplifier, the current output is very closely proportional to the voltage applied to the grid of the first tube. This linearity of response is due to the constancy of the parameters within the amplifier. Any departure from constancy of these parameters affects the relation between input and output directly.

The servo-mechanism differs from the simple amplifier in that the responsibility for the functional relation is not placed directly on the amplifying element of the servo. Here it will be necessary to distinguish between the input to the servo-mechanism, which is the indication of the measuring instrument, and the input to the amplifier element in the servo which is something different. The output of the servo amplifier element can be considered as the output of the servo, however. In a servo-mechanism, the input to the servo amplifier element is connected to the *difference* between the servo input and output. When this difference is finite the servo output is driven in such a manner as to tend to make this difference zero. Thus the only function of the servo am-

---

*In a companion paper, the design and test of a high performance servo-mechanism, designed on the basis of the analysis in the present paper, is given.

†For a description of an ingenious and successful mechanical torque (and hence power) amplifier developed by Mr. Neiman of the Bethlehem Steel Corporation see reference 29, bibliography.

plifier element is to apply sufficient force to the servo output to bring it rapidly to correspondence with the servo input.   Such an amplifier element can be a relatively crude affair.   In fact it may consist merely of a suitable relay or switch controlling an electric motor.   An illustration of this action is furnished by a servo-mechanism used on an early model of a machine for solving differential equations.[30]   The servo input was the angle of a rotating shaft, the output was the angle of another rotating shaft.   When these two angles differed, a contact started the electric servo-motor which drove the output shaft in the direction to restore coincidence of the angles.

A servo-mechanism *may thus be defined as a power-amplifying device in which the amplifier element driving the output is actuated by the difference between the input to the servo and its output.*   An ideal servo-mechanism is one in which the input and output indications (expressed in common units) are equal at all instants of time.   Although the ideal servo is never realized in practice, its operation furnishes the standard by which the operation of actual servos is judged.

When compared to the ideal, an actual servo-mechanism is subject to two principal defects, oscillation and lag.   Oscillation is a periodic deviation of output from input.   Lag is an average or unidirectional deviation.   Either or both of these forms of deviation are always present in some degree and it is the purpose of design to reduce these deviations to a magnitude which is negligible in any particular application.

A servo-mechanism is by nature the type of device in which oscillation would be expected to occur if definite preventative means were not employed.   To simplify the discussion, assume that the input indication is a constant.   The output is acted upon by a force tending to return it to coincidence with the input.   In the absence of damping forces such a system would oscillate indefinitely at any amplitude at which it was started.   Suitable damping reduces these oscillations to a small amplitude or prevents them altogether, depending on the type of servo.   In certain applications, small oscillations may be useful in minimizing the effects of friction by making the dynamic rather than the larger static coefficient effective.   At best this scheme reduces one evil at the cost of

introducing another.   The friction evil might better be treated
by increasing the ratio of driving torque to friction torque.
Continuous oscillation necessarily introduces relatively large
wear and tear in moving parts.   Oscillation, then, is to be
tolerated only as an evil justified by cost or by being the lesser
of two optional evils.

Lag occurs in some degree in practically all servo-mechanisms when the input has motion.   That is, the output indication at a given instant corresponds to that of the input a
short time before.   In some applications the effect of lag
may be negligible while in others it may be very serious.[30]
It does no harm, for example, in a recording-instrument servo
provided the magnitude of the lag is less than the permissible
error, and provided this source of error is taken into account
in the design of the instrument.   In a closed-cycle control
system, however, lag in the servo-mechanism frequently introduces negative damping into the system as a whole, which
must be overbalanced by positive damping from some other
source if the system is to be stable.*   Consequently it is an
effect that must be carefully considered.

The quantitative study of oscillation and lag requires a
separate analysis for each type of servo.   Some servos are
inherently oscillatory while others can be made non-oscillatory
by suitable design.   Three main types in common use are
treated in this paper and will be described briefly before
proceeding to the mathematical analysis.

The first type of servo-mechanism and one that is widely
used because of its simplicity, may be called a relay servo,
because of the essentially "off" and "on" nature of the forces
acting on the output element.   Several forms of gyro-compass
repeaters, automatic pilots, and gyro-stabilizers for ships use
this type of servo.[6, 13, 14, 15, 18, 20, 21, 27]   In this type the restoring
force applied to the output element is usually substantially
constant in magnitude, while operative.   Ideally, this force
would be brought into operation by an infinitesimal deviation
of the output from the input but practically it is usually
necessary to have a small range of deviation over which the
restoring force is inactive.   If the inactive deviation range
were infinitesimal, and if there were no time lag in the applica-

---

* For example see ref. 4 (1922), pp. 305–8, and ref. 30.

tion of the restoring force, this servo could be made to operate with an infinitesimal amplitude of oscillation, an infinite frequency of oscillation, and an infinitesimal lag error. It is quite evident that such operation represents a limiting case which at best can only be roughly approached with actual physical apparatus. Practically the amplitude and frequency of oscillation are finite, and in most cases lag error is present. The limiting case is of interest however from the point of view of analysis and from its significance as an ideal.

This type of servo has certain disadvantages. The rather sudden application of the entire available driving torque, first in one direction and then in the other is conducive to large wear and tear of the entire mechanism. In practice the restoring force is usually initiated by electric contacts which may be somewhat troublesome especially when only very small forces are available for their operation. This type of servo has the asset of simplicity, however, and may be useful where static friction in the mechanism is troublesome, and a relatively crude type of control suffices.

The second type of servo-mechanism is one in which the correction of the output is made in finite steps at definite time intervals. This type is extensively used in recording instruments and controllers for quantities that vary relatively slowly.[1, 2, 3, 9, 10] It is well illustrated by a temperature recorder in which the position of the pen is periodically tested to ascertain whether or not it corresponds to the indicated temperature. If in error, the pen position is given a small finite change in the correct direction and a short time later its position is again tested, and so on. In this scheme, the magnitude of the step correction is usually made approximately proportional to the deviation between output and input. When properly designed and adjusted this type of servo is non-oscillatory. The necessary conditions for non-oscillation are: first, that any correction and the indication of it shall be substantially completed before a new test is made; second, that the correction applied shall not be greater than that required to reduce the deviation approximately to zero; and third, that the inactive range over which the output can vary without causing a correction to be made shall be at least as large as the smallest possible correction step.

A modification of this type is used in an automatic steering gear suitable for small craft.[8]   The rudder, which is the output of the servo, is periodically displaced from mid position an amount approximately proportional to the error in the ship's heading, and then allowed to trail back to mid position before the heading is again tested and another temporary displacement is given to the rudder.

Evidently the finite-step or impulse type of servo-mechanism can be non-oscillatory.   However it does have a persistent lag error when the input is so varying that corrections must be applied continually in one direction.   It is also inherently somewhat slower than the other types of servos, since in order to be non-oscillatory, the driving force can be effective during only a fraction of the cycle of operation.

The third type of servo considered here is one in which the restoring force, acting continuously on the output element, is approximately proportional to the deviation of the output.

This type is rapidly coming into use and is now employed in a number of devices such as certain gyro-compass followers,[6, 7] and various automatic recorders.[22, 23, 24, 25, 26]   By the use of suitable damping, its action can be rendered aperiodic or oscillatory to any desired degree.   Where a high speed of response, high sensitivity, and freedom from hunting are desired this type undoubtedly has the greatest possibilities of the three.   Small deviations call into play only small restoring forces, hence its operation lacks the somewhat violent nature characteristic of relay servos.   Moreover, these small forces can be called into play promptly and large deviations are thereby avoided.   This type can be built to have very rapid response.

A servo-mechanism of this third type, described in a companion paper, capable of operating on the output of a photo-electric cell and of delivering about one-tenth of a horse power has been built which substantially completes the correction of a small deviation in about one-twentieth of a second. If desired this speed could undoubtedly be improved by a factor of five using the same basic design.   This servo-mechanism as normally operated is aperiodic in response.*

---

* See footnote on p. 282.

Having discussed the general characteristics of servo-mechanisms, and the method of operation of three of the most important types, attention is turned to the analysis. In what follows only a very simple dynamic system is treated in detail, one in which friction, inertia, and a restoring force are associated with the output element. This representation is very close to the truth for many servos. For example, in the case of the continuous-control servo referred to above, test and calculation agreed within a small experimental error. In other cases, the idealizations made in order to make a usefully simple analytical treatment possible may depart somewhat more from the facts. Nevertheless, the analysis of an idealized case gives a real insight into the characteristics of a given servo. In most specific cases in which a general analytical solution is too cumbersome to be useful, a restricted analytical or a numerical solution can be readily made, taking into account departures from more idealized conditions. The methods of treating these more involved cases will be outlined briefly at suitable points in the analysis. The first type to be considered is the relay-servo.

RELAY-TYPE SERVO-MECHANISMS.

The relay type of servo-mechanism is characterized primarily by a constant-magnitude restoring force brought into play when the output deviates by some predetermined amount from the input. This characteristic alone does not fix the performance however as the effects of a number of other factors are important. Among these others are included: first, the type of dynamic system to which this force is applied, which includes one or more inertia and friction parameters, and possibly elastance parameters; second, the nature of the friction parameters, whether the frictional force is independent of, or a function of, the output velocity; third, the inactive range, or the limits of the output deviation within which the restoring force is not in action; fourth, the time delay between the indication that the restoring force is to be applied or removed, and the actual application or removal of this force; fifth, the nature of variation of the input, whether this variation is relatively slow so that the servo maintains the output at a substantially constant value, or is rapid and unidirectional

so that the servo must move the output more or less continuously in one direction. Each of these factors must be considered in the analysis of any particular case. A compre-
hensive treatment of all cases is evidently beyond the scope of
this paper, but a few significant cases will be analyzed and
from the results obtained deductions can be made covering
other cases.

In this section an analysis is made of the cases indicated
in the following outline:

Relay Servo-Mechanisms.

A. No inactive zone.
    1. Friction force proportional to velocity (viscous
       friction).
    2. Friction force independent of velocity (Cou-
       lomb friction).
B. Finite inactive zone.
    1. Friction force proportional to velocity.
    2. Friction force independent of velocity.
C. Cases *A* and *B* with time delay.

The properties of particular interest are the amplitude of
steady-state oscillation, and the magnitude of the average lag
error as a function of the input speed. These quantities will
be determined in what follows.

Because the output of many servos is in the form of mechanical rotation the analysis is given in these terms. However, the equations can be used as given for certain other
systems, by merely redefining the symbols in terms of the
analogous parameters of other types of motion such as
mechanical translation or electric current. A more specific
discussion of this subject is given further on in the paper.

Case A1. Relay-type servo having no inactive zone and
output-element characterized by inertia and viscous friction.
Three different conditions of operation of this servo will be
considered: first, stationary input and zero time lag in application of restoring force; second, stationary input, and finite
time lag; third, uniformly varying input, zero time lag. For
each of these conditions the amplitude of steady-state oscillation and the lag error will be determined.

For the first condition consider the action of the physical
system shown schematically in Fig. 1.

FIG. 1.

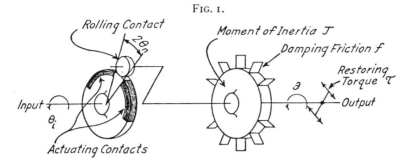

Schematic diagram of the significant dynamic elements of the relay-type servo-mechanism.

Let θ_i = angle of servo input (radians),
   θ = angle of servo output (radians),
   $2\theta_n$ = inactive range (radians) (= 0 for case A),
   J = moment of inertia of servo referred to output (gm. cm.²),
   f = friction or damping torque per unit angular velocity of output (dyne cm. per radian per sec.),
   $\pm \tau$ = restoring torque acting on output (dyne cm.),

$$p = \frac{d}{dt}, \text{ the time derivative operator,}$$

$$\omega = p\theta = \text{angular velocity (radians per sec.).}$$

The restoring torque τ is normally of sign opposite to that of the quantity $\theta_i - \theta$.   A sketch of θ for one cycle of steady-state operation is shown in Fig. 2.   To find the amplitude of

FIG. 2.

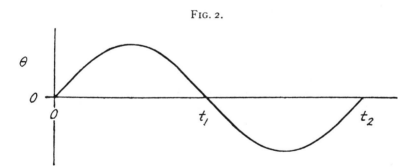

steady-state oscillation it is necessary merely to solve for the motion of the output over the range $0 < t < t_1$ in terms of

ω_0, the value of $p\theta$ at $t = 0$, and to make $p\theta$ at t_1 equal to $-\omega_0$. The resulting equation in ω_0 and the system parameters determines ω_0 for steady-state operation.

For the interval $0 < t < t_1$ the differential equation of motion is

$$- \tau - fp\theta = Jp^2\theta. \tag{1}$$

The known terminal conditions are

$$
\begin{array}{c}
t = 0 \\
\theta = 0 \\
\omega = \omega_0
\end{array}\Bigg\}, \quad (2)
\qquad\qquad
\begin{array}{c}
t = t_1 \\
\theta = 0 \\
\omega = -\omega_0
\end{array}\Bigg\} \quad (3)
$$

Integrating (1) and inserting conditions (2) there results

$$\omega = -\omega_s + (\omega_s + \omega_0)\epsilon^{-t/T}, \tag{4}$$

$$\theta = -\omega_s t + T(\omega_s + \omega_0)(1 - \epsilon^{-t/T}), \tag{5}$$

in which

$$\omega_s = \frac{\tau}{f} = \text{runaway velocity of output,}$$

$$T = \frac{J}{f}.$$

Solving (4) for t_1 with $\omega = -\omega_0$ and using this value of t_1 in (5) for which $\theta = 0$, there results

$$\ln \frac{1 - \dfrac{\omega_0}{\omega_s}}{1 + \dfrac{\omega_0}{\omega_s}} + 2\frac{\omega_0}{\omega_s} = 0. \tag{6}$$

Equation (6) is satisfied only for vanishingly small values of ω_0/ω_s. Since ω_s is finite, ω_0 must be vanishingly small. From this result it is seen that the servo output in this case oscillates, but at an infinitesimal amplitude and an infinite frequency. Physically this result is absurd for any actual servo, hence the necessary conclusion is that this case is too greatly idealized to represent the facts. It is of interest, however, as a limiting case.

Because (6) is very nearly satisfied for small finite values of ω_0/ω_s, it is to be inferred that only a slight departure from the assumed conditions could result in a finite amplitude and frequency of oscillation.

By taking time lag in the change of the restoring force into account, a practical case results. As the second condition then the results of time lag will be considered.

This condition differs from the one just treated only in that the restoring torque is changed in sign, not when $\theta = 0$ but t_1 seconds later. A cycle of output angle and restoring torque is shown in Fig. 3a. The plan of solution is similar to that just used.

FIG. 3.

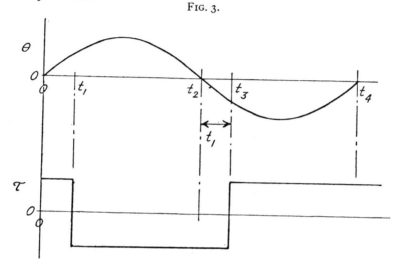

Restoring torque and output angle for case A1 with time lag present. Input stationary.

For the interval, $0 < t < t_1$, the differential equation of motion is

$$\tau - fp\theta = Jp^2\theta. \tag{7}$$

For the interval $t_1 < t < t_3$ this changes to (1). The known terminal conditions are:

$$
\left.
\begin{array}{l}
t = 0 \\
\omega = \omega_0 \\
\theta = 0
\end{array}
\right\} \quad (8)
\qquad
\left.
\begin{array}{l}
t = t_2 \\
\omega = \omega_2 = -\omega_0 \text{ for steady state} \\
\theta = 0
\end{array}
\right\} \quad (9)
$$

and that at $t = t_1$, ω and θ are continuous. Integrating (7) and (1), inserting the terminal conditions (8) and (9), and equating the two resulting expressions for ω at $t = t_1$, there results

$$\epsilon^{-t_1/T} + \epsilon^{(t_2-t_1)/T} = \frac{2\omega_s}{\omega_s - \omega_0} . \qquad (10)$$

Equating the two expressions for θ at $t = t_1$,

$$\omega_s(2t_1 - t_2) + T(\omega_0 - \omega_s)(2 - \epsilon^{-t_1/T} - \epsilon^{(t_2-t_1)/T}) = 0. \qquad (11)$$

Substituting (10) in (11), solving the result for t_2 and putting this in (10), gives, after some algebraic reduction, the following equation in t_1 and ω_0:

$$a^2(1 - \Omega_0)\epsilon^{2\Omega_0} - 2a + (1 - \Omega_0) = 0, \qquad (12)$$

in which

$$a = \epsilon^{t_1/T},$$

$$\Omega_0 = \frac{\omega_0}{\omega_s} .$$

This relation between ω_0 and t_1 plotted in dimensionless form as Ω_0 vs. t_1/T is shown in Fig. 4. The conclusion drawn by

FIG. 4.

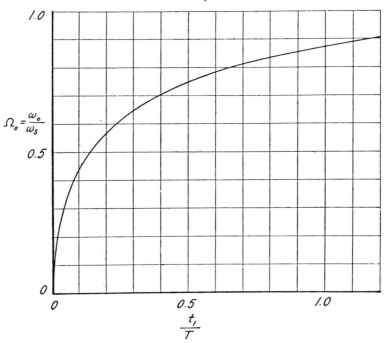

Amplitude of steady-state oscillation as a function of time lag for case AI.   Input stationary.

inference from (6) above that ω_0 might acquire a relatively large finite amplitude for even a small departure from the assumed ideal conditions is well borne out by Fig. 4. Evidently if a small amplitude is desired from this type of servo, it is very important to make the time lag in the application of the restoring-torque change very small. This lag is the effective lag due to all causes such as the time required to actuate relays and that required to establish a torque after the necessary contacts, or valves, etc., have been operated.

The third condition will next be considered. This condition is similar to the first, i.e., time lag is not considered, but the input instead of being stationary is moving at a constant speed,

$$\left.\begin{array}{c} p\theta_i = \omega_a \\ \theta_i = \omega_a t \end{array}\right\}, \tag{13}$$

where ω_a is constant. A curve of the input angle θ_i and the output angle θ for a sample cycle beginning at $t = 0$ is shown in Fig. 5. The differential equations of motion are

FIG. 5.

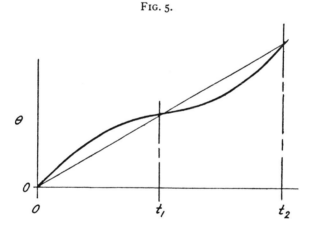

$$-\tau - fp\theta = Jp^2\theta \quad \text{for} \quad 0 < t < t_1 \tag{14}$$

and

$$\tau - fp\theta = Jp^2\theta \quad \text{for} \quad t_1 < t < t_2. \tag{15}$$

For steady-state operation the known terminal conditions are

$$
\left.\begin{array}{l} t = 0 \\ \omega = \omega_0 \\ \theta = 0 \end{array}\right\}, \quad (16) \qquad
\left.\begin{array}{l} t = t_1 \\ \omega = \omega_1 \\ \theta = \omega_a t_1 \end{array}\right\}, \quad (17) \qquad
\left.\begin{array}{l} t = t_2 \\ \omega = \omega_0 \\ \theta = \omega_a t_2 \end{array}\right\} \quad (18)
$$

Intergrating (14) and using the terminal conditions (16), there results for $0 < t < t_1$:

$$
\omega = -\omega_s + (\omega_s + \omega_0)\epsilon^{-t/T}, \tag{19}
$$

$$
\theta = -\omega_s t + T(\omega_s + \omega_0)(1 - \epsilon^{-t/T}). \tag{20}
$$

Fcr the interval $t_1 < t < t_2$ the corresponding expressions as obtained from (15) and (17) are

$$
\omega = \omega_s - (\omega_s - \omega_1)\epsilon^{-(t-t_1)/T}, \tag{21}
$$

$$
\theta = \omega_s(t - t_1) + \omega_a t_1 - T(\omega_s - \omega_1)(1 - \epsilon^{-(t-t_1)/T}). \quad (22)
$$

From the relations (17) to (22) the steady-state amplitude of oscillation can be obtained in terms of the relation of ω_0 and ω_1 to ω_a, that is in terms of the relation of the maximum and minimum angular velocities to the average angular velocity of the output. Let

$$
\left.\begin{array}{l} \dfrac{\omega_a}{\omega_s} = \Omega_a \\[2mm] \dfrac{\omega_0}{\omega_s} = \Omega_0 \\[2mm] \dfrac{\omega_1}{\omega_s} = \Omega_1 \end{array}\right\}. \tag{23}
$$

Using (17) in (19) and expressing the result in terms of (23),

$$
\epsilon^{-t_1/T} = \frac{1 + \Omega_1}{1 + \Omega_0}. \tag{24}
$$

(17) in (20) gives

$$
\frac{t_1}{T} = \frac{1 + \Omega_0}{1 + \Omega_a}(1 - \epsilon^{-t_1/T}). \tag{25}
$$

Using (24) in (25) and solving the result for t_1/T, and substituting this in (24) gives one equation relating Ω_a, Ω_0 and Ω_1 thus:

$$\frac{1 + \Omega_1}{1 + \Omega_0} = \epsilon^{-(\Omega_0 - \Omega_1)/(1 + \Omega_a)}. \tag{26}$$

Another expression relating these same quantities can be obtained from (21) and (22) with (18). Using (18) in (21),

$$\epsilon^{-(t_2 - t_1)/T} = \frac{1 - \Omega_0}{1 - \Omega_1}. \tag{27}$$

(18) in (22) gives

$$\frac{t_2 - t_1}{T} = \frac{1 - \Omega_1}{1 - \Omega_a}(1 - \epsilon^{-(t_2 - t_1)/T}). \tag{28}$$

Using (27) in (28) and solving the result for $(t_2 - t_1)/T$ and substituting this in (27) gives the second equation relating Ω_0 and Ω_1 to Ω_a. This equation is

$$\frac{1 - \Omega_0}{1 - \Omega_1} = \epsilon^{-(\Omega_0 - \Omega_1)/(1 - \Omega_a)}. \tag{29}$$

Equations (26) and (29) taken simultaneously suffice to determine Ω_0 and Ω_1 in terms of Ω_a. No general analytical solution can be obtained, at least by elementary means. However from the result obtained for the previous case, i.e., for $\Omega_a = 0$ it is worth trying to see if a small amplitude of oscillation satisfies (26) and (29). To do this let

$$\Omega_0 = (1 + \rho_0)\Omega_a,$$
$$\Omega_1 = (1 - \rho_1)\Omega_a, \tag{30}$$

where ρ_0 and ρ_1 are small positive quantities. Furthermore if ρ_0 and ρ_1 are small, they are presumably approximately equal. If a solution to (26) and (29) can be obtained on the assumption that they are equal, this assumption will be justified. Assume tentatively then that

$$\rho_0 = \rho_1 = \rho. \tag{31}$$

Using (30) and (31) in (26) and (29), the following expressions are obtained:

$$\frac{1 - m\rho}{1 + m\rho} = \epsilon^{-2m\rho}, \tag{32}$$

$$\frac{1 - n\rho}{1 + n\rho} = \epsilon^{-2n\rho}, \tag{33}$$

in which

$$m = \frac{\Omega_a}{1 + \Omega_a}, \qquad n = \frac{\Omega_a}{1 - \Omega_a}.$$

Expanding the left side of (32) by division and the right side by the exponential series,

$$1 - 2m\rho + 2m^2\rho^2 - 2m^3\rho^3 + \cdots$$
$$= 1 - 2m\rho + \frac{4m^2\rho^2}{2} - \frac{8m^3\rho^3}{6} + \cdots. \tag{34}$$

Doing the same with (33),

$$1 - 2n\rho + 2n^2\rho^2 - 2n^3\rho^3 + \cdots$$
$$= 1 - 2n\rho + \frac{4n^2\rho^2}{2} - \frac{8n^3\rho^3}{6} + \cdots. \tag{35}$$

(34) and (35) are satisfied for vanishingly small values of $m\rho$ and $n\rho$ and are approximately satisfied for values of these quantities for which their cubes are negligible in comparison with unity. Barring the limiting case in which $\Omega_a = 1$, i.e., $\omega_a = \omega_s$, (34) and (35) are satisfied by vanishingly small values of ρ. Since ρ is the maximum fractional deviation of the instantaneous output angular velocity from the average or input velocity, it is seen that the oscillations are of vanishingly small amplitude and of infinite frequency.

Although as in the first condition, this is an idealized case that cannot be realized physically, the result has interest and significance. It shows that an ideal relay-type servo is capable of following a constant input velocity with precision, i.e., without finite deviation in the nature of lag or oscillation. That this condition could be approached even under ideal conditions is interesting.

The inevitable presence of time lag in the change of the restoring torque causes a finite amplitude and frequency in any actual servo. This fact could be demonstrated by explicit formulation but the analysis will not be given here. Instead the effect of such time lag in this and other cases in which the analysis is unwieldy is discussed further on in the paper.

Case A2. In the case just treated, the friction or damping force was assumed to be proportional to the velocity. In the present case, the frictional force is assumed to follow the Coulomb law, that is, to be independent of velocity. Otherwise the two cases are identical. As was done above, the response of the servo output for the condition of a stationary input will be treated first. As a second condition the effect of time lag is considered with the input stationary. A third condition with constant input velocity and zero time lag is considered last.

In Fig. 6 the variations of the torques, the output angle

FIG. 6.

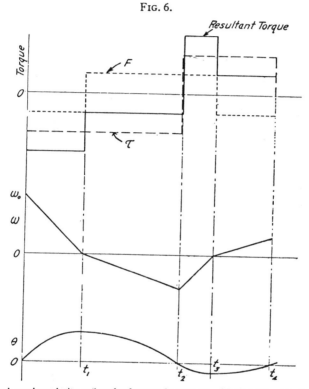

Torques and angular velocity and angle of output for case A2. No time lag. Input stationary.

and the output velocity during a cycle of operation for the first condition are sketched. As before, the amplitude of steady-state oscillation and the output lag are the quantities of interest.

Let

$$\pm F = \text{friction force, sign opposite to velocity.}$$

The other quantities are as previously defined. The differential equations of motion are

$$-\tau - F = Jp^2\theta = Jp\omega \quad \text{for} \quad 0 < t < t_1 \qquad .(36)$$

and

$$-\tau + F = Jp^2\theta = Jp\omega \quad \text{for} \quad t_1 < t < t_2. \qquad (36a)$$

The known terminal conditions are

$$\left. \begin{array}{l} t = 0 \\ \omega = \omega_0 \\ \theta = 0 \end{array} \right\}, \quad (37) \qquad \left. \begin{array}{l} t = t_1 \\ \omega = 0 \\ \theta = \theta_1 \end{array} \right\}, \quad (38) \qquad \left. \begin{array}{l} t = t_2 \\ \omega = \omega_2 \\ \theta = 0 \end{array} \right\}. \quad (39)$$

Integrating (36) and using (37) to determine the constants of integration give, for the interval $0 < t < t_1$,

$$\omega = -\alpha_1 t + \omega_0 \qquad (40)$$

and

$$\theta = -\frac{\alpha_1}{2} t^2 + \omega_0 t, \qquad (41)$$

where

$$\alpha_1 = \frac{\tau + F}{J}. \qquad (42)$$

Using (38) in (40) to obtain t_1, and putting this value of t_1 in (41) gives

$$t_1 = \frac{\omega_0}{\alpha_1},$$

$$\qquad (43)$$

$$\theta_1 = \frac{\omega_0^2}{2\alpha_1}.$$

Integrating (36a) and evaluating the constants of integration by (38) and (43) give

$$\omega = -\alpha_2(t - t_1), \qquad (44)$$

$$\theta = -\frac{\alpha_2}{2} t^2 + \frac{\alpha_2}{\alpha_1} \omega_0 t + \frac{\omega_0^2}{2\alpha_1} - \frac{\alpha_2\omega_0^2}{2\alpha_1}, \qquad (45)$$

where

$$\alpha_2 = \frac{\tau - F}{J}.$$

(39) in (45) gives

$$t_2 = t_1 \left(1 \pm \sqrt{\frac{\alpha_1}{\alpha_2}} \right), \qquad (46)$$

in which the plus sign alone is of interest since $\alpha_1 > \alpha_2$. Using (46) in (40) it is found that

$$\frac{\omega_2}{\omega_0} = -\sqrt{\frac{\alpha_2}{\alpha_1}}. \tag{47}$$

Since the second half of the cycle is identical in form with the first half, the ratio of initial to final velocity will be the same in both cases. The value ω_4 of ω at $t = t_4$ is then given by

$$\frac{\omega_4}{\omega_2} = \frac{\omega_2}{\omega_0} = -\sqrt{\frac{\alpha_2}{\alpha_1}}$$

and

$$\frac{\omega_4}{\omega_0} = \frac{\alpha_2}{\alpha_1}. \tag{48}$$

Equation (48) shows that if the servo starts with a finite amplitude of oscillation, this amplitude is reduced by a constant factor each cycle, and thus the amplitude approaches zero asymptotically. In the steady-state, therefore, this ideal servo, like the previous one, oscillates with zero amplitude and infinite frequency when the input has a constant value.

Although this condition if of interest as a limiting case, the effect of time lag is always present practically. As the second condition then, the effect of a definite time lag of t_1 seconds in the reversal of the restoring torque is considered. The nature of the variations of angle, velocity, and torque with time in this condition is sketched in Fig. 7. Remembering that during each interval when the resultant torque is constant, the motion is that of a uniformly accelerated body and using the following terminal conditions,

$$t = 0 \atop \left. \begin{matrix} \omega = \omega_0 \\ \theta = 0 \end{matrix} \right\}, \; (49) \qquad t = t_2 \atop \left. \begin{matrix} \omega = 0 \\ \theta = \theta_2 \end{matrix} \right\}, \; (50) \qquad t = t_3 \atop \left. \begin{matrix} \omega = -\omega_0 \text{ for} \\ \text{steady state} \\ \theta = 0 \end{matrix} \right\}, \; (51)$$

the following expression is obtained for θ_2 starting from the conditions at $t = 0$:

$$\theta_2 = \omega_0 t_1 + \frac{\alpha_2}{2} t_1{}^2 + \frac{1}{2\alpha_1}(\omega_0 + \alpha_2 t_1)^2. \tag{52}$$

FIG. 7.

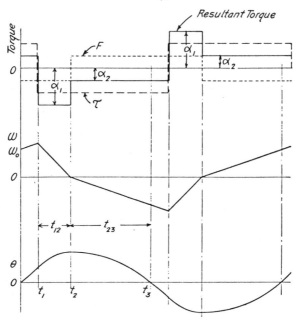

Same as Fig. 6 except that time lag is present.

Using (51) and writing an expression for θ_2 starting from $t = t_3$,

$$\theta_2 = \frac{\omega_0{}^2}{2\alpha_2}. \tag{53}$$

Equating (52) to (53) and solving the result for ω_0 there results for steady-state operation:

$$\omega_0 = \alpha_2 t_1 \frac{\alpha_1 + \alpha_2 \pm \sqrt{2\alpha_1(\alpha_1 + \alpha_2)}}{\alpha_1 - \alpha_2}. \tag{54}$$

Equations (54) and (53) give the interesting result that when the friction obeys the Coulomb law, the relay servo with a stationary input oscillates with an amplitude proportional to the square of the time lag in the restoring force change. In contrast to the previous case, similar except that the damping was due to viscous friction instead of Coulomb friction, the amplitude due to a small time lag is very small.   From the point of view then of making the amplitude of oscillation small, a servo with Coulomb friction acting on the output is preferable to one with viscous friction. This statement applies, as will be seen below, only to the condition with a substantially stationary input.

The third condition is similar to the first for this case, except that the input, instead of being stationary, has a constant angular velocity ω_a.

In Fig. 8 are shown the general relations between the

FIG. 8.

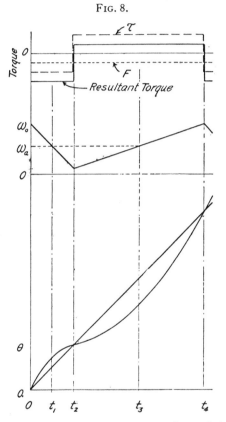

Similar to Fig. 6 except that input has constant input velocity ω_a.

torques, velocities, and angles, and time for this condition. It will be assumed that the output velocity is always positive, i.e., that although the output velocity varies above and below the average or input velocity, it never becomes negative, and that as a consequence the frictional force always acts in the same direction. The implications of this assumption become evident in the analysis.

The differential equations of motion are:

$$- \tau - F = Jp^2\theta = Jp\omega \quad \text{for} \quad 0 < t < t_2, \tag{55}$$

$$\tau - F = Jp^2\theta = Jp\omega \quad \text{for} \quad t_2 < t < t_4. \tag{56}$$

The known terminal conditions are:

$$\left.\begin{matrix} t = 0 \\ \omega = \omega_0 \\ \theta = 0 \end{matrix}\right\}, \quad (57) \qquad \left.\begin{matrix} t = t_2 \\ \omega = \omega_2 \\ \vartheta = \omega_a t_2 \end{matrix}\right\}, \quad (58) \qquad \left.\begin{matrix} t = t_4 \\ \omega = \omega_4 \\ \theta = \omega_a t \end{matrix}\right\}. \quad (59)$$

From (55) and the terminal conditions (57) the expressions for output velocity and angle for the interval $0 < t < t_2$ are found to be

$$\omega = -\alpha_1 t + \omega_0, \tag{60}$$

$$\theta = -\frac{\alpha_1}{2} t^2 + \omega_0 t \tag{61}$$

as before. From the value of θ at $t = t_2$ in (58) substituted in (61), the value of t_2 is found to be

$$t_2 = 0 \text{ or } 2\left(\frac{\omega_0 - \omega_a}{\alpha_1}\right), \tag{62}$$

the second value only being of interest. Putting this value in (60) gives

$$\omega_2 = 2\omega_a - \omega_0. \tag{63}$$

Integrating (56) and evaluating the constants by (58) and (63), the following expressions for velocity and angle during the interval $t_2 < t < t_4$ are obtained:

$$\omega = \alpha_2 t + 2\omega_a\left(1 + \frac{\alpha_2}{\alpha_1}\right) - \omega_0\left(1 + 2\frac{\alpha_2}{\alpha_1}\right), \tag{64}$$

$$\theta = \frac{\alpha_2}{2}(t^2 - t_2{}^2) - (\omega_a - \omega_0)\left(1 + 2\frac{\alpha_2}{\alpha_1}\right)t_2$$

$$+ 2\omega_a\left(1 + \frac{\alpha_2}{\alpha_1}\right)t - \omega_0\left(1 + 2\frac{\alpha_2}{\alpha_1}\right)t. \quad (65)$$

The value of t_4 in terms of known quantities can be found by using the value of θ given in (59) in (65) and is

$$t_4 = t_2 \quad \text{or} \quad 2\frac{(\alpha_1 + \alpha_2)}{\alpha_1\alpha_2}(\omega_0 - \omega_a). \quad (66)$$

The second value is, of course, the only one of interest. Putting this value of t_4 in (64) gives the angular velocity at the end of the cycle in terms of the velocity at the beginning. This value is

$$\omega_4 = \omega_0. \quad (67)$$

This result is surprising at first glance for it shows that this servo, under the given conditions, continues to oscillate with any finite amplitude at which it may happen to be started. That this result is sound physically may be seen by considering the curve of resultant torque in Fig. 8. Even though friction is present, when it always acts in the same direction and is of constant magnitude, its effect is merely to increase the magnitude of the restoring force during the first portion of the cycle, and to decrease it during the second portion. Thus no damping action whatever is produced. This is the first case in which the relay-type of servo has been found to have the possibility of a finite amplitude of oscillation in steady-state operation under the ideal condition of zero time lag.

This result, of course, holds only for the postulated condition that the output velocity is unidirectional. If the amplitude of velocity oscillation is sufficiently large in comparison with the input or average output velocity so that reversal of the output velocity occurs, the action is quite different. Due to reversal of the output velocity and therefore of the direction of the friction force, a damping action is produced and the amplitude of oscillation is reduced. This reduction continues until the output velocity just fails to reverse, when the damping action ceases. Thereafter the output oscillations continue indefinitely at such amplitude that reversal of the output velocity just fails to occur. Time lag aggravates this condition in a way that is discussed further on.

Evidently this type of servo-mechanism with constant-magnitude restoring and friction torques is unsuited for use with other than a fixed or very slowly varying input, because of the absence of useful damping when the input has appreciable velocity.

In the cases treated thus far the restoring torque has been acting at all instants either in one direction or the other. Actually there is always some inactive period which may be so small, however, in many instances that its effect is insignificant. In other instances, a finite inactive range, over which the deviation of input from output may vary without bringing the restoring torque into action, is purposely provided. The analysis of the performance of servos with such an inactive range is therefore important and will be considered next.

Case B1. In this case is considered a servo-mechanism having a constant-magnitude restoring torque and a viscous-friction torque acting on the output element.

It differs from the servo with these properties treated under A1 in that the restoring torque is not brought into action until the output differs in magnitude from the input by some predetermined angle θ_n. The analysis of this servo, while straight-forward and simple when carried out numerically for a particular case, is rather clumsy analytically because of the large number of time intervals into which a cycle of operation must be subdivided. For each of these subdivisions a different differential equation applies. Each integration to obtain an expression for angle involves two constants determined by conditions at the beginning of the time interval. Since at least five such time intervals appear in one cycle the analytical expressions become rather cumbersome by the end of the cycle.

Certain conclusions of value can be drawn without explicit formulation, however. If this servo has a stationary or slowly moving input, the inactive zone allows the servo-mechanism to remain inoperative much of the time, since the restoring torque would be called into action only when the deviation exceeded the inactive or tolerance range θ_n. It is readily seen that under these conditions the output angle is uncertain within the range $\pm \theta_n$ from the input.

This servo will not oscillate because all of the damping forces exist that were present with no inactive zone. Consequently any initial oscillation will damp down to an amplitude that lies within the inactive zone, and then cease entirely. The effect of time lag in the application or removal of the restoring torque is discussed further on.

When the input of such a servo has an appreciable velocity, conclusions cannot be so readily drawn without a mathematical formulation, which as seen above is cumbersome. Numerical computation is perhaps as direct a method as any for obtaining an explicit result in a particular case. For a general study covering all cases, mechanical methods are the most attractive. Thus on a device such as the differential analyzer [32] many particular solutions could be run off with relative ease. These solutions could take the form of curves relating the dimensionless variables θ/θ_n and t/T, with ω_0/ω_a and ω_s/ω_a as dimensionless parameters whose values would be different for each particular solution. All possible solutions could be represented on a series of curve sheets, each sheet containing a nest of curves of θ/θ_n vs. t/T with ω_0/ω_a as a parameter, all for a given value of ω_s/ω_a. Each other sheet would be similar except that it would be for a different value of ω_s/ω_a.

The case under discussion has not sufficient interest to warrant the presentation of such a set of solutions in this paper. However, the operation of this servo-mechanism will be illustrated by a numerical example using the following values:

$$\omega_s = 2,$$
$$\omega_a = 0.5,$$
$$\omega_0 = 1.5,$$
$$\theta_n = 0.1,$$
$$T = 1.$$

Any consistent set of units can, of course, be used. The results of this calculation are plotted in Fig. 9 which shows a rather rapid damping out of oscillations. In the steady state the output oscillates with an infinitesimal amplitude about the lagging limit, i.e., the limit at which the positive-direction restoring torque comes into action, or at an output angle

$$\theta = \theta_i - \theta_n.$$

This servo then has a definite lag angle θ_n which is half the inactive range, assuming that this range is centered on the output position for exact correspondence between input and output. From these results it appears that an inactive range has little usefulness from the theoretical point of view when the input is moving unidirectionally. Practically it may be necessary because of electrical contact difficulties or the like.

That the conclusions drawn from Fig. 9 for a constant

FIG. 9.

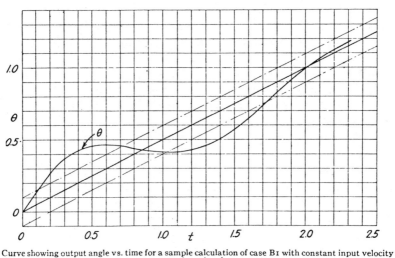

Curve showing output angle vs. time for a sample calculation of case B1 with constant input velocity ω_a, and no time lag.

input velocity are of rather general application can be seen from the following considerations. Suppose the initial amplitude of oscillation is very large in comparison with the width of the inactive zone. Then the analysis given under case A1 will apply approximately, leading to a smaller amplitude, when conditions will become similar to those shown in Fig. 9. If the initial amplitude is very small, the steady-state will be reached in less time than that required for the initial conditions shown in Fig. 9.

Case B2.   In the case just considered the damping was due to viscous friction.   In the present case Coulomb friction is assumed.   The inactive zone is the same as assumed above, i.e., the output has a permissible deviation of $\pm \theta_n$ from the input before the restoring torques are effective.   Except for the presence of the inactive range this case is identical with case A2.

Considering first that the input is moving slowly or not at all, a consideration of the motion of the output and the forces acting upon it will show that oscillations damp out relatively rapidly, and that the servo may remain inactive much of the time as in the foregoing case.   The same uncertainty as to output position also exists.

When a constant velocity input is considered, the prediction of behavior is somewhat more involved.   First assume that the initial amplitude of output oscillation is not so large as to cause reversal of the output velocity.   A cycle of operation can then be represented as shown in Fig. 10.   It can be shown readily that steady-state operation of this servomechanism can occur with any amplitude of oscillation for which the output velocity is never negative as assumed above. This is done as follows: From the differential equations and terminal conditions, or by inspection, remembering that only constant accelerations are encountered, the following relations are obtained:

$$\omega_0 - \omega_a = \omega_a - \omega_1, \tag{68}$$

$$\omega_a - \omega_2 = \omega_3 - \omega_a, \tag{69}$$

$$\omega_2 = \omega_1 - \alpha_0 t_{12}, \tag{70}$$

$$\theta_2 = \theta_1 + \omega_1 t_{12} - \frac{\alpha_0}{2} t_{34}{}^2 = \theta_1 + \omega_a t_{12} - 2\theta_n, \tag{71}$$

$$\omega_4 = \omega_3 - \alpha_0 t_{34}, \tag{72}$$

$$\theta_4 = \theta_2 + \omega_3 t_{34} - \frac{\alpha_0}{2} t_{34}{}^2 = \theta_3 + \omega_a t_{34} + 2\theta_n. \tag{73}$$

The significance of the subscripts is indicated in Fig. 10. Equations (68) and (69) express the fact that, in uniformly

FIG. 10.

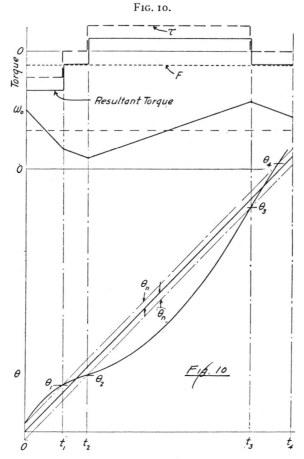

Torques, and angular velocity and angle of output for case B2 with constant input velocity ω_a and no time lag.

accelerated motion, the average velocity is equal to the average of the initial and final velocities. From these six equations it can be shown that

$$\omega_4 = \omega_0 \qquad (74)$$

which will prove the statement made above. To demonstrate (74) substitute ω_1 from (68) in (70), substitute the resulting expression for ω_2 in (69) and solve for ω_4. This result substituted in (72) gives the following relation between ω_4 and ω_0:

$$\omega_4 = \omega_0 + \alpha_0(t_{12} - t_{34}).$$

To prove (74) it is therefore sufficient to show that

$$t_{12} = t_{34}.$$

This can be done by solving (71) and (73) for t_{12} and t_{34} respectively, and expressing ω_1 and ω_3 in the results in terms of ω_0.

It follows then that the introduction of an inactive range in the constant-friction relay servo does not damp out oscillations when the input has a constant velocity and the output velocity is unidirectional. These conditions are representative of normal operation, hence this servo is inherently unsatisfactory. To damp out oscillations under these conditions, the frictional force must have a component that varies in magnitude as some positive power of the velocity.

This completes the analysis of the various cases of relay-type servo-mechanisms considered in this paper. As this analysis has involved a considerable amount of detail, it is well to summarize the important results. This is done in Table I, which is self-explanatory except for certain of the effects of time lag discussed below. These results apply to a servo-mechanism in which the output element has inertia, and is acted upon by constant magnitude restoring forces, and a frictional force of some type.

C. Effect of Time Lag in Change of Restoring Torque.—Although in two cases, the effect of time lag in the application and removal of restoring torques was treated mathematically, a treatment of its effect in other cases was postponed for a general discussion which will be given here.

Time lag, as stated before, is the delay between (*a*) the time at which the restoring torque would be applied or removed were the relay to act and the resulting torque to build up instantaneously when the need for a change is indicated by a predetermined discrepancy, between input and output and (*b*) the time when the torque actually becomes effective. This time lag may be different for the application and the removal of the torque when there is an inactive zone, but it suffices for this discussion to consider the combined effect of these two time lags. Actually there is always some delay or time, lag. In some cases it may be so small as to have an insignificant

TABLE I.

	Stationary Input.			Constant Velocity Input.		
	Steady-state Oscillation.		Average Lag Error.	Steady-state Oscillation.		Average Lag Error. No Time Lag.
	No Time Lag.	Time Lag.		No Time Lag.	Time Lag.	
A. No inactive zone 1. Viscous friction	Zero amplitude infinite frequency	Finite and function of time lag	Zero	Zero amplitude infinite frequency	Finite amplitude	Zero
2. Coulomb friction	Zero amplitude infinite frequency	Finite and proportional to square of time lag	Zero	Indeterminate	Amplitude increases until postulated conditions fail to hold	Indeterminate
B. Finite inactive zone 1. Viscous friction	None	May or may not oscillate depending on parameters	± half of inactive range with no time lag	Zero amplitude infinite frequency	Finite amplitude	± half of inactive range
2. Coulomb friction	None			Indeterminate	Amplitude increases until postulated conditions fail to hold	Indeterminate

effect upon the action of the servo.   In many cases, however, its effect is significant.   This is demonstrated by the great sensitiveness of the viscous-friction damped relay servo to time lag, where a small time lag results in a relatively large amplitude of steady-state oscillation.   On the other hand, a relay servo having Coulomb friction is relatively insensitive to small values of time lag.   These facts have been demonstrated for a servo with an approximately constant input and no inactive range.

When the input has appreciable velocity, or a finite inactive zone, the formal analysis, while straightforward, is somewhat cumbersome because of the number of points during a cycle at which the differential equation of motion changes and constants of integration must be reëvaluated.   In any particular case this process can be carried through numerically with relative ease.

A qualitative analysis of the effect of time lag in these cases is readily made however.   In general the result of a time lag in changing the restoring forces is to continue to increase the output velocity in a given direction after the instant at which this velocity should be decreased.   Thus in an ideal relay servo with no inactive zone, if the output is ahead of the input, but is moving toward the input, the restoring torque should reverse in sign at the instant when the input and output coincide.   From this instant the output is accelerated toward the input.   When time lag is present, the restoring torque acting prior to this instant continues to act for the time-lag interval after this instant.   During this interval the restoring torque is tending to drive the output away from the input rather than toward it.   As a result, the overshoot is greater than when no time lag is present, and the relative velocity between input and output is greater when they again are in coincidence than it would have been in the absence of time lag.   At the time of this second coincidence the time lag in changing the sign of the restoring torque again causes an increment of output velocity in the wrong direction.   The net result, as is seen, is to tend to increase the amplitude of oscillation.   Looked at in a slightly different way, the impressed torque, when time lag is present, may be resolved into two components, one with no time lag, the effect of which has been studied in each case, and a second which gives an impulse to the output in the direction of its motion, at a time when this motion should be retarded. This impulse evidently adds energy to the oscillation, a fact demonstrated by the results in the two cases analyzed with time lag present.

The point of real interest is the effect of this increment of energy on the steady-state amplitude of oscillation. The cases previously considered are divided into two groups with respect to this effect. The first group includes the cases where positive damping exists as indicated by a zero amplitude of steady-state oscillation. The second includes the cases where no damping exists as indicated by the indefinite persistence of any existing finite amplitude.

The effect of time lag on the first group is to increase the amplitude of oscillation until the friction or damping work per cycle due to increased amplitude is equal to the work per cycle put into the output as a result of the time lag. This amplitude is finite for the cases considered in the first group.

In the second group time lag results in an amplitude of oscillation that increases until the postulated conditions of operation cease to hold. It has been seen that in the Coulomb-friction relay servo there is no damping so long as the output moves unidirectionally. When the amplitude is so large as to cause reversal of the output during some portion of the cycle, however, net work is done against friction, and positive damping is thereby introduced. Due to time lag, then, the amplitude of oscillation will increase until the work per cycle resulting from time lag is equal to the net friction work per cycle resulting from reversal of the output unless something breaks before this limit is reached. Evidently, unless the input velocity is so small that sufficient positive damping to overcome the negative damping effect of any time lag present can be introduced by a reasonable amplitude of oscillation of the output, the second group of servos is quite useless for practical purposes.

The analysis of the operation of relay servos given in the foregoing pages has assumed that the dynamic elements of the system under consideration can be reduced to an output member that has inertia, and that only frictional and constant-magnitude restoring forces act upon this member.

This is the simplest type of relay servo-mechanism but one which includes many practical cases.

This concludes the discussion of relay-type servo-mechanisms given here, which could, of course, be almost indefinitely extended. Attention will now be turned to the second, or definite-correction type.

DEFINITE-CORRECTION SERVO-MECHANISM.

Because of the relative simplicity of the operating characteristics of this type of servo-mechanism no extensive mathematical analysis will be given. The criteria for nonoscillatory response were given under the discussion of general characteristics of servo-mechanisms. Assuming that these criteria are satisfied, the other important operating characteristics are the maximum input speed which this servo can follow and its lag error. The maximum input speed is easily found.

In the definite-correction servo-mechanism, the value of the output is periodically measured to determine whether or not it differs by more than half the inactive range from the value of the input. The time interval between successive measurements is a definite quantity, independent of the manner in which the input may be varying. As a result of a measurement, a definite correction is applied to the output. If the time interval is Δt and the maximum correction that can be applied to the output in one interval is $\Delta\theta_m$, the maximum input speed ω_{im} that this servo can follow is evidently

$$\omega_{im} = \frac{\Delta\theta_m}{\Delta t}.$$

If this servo is to follow a high-speed input, either $\Delta\theta_m$ must be large or Δt small, or both. Δt is fixed by the time required for the correction to be substantially completed and for the measuring device to take up an indication substantially in accordance with the corrected value of the output. The lower limit for Δt is then established by design considerations relating to the time required for the correction to be effected, and to the time constant of the measuring device.

If only one size correction $\Delta\theta$ can be made, then $\Delta\theta$ is fixed by the permissible error, since, to prevent oscillation, $\Delta\theta$ can be no larger than twice the range of deviation of the out-

put from the input over which no correction is applied.   This
limitation can be mitigated by grading the size of correction,
making the smallest size satisfy the error condition, and the
largest satisfy the maximum input-speed condition.   Of
course the correction should be proportional to the measured
deviation, and just sufficient to restore the deviation to zero.
Nearly all definite-correction servos have this proportional
correction feature.

One limitation to which this type of servo is subject should
be noted.   It can follow satisfactorily only those components
of input velocity that are sufficiently continuous in a mathematical sense to make the input velocity approximately constant over at least two successive time intervals Δt.   That is,
the output varies according to a sort of step function in which
the size of a given step is determined largely by the average
input velocity during the preceding interval Δt.   If the input
velocity is sufficiently continuous, this output step function
will be a good approximation to the input function.   In this
type of servo, however, the output always lags behind the
input, because a deviation is corrected only after it has occurred.   The input is always one jump ahead.

The definite-correction servo has a number of advangages
for certain classes of work.   When properly adjusted it is
non-oscillatory.   When the input is constant or varies only
slowly, the output drive may be inactive for considerable
periods, thus saving wear and tear.   A malfunctioning of the
servo is more likely to result in inaction than in a racing away
of the output toward infinity or the limit stop.   The relay
servo is always hunting a balance position, whereas the definite-correction servo moves only when balance is disturbed.
In a loose way it may be said that the relay servo keeps the
output in equilibrium only by continuous juggling, while the
output of the definite-correction servo is in static equilibrium.
This fact makes for long life of the latter.   The definite-correction servo is, however, somewhat slower than other types
because of the relatively small fraction of the time during
which the restoring force can be effective.   This slowness
imposes a restriction upon the field of application rather than
upon its effectiveness in any application for which it is suitable.
The definite-correction servo is widely used, especially for re-

cording and control in industrial processes where it is normally highly successful.[1, 2, 3, 9, 10, 11, 12]

CONTINUOUS-CONTROL SERVO-MECHANISM

The third type of servo-mechanism, in which the indicating element continuously controls the restoring force acting on the output element in both magnitude and direction, is perhaps the most interesting of the three. General considerations indicate that this smooth torque control should be superior to the somewhat crude "off-on" control of the relay servo. Also the continuous use of the input indications should be superior to the occasional use of these indications characteristic of the definite-correction servo. Thus the continuous-control type appears on casual inspection to have inherent advantages over the other two types, especially where accurate, rapid following is required.

The continuous-control type is interesting for another reason. Because of their nature, the forces acting in this servo are easily expressed mathematical functions of displacement and velocity. Consequently a thorough-going analysis of the performance of this servo is more readily made than of the two previous types.

The analysis given here applies to a servo in which an output member, which has inertia, is acted upon by restoring and damping torques, the restoring torque being a function of the deviation between input and output and its derivatives, and the damping torque proportional to the output velocity. Most continuous-control servos fall into this category. Especially in high-speed servos the forces obeying the Coulomb friction law are likely to be quite negligible when compared with the forces having the effect of viscous friction.

Two cases are considered in what follows. In the first, the restoring torque is proportional to the input-output deviation and the necessary damping is secured by viscous-friction-law torques. In the second, the restoring force is a linear function of the first and second time derivatives of the deviation as well as of the deviation itself. A third case, not considered in detail here because of certain practical limitations involved in its utilization in servo-mechanisms, but of interest in other similar dynamic systems, is that in which the first

time derivative of the restoring torque is a linear function of
the deviation and its first two time derivatives.   Minorsky [4]
treats these last two cases as applied to the rudder-hull dynamic system of the ship-steering problem.   The reason that
the third case will not be generally useful in servo-mechanisms
is that this type of control of the restoring torque relies largely
on the derivatives of the deviation and these derivatives in the
usual system utilizing a servo-mechanism are not sufficiently
continuous to provide smooth operation of the servo.   In
case a servo is used in connection with a large vessel or other
large inertia system, such a control of servo restoring torques
might be attractive.   However, the second-case type of control has not yet been exploited in servos and until this has been
done that of the third case can wait.   Analysis of the third
case is not difficult and has been carried out to the point of
determining the conditions for stability and the nature of the
steady-state deviations by Minorsky for the rudder-hull
dynamic system. [4]   A general solution involves the solution of
a cubic or higher degree equation making the results cumbersome at best.   Again, however, numerical solutions for particular cases are quite straightforward.

With this preliminary discussion, attention is now turned
to a mathematical consideration of the first case, in which the
servo restoring-force control is a linear function of the input-
output deviation.   Expressions for the steady-state and transient behavior are derived and a criterion of merit for such a
servo is developed.

(a) Restoring Torque Proportional to Deviation.

For this analysis consider the physical system shown in
Fig. 11.   This system correctly represents the dynamic elements of a continuous-control servo on the condition that the
effect of all inertia and damping torques can be assumed to
be concentrated at the output shaft.   For many continuous-
control servos these assumptions correctly represent the facts.

The amplifier shown in Fig. 8 has the effect of reducing
the reaction on the input of the resilience torque acting on the
output by the amplification factor.   Such amplification in
some form is nearly always present in this type of servo.   It
may take electrical, mechanical or other form but the net

FIG. 11.

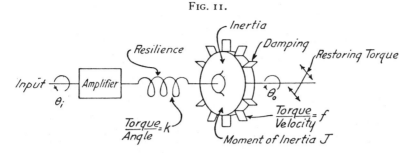

Schematic diagram of the significant dynamic elements of the continuous-control type servo-mechanism.

effect is the same in any case. It should be noted that the amplification factor of the amplifier affects the resilience constant, and to that extent the amplifier and resilience element are not independent as shown in Fig. 11.

In the analysis which follows a number of physical quantities not previously used appear, hence the notation for this analysis will be entirely redefined. Let

θ_i = input angle (radians),
θ_0 = output angle (radians),
$\theta = \theta_i - \theta_0 =$ lag of output with respect to input (radians),
k = resilience constant (dyne-cm. per radian),
J = moment of inertia of servo referred to output shaft (gm.-cm.²),
f = damping constant (dyne-cm. per radian per sec.),
t = time (seconds),
1 = Heaviside's unit function,
$p = \dfrac{d}{dt}$,
$\omega + p\theta$.

c.g.s. units are indicated in parentheses but of course any other consistent set is equally suitable since no empirical factors are used.

The equations governing the motion of the elements shown in Fig. 11 are:

$$k(\theta_i - \theta_0) - fp\theta_0 = Jp^2\theta_0, \qquad (75)$$

$$\theta_i = \Phi(t). \qquad (76)$$

$\Phi(t)$ is some arbitrary function of time representing the input motion. These equations may be solved in terms of any of the angle variables. The deviation or error angle θ is of primary interest, and from it and the given θ_i, the output angle is readily obtained. Therefore, the solution will be made in terms of θ.

Substituting

$$\theta_0 = \theta_i - \theta$$

in (75) gives

$$(k + fp + Jp^2)\theta - (fp + Jp^2)\theta_i = 0$$

or

$$\theta = \frac{f + Jp}{k + fp + Jp^2} p\theta_i. \tag{77}$$

To obtain the error angle θ as an explicit function of time some $\theta_i = \Phi(t)$ must be chosen for which the error is to be determined. An interesting case and one that imposes a severe test on a servo consists in suddenly applying a constant velocity to the input, previously at rest. Furthermore, from the response to this particular time function, the response to any $\Phi(t)$ can be determined by the use of the superposition integral.[31] If the magnitude of the suddenly-applied velocity is ω_1,

$$p\theta_i = \omega_1 1. \tag{78}$$

This expression in (77) gives

$$\theta = \frac{p + \dfrac{f}{J}}{p^2 + \dfrac{f}{J}p + \dfrac{k}{J}} \omega_1 1. \tag{79}$$

Equation (79) can be evaluated directly from a table of operational formulas.[31] To do this the denominator of (79) should be written in the form,

$$(p + \alpha - \beta)(p + \alpha + \beta),$$

in which

$$\alpha \pm \beta = \frac{f}{2J} \pm \sqrt{\frac{f^2}{4J^2} - \frac{k}{J}}. \tag{80}$$

Three cases of solution of (79) exist corresponding to positive, zero, and negative values of the quantity under the radical of (80). Two of these cases are of particular interest in this problem, the first or critically-damped case for which

$$\beta = 0 \tag{81}$$

and the second or oscillatory case for which

$$\beta = j\sqrt{\frac{k}{J} - \frac{f^2}{4J^2}}$$
$$= j\phi, \tag{82}$$

ϕ being a real number. The third case for which β is real, corresponds physically to overdamping. This case is of relatively minor interest from the design point of view because the steady-state lag error is greater and the speed of response less than in the critically-damped case, and there are no compensating advantages in the use of over-damping.

Although (81) is a limiting case of (82), the solution of (79) using (81) is somewhat simpler than that for the oscillatory case and is therefore given first. Putting (80) and (81) in (79) there results

$$\theta = \left[\frac{p}{(p + \alpha)^2} + \frac{2\alpha}{(p + \alpha)^2}\right]\omega_1 1. \tag{83}$$

From a table of operational formulas,* the time function corresponding to (83) is found to be

$$\theta = \omega_1\left[\frac{2}{\alpha} - \left(t + \frac{2}{\alpha}\right)\epsilon^{-\alpha t}\right].$$

Substituting the time constant T for $1/\alpha$ and dividing the resulting expression by $\omega_1 T$ to obtain a dimensionless equation,

$$\frac{\theta}{T\omega_1} = 2 - \left(\frac{t}{T} + 2\right)\epsilon^{-(t/T)}. \tag{84}$$

This equation is plotted as curve 1 of Fig. 12 in terms of the dimensionless variables $\theta/T\omega_1$ and t/T. These results are discussed below in connection with those which will now be obtained.

---

* Ref. 31 of Bibl., Formulas (7) and (8) of Appendix C.

When ϕ is finite and real, the response is oscillatory. A solution for this case is important. Here (79) can be written in the form,

$$\theta = \frac{p + 2\alpha}{(p + \alpha - j\phi)(p + \alpha + j\phi)} \omega_1 1. \qquad (85)$$

Equation (85) integrates into the form,

$$\theta = \psi_1(t) + 2\alpha\psi_2(t). \qquad (86)$$

By formula * the first and second terms of (86) are found to be respectively:

$$\psi_1(t) = \frac{\omega_1}{\phi} \epsilon^{-\alpha t} \sin \phi t \qquad (87)$$

and

$$\psi_2(t) = \frac{\omega_1}{\alpha^2 + \phi^2}\left[1 - \epsilon^{-\alpha t}\left\{ \frac{\alpha}{\phi}\sin \phi t + \cos \phi t \right\} \right]. \qquad (88)$$

The latter is obtained after some algebraic manipulation. Substituting (87) and (88) in (86), replacing $1/\alpha$ by the time constant T, and dividing by $\omega_1 T$ to make the expression dimensionless, there results

$$\frac{\theta}{\omega_1 T} = \frac{2}{1 + T^2\phi^2}$$
$$\times\left[1 - \epsilon^{-(t/T)}\left\{ \cos \phi t + \left(\frac{1 - T^2\phi^2}{2T\phi} \right) \sin \phi t \right\} \right] \qquad (89)$$

or in polar form

$$\frac{\theta}{\omega_1 T} = \frac{2}{1 + T^2\phi^2} - \frac{1}{T\phi}\epsilon^{-(t/T)} \sin (\phi t + \xi), \qquad (90)$$

where

$$\xi = \arctan \frac{2T\phi}{1 - T^2\phi^2}. \qquad (91)$$

In Fig. 12 (89) is plotted for various values of $T\phi$ as indicated on the curves 2, 3, and 4.

---

* Bibl. 31, formulas 17 and 11 of Appendix C.

FIG. 12.

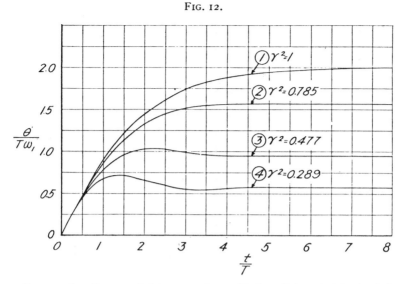

Response of continuous-control type servo to a suddenly-applied constant input velocity. Curves of input-output deviation as a function of time with the relative damping γ^2 as a parameter. Dimensionless variables are used making these curves applicable to a servo with any constants.

Before discussing the curves of Fig. 12, it is of interest to have a single quantity which characterizes the degree of oscillation in the servo response. One such quantity, which may be called the relative damping factor, is defined by

$$\gamma^2 = \frac{f^2}{4Jk}.\qquad(92)$$

From the values of T and ϕ in terms of f, k, and J it is easily shown that

$$\gamma^2 = \frac{1}{1 + T^2\phi^2}.\qquad(93)$$

From the condition for critical damping, that $\phi = 0$, it will be seen that $\gamma^2 = 1$ characterizes this case, while for less than critical damping

$$\gamma^2 < 1.$$

For large values of t/T, that is when the steady-state has been reached, the substitution of (93) in (89) or (90) yields the following very simple relation between the steady state lag angle θ_s and the impressed velocity ω_1:

$$\theta_s = 2\gamma^2 T\omega_1. \tag{94}$$

The implications of equations (84), (89) and (94) will now be considered with the aid of Fig. 12. Several interesting facts can be deduced.

In the first place, the steady-state lag error is shown to be proportional to the time constant T, to the angular velocity ω_1 of the input, and to the relative damping factor γ^2. This shows that a servo with a high speed of response, i.e., a small time constant T, is inherently more accurate than a slower servo. This is true not only for the particular input function investigated here but for any form of $\theta_i = \Phi(t)$ because the factor T is carried through the evaluation of the superposition integral used to obtain the error for any $\Phi(t)$.

The significance of the relative damping factor γ^2 can perhaps best be seen by reference to Fig. 12. γ^2 characterizes the form of the deviation or error curve, that is, the amount of overshoot and persistence of oscillation. A small γ^2 correponds to a large overshoot and subsequent oscillation. A small γ^2 also results in a small steady-state error. In designing this type of servo for small error, it is evident that some compromise must be made between the tolerated error and the amount of oscillation permitted. Fortunately the steady-state error can be very materially reduced by using a small γ^2 without introducing a very large overshoot. In certain applications, however, where the input may contain periodic components of a period comparable to that of the natural period of the servo, an aperiodic adjustment, i.e., $\gamma^2 = 1$, may be necessary.

From the design point of view perhaps the most important deduction from these equations follows from the establishment of a figure of merit for servos. Suppose this figure of merit is arbitrarily taken as the product of the numerical measure of two desirable properties of a servo-mechanism. One of these factors is taken as the maximum attainable speed ω_m. The other is taken as the smallness $1/\theta_s$ of the steady-state error θ expressed as a fraction of the total angle ω_1 turned per unit time, or ω_1/θ_s. Let the product of these two factors be the figure of merit M thus:

$$M = \omega_m \frac{\omega_1}{\theta_s}.$$

It has been seen that θ is proportional to ω in the steady-state, hence values corresponding to the maximum speed may be used and

$$M = \frac{\omega_m{}^2}{\theta_m}, \tag{95}$$

where θ_m is the steady-state error at the speed ω_m. The expression for M can be written in terms of the design constants k, f, and J as follows: From (94) using $\omega_1 = \omega_m$,

$$\frac{\omega_m}{\theta_m} = \frac{1}{2\gamma^2 T},$$

$$\frac{1}{T} = \alpha = \frac{f}{2J} = \gamma\sqrt{\frac{k}{J}}$$

and

$$M = \frac{\omega_m}{2\gamma}\sqrt{\frac{k}{J}} = \frac{k\theta_m}{4\gamma^2 J} = \frac{\tau_m}{4\gamma^2 J}, \tag{96}$$

where τ_m is the maximum torque that can be applied to the output element. Taking (95) as a measure of the desired performance of a servo-mechanism, equation (96) shows that in design, the ratio of the maximum torque τ_m to the moment of inertia J should be as large as practicable. It also shows that the relative damping should be as small as can be permitted. These are important results as they give a very definite guide for design.

In the companion paper above referred to this criterion is used as a basis for design. Tests of the servo thus designed show very fast response and an excellent agreement between test and calculation.

(b) Restoring Torque Proportional to Deviation and its First Two Time Derivatives.

The second case of the continuous-control servo will now be considered, the case in which the restoring torque is a linear function not only of the deviation θ but of its first two time derivatives. Let the restoring torque τ be given by the following relation:

$$\tau = (k + lp + mp^2)(\theta_i - \theta_0), \tag{97}$$

where l and m are constants of proportionality.

Consider the response to a suddenly-applied input velocity ω_1 just as in the first case. Using (97) in the differential equation of motion, the operational expression for the deviation or error angle θ is

$$\theta = \frac{\dfrac{J}{J_1}p + \dfrac{f}{J_1}}{p^2 + \dfrac{f_1}{J_1}p + \dfrac{k}{J_1}} \omega_1 l, \tag{98}$$

where

$$f_1 = l + f,$$
$$J_1 = m + J. \tag{99}$$

Redefining α and ϕ in terms of the parameters of the denominator of (98),

$$\alpha \pm j\phi = \frac{f_1}{2J_1} \pm \sqrt{\frac{k}{J_1} - \frac{f_1^2}{4J_1^2}} \tag{100}$$

and evaluating (98) as an explicit function of time in the same way that (85) was evaluated gives the following expression for the deviation or error angle θ:

$$\theta = \frac{\omega_1 f}{J_1 \alpha^2} \left[1 - \epsilon^{-\alpha t} \left\{ 1 + \alpha t - \frac{\alpha^2 Jt}{f} \right\} \right] \tag{101}$$

for the critically-damped case, i.e., for $\phi = 0$, or

$$\theta = \frac{\omega_1 f}{J_1(\alpha^2 + \phi^2)}$$
$$\times \left[1 - \epsilon^{-\alpha t} \left\{ \cos \phi t - \frac{J(\alpha^2 + \phi^2) - \alpha f}{f\phi} \sin \phi t \right\} \right] \tag{102}$$

for the oscillatory case, i.e., for ϕ finite and real.

These are very interesting results for they show that the steady-state error of this servo can be made zero by making the factor f zero. f is the coefficient of damping for the viscous-friction torque acting on the output. Although this damping is made zero, the servo operation can be made aperiodic or oscillatory in any desired degree by the damping effect introduced by the component of restoring torque depending on the first derivative of θ. By this method the damping effect is due not to the velocity $p\theta_0$ of the output but to the relative velocity $p\theta$ of the output with respect to the input. The magnitude of this damping is proportional to the coefficient l in (97). Physically the torque corresponding to the term $lp\theta$ in (97) is readily introduced, when a d-c. vacuum-tube amplifier is used, by an inductance component of inter-stage coupling.

The effect of introducing a component of restoring torque proportional to the relative acceleration of output and input, i.e., of introducing the effect represented by the term $mp^2\theta$ of (97) is only to alter the equivalent output inertia from the actual inertia J to an equivalent value $J_1 = J + m$. By making m negative numerically, the equivalent inertia can be given any value positive or negative. Negative values evidently are barred practically because the coefficient of the exponent in the transient-term exponential would be positive and the servo would be unstable. So long as m $> - J$, however, the system will be stable assuming of course that

$$f_1 = f + l > 0.$$

Subject to the above conditions for dynamic stability, a given servo with adjustable coefficients l and m can be made arbitrarily accurate, and fast in the sense that transient effects disappear rapidly, by giving suitable values to l and m. This is a very important result.

Physically the effect of a coefficient m can be introduced in a d-c. vacuum-tube amplifier by the use of two successive inductive interstage couplings. A negative m is secured by reversing the sense of one of these couplings. Mechanically these coefficients l and m could presumably be introduced by suitable viscous friction and inertia couplings respectively. Gyroscopic methods could also be applied to this end.[4]

The results of the analysis of this last case open up a great opportunity for improvement in the performance of servo-mechanisms of the continuous-control type. At sufficiently high speeds of response, however, the dynamic system of a simple servo-mechanism fails to reduce to the system postulated in this analysis and the analysis no longer applies. This fact sets one limitation upon indefinite improvement. Another important practical factor is that of stability of operation which may fix a limit below which the effective inertia cannot be reduced. Nevertheless, very real gains in the performance of even present high speed response servos using deviation control should be possible by the addition of deviation derivative control.

SIMILAR DYNAMIC SYSTEMS.

Earlier in the paper it was stated that although the analyses were made in terms of rotation, they could be applied without change to a dynamically similar system involving translation, electric currents, etc. As an example of this parallelism an electric circuit which is equivalent to the continuous-control servo analyzed is shown in Fig. 13. Using

FIG. 13.

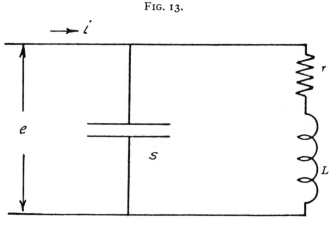

Electric circuit equivalent to continuous-control servo-mechanism.

the parallelism shown in Table II it is seen that the charge on the susceptance S is analogous to the servo error.
Thus, writing the operational expression for q_s, the charge on the condenser, there results.[31]

$$q_s = \frac{e}{S} = \frac{iZ(p)}{S}$$

$$= \frac{1}{S} \frac{(r + pL)\dfrac{S}{p}}{r + pL + \dfrac{S}{p}} i$$

TABLE II.

Quantity.	Rotation.	Translation.	Electric Circuit.
Displacement	Angle θ	Distance x	Charge q
Velocity	$p\theta$	px	Current i
Acceleration	$p^2\theta$	p^2x	pi
Force	Torque τ	Force F	e.m.f. e
Inertia	Moment of inertia J	Mass M_x	Inductance L
Resilience	Spring constant K	Spring constant K_x	Elastance $S = \dfrac{1}{\text{capacitance}}$
Loss constant	Viscous friction f	Viscous friction f_x	Resistance r
Time	t	t	t

$$= \frac{r + pL}{S + rp + Lp^2} i. \tag{103}$$

The resemblance of (103) to (77) is evident, remembering that $pq = i$. If the current i is taken as

$$i = I1, \tag{104}$$

where I is a constant, (103) can be written

$$q_s = \frac{p + \dfrac{r}{L}}{p^2 + \dfrac{r}{L}p + \dfrac{S}{L}} I1. \tag{105}$$

Equation (105) has solutions of exactly the same form as (84) and (89) but in terms of the electrical quantities corresponding to rotational quantities as shown in Table II.

The translational problem is strictly analogous to the rotational problem so it is unnecessary to go into details of the analogy here.

As mentioned before, the rudder-hull dynamic system involved in the automatic steering of ships is similar to the servo-mechanisms here analyzed. These examples suffice to illustrate the correspondence between analogous systems.

ANALYSIS OF OTHER SERVO-MECHANISMS.

In the foregoing analysis of servo-mechanisms, a dynamically simple system was assumed both because most servo-mechanisms are adequately represented by this system and because this system is readily analyzed. This system consists of an output element having inertia, and acted upon by a frictional force of some type and a restoring torque which is some function of the input-output deviation.

For servo-mechanisms involving additional dynamically-significant elements, the analysis will be somewhat more complex. The differential equations of motion will, in general, be of higher order than the second and hence somewhat more lengthy algebra will be involved in their solution. The method of formulation is, however, essentially the same as that used here.

In the case of the continuous-control type of servo; the procedure is easily stated. By writing the differential equations of motion, a relation between the deviation angle θ (using rotational terms as illustrative), the input angle θ_i and the time derivative operator p of the form,

$$\theta = F(\theta_i, p),$$

can be obtained. Equations (77) and (98) are of this form. Operational methods of evaluating such expressions as explicit time functions such as (89) and (102), due to Heaviside and later workers,* are sufficiently well developed so that the process is one of routine algebra for equations resulting from lumped-parameter systems. Practically all servo-mechanisms will be included in this class. Numerous distributed parameter systems can also be treated but may involve more than algebraic work. For lumped-parameter systems the greatest difficulty is likely to be associated with the determination of the roots of higher degree algebraic equations in one variable, a cumbersome but straightforward process numerically. Such an analysis presupposes constant parameters, i.e., a linear system, an assumption widely justified particularly for the small variations which are usually of primary importance.

It should also be mentioned that entire closed-cycle control systems are dynamically similar to servo-mechanisms and their operation is investigated by the same methods. Often a

---

* See ref. 31 for these methods and references to other works on the subject.

closed-cycle control contains a servo-mechanism. If the time constants of the servo and of the system being controlled are widely different, it is often possible to simplify the problem considerably by analyzing the two parts separately. Otherwise the equations of the entire system must be solved as a unit.

CONCLUSIONS.

In this paper the performance of three important types of servo-mechanisms has been analyzed. These types include the relay servo, the definite-correction servo, and the continuous-control servo.

The first or relay type is always oscillatory in response to a varying input but under certain conditions the amplitude of oscillation and lag error can be made small. If unidirectional motion occurs, the presence of Coulomb friction alone will not damp out an initial amplitude of oscillation. Time lag in the application of the restoring forces tends to increase the amplitude of oscillation.

The second or definite-correction type is aperiodic in operation when properly adjusted and is quite suitable for use with slowly varying quantities. There is a slight lag error.

The third or continuous-control type is probably the best type where high-speed response and smoothness of control are required. This type can be made to have a response which is aperiodic or oscillatory with any given decrement, by suitable design and adjustment. By using the first and second derivatives of the deviation of the output from the input, as well as the deviation itself, to control the restoring force applied to the output, a very high rate of response and a very small steady-state deviation should be attainable. This type of servo has the advantage of being susceptible to rather easy and complete analysis. Tests on a high-speed of response servo of this type with deviation control alone show an excellent check with the theory. The design and test of this unusually fast servo are given in a companion paper.

ACKNOWLEDGMENTS.

The writer is indebted to Dr. V. Bush who suggested the nature and scope of this study, and who critically reviewed the manuscript. He also wishes to thank those of the Electrical Engineering staff at the Massachusetts Institute of Technology who have contributed helpful suggestions.

BIBLIOGRAPHY.

1. Stein, I. M.: "Precision Industrial Recorders and Controllers," *J. F. I.*, Vol. 209, No. 2, pp. 201–228, Feb. 1930.

2. Harrison, T. R.: "The New Brown Potentiometer Recorder," *Rev. of Sci. Inst.*, Vol. 2, No. 10, pp. 618–625, Oct. 1931.

3. "Kent Recording and Controlling Apparatus," *Engineering* (London), Vol. 132, No. 3428, pp. 407–408, Sept. 25, 1931.

4. Minorsky, N.: "Directional Stability of Automatically Steered Bodies," *Jour. Am. Soc. of Naval Engrs.*, Vol. 34, No. 2, pp. 280–309, May 1922.
   "Automatic Steering Tests," *Jour. Am. Soc. of Naval Engrs.*, Vol. 42, No. 2, pp. 285–310, May 1930.

5. Sauvaire-Jourdan: "Comment On Peut Faire Gouverner Automatiquement un Navire," *La Nature* (Paris), No. 2848, pp. 355–358, April 15, 1931.

6. Ferry, E. S.: "Applied Gyro-Dynamics," John Wiley and Sons, Inc., 1932.

7. "New Compass and Path Indicator," *Shipbldg. and Shipping Rec.*, p. 684, Dec. 5, 1929.

8. "Automatic Yacht Steerer," *Sci. Amer.*, p. 238, Apr. 1933.

9. "Hydraulic Turbine Governors and Frequency Control," *Proc. N. E. L. A.*, Vol. 88, pp. 542–572, 1931. Also as Pub. No. 13. Report Hydraulic Power Committee. (Contains extensive Bibliography.)

10. McCrea, H. A.: "Automatic Control of Frequency and Load," *G. E. Rev.*, Vol. 32, No. 6, pp. 309–313, June 1929.

11. Jones, D. M.: "Controlling Load, Maintaining Frequency," *El. Wld.*, Vol. 95, pp. 1072–1076, May 31, 1930.

12. Sporn and Marquis: "Frequency, Time and Load Control on Interconnected Systems," *El. Wld.*, Vol. 99, pp. 618–624, April 2, 1932.

13. Chalmers, T. W.: "The Automatic Stabilization of Ships," Chapman and Hall, Ltd., London, 1931. This material also appeared in *The Engineer*, London, Apr. 4 to June 27, 1930.

14. "The Gyroscopic Stabilizing Equipment of the Lloyd Sabado Liner Conte de Savoia," *The Engineer* (London), Vol. 153, I, pp. 32–35, 45, Jan. 1, 1932; II, pp. 62–65, Jan. 8, 1932.

15. Frisch, E.: "Controlling a Gyro Stabilizer," *El. Jl.*, Vol. 29, No. 7, pp. 408–410, July 1931.

16. Schilovsky, P. P.: "The Gyroscopic Stabilization of Ships," *Engineering* (London), Vol. 134, No. 3491, pp. 689–690, Dec. 9, 1932; Discussion, Vol. 135, pp. 53–54, Jan. 13, 1933; p. 107, Jan. 27, 1933.

17. Viterbo, F.: "Girostabilizzàtori antirullanti (Anti-Rolling Gyro Stabilizer)," *Ingegnere*, Vol. 5, No. 3, pp. 140–152, Mar. 1931.

18. Haus, F.: "Stabilité automatique des Avions," *L'Aéronautique*, Vol. 14, Nos. 156–159, May, June, July and August 1932. English translation, "Automatic Stability of Airplanes," N.A.C.A. Tech. Memo. No. 695, Dec. 1932.

19. Green and Becker: "Radio Aids to Air Navigation," *Elect. Engrg.*, Vol. 52, No. 5, pp. 307–312, May 1933.

20. SPERRY, E. A., JR.: "Description of the Sperry Automatic Pilot," *Aviation Eng.*, pp. 16–18, Jan. 1932.

21. HUGGINS, M.: "Gyropilot Goes Cross-Country," *Aero Digest*, Vol. 17, No. 1, pp. 51–52, July 1930.

22. LA PIERRE, C. W.: "An Improved Photoelectric Recorder," *G. E. Rev.*, Vol. 36, No. 6, pp. 271–274, June 1933.

23. BERNARDE AND LUNAS: "A Recorder for Minute Quantities," *El. Jl.*, Vol. 30, No. 3, pp. 108–109, March 1933.

"A new Electronic Recorder," *Elect. Engrg.*, Vol. 52, No. 3, pp. 168–170, March 1933.

24. GULLIKSEN, F. H.: "Recent Developments in Electronic Devices for Industrial Control," A. I. E. E., Paper No. 33–24, 1933.

25. ALFRIEND, J. V.: "Light Sensitive Process Control," A. I. E. E. Paper No. 33–45, 1933.

26. HARDY, A. C.: "A Recording Photoelectric Color Analyzer," *Jour. Opt. Soc. Am.*, Vol. 18, No. 2, pp. 96–117, Feb. 1929.

27. BAYLE, M. A.: "Un Héliostat Sans Mouvement d'Horlogerie," *Rev. d'Optique*, Vol. 10, No. 12, pp. 495–511, Dec. 1931.

28. OSBOURNE, A.: "Teaching the Robot to Think Ahead," *Chem. and Met. Engr.*, Vol. 38, No. 4, pp. 245–247, April 1931.

29. NEIMAN, *American Machinist*, Vol. 66, pp. 895–897, May 26, 1927.

30. BUSH AND HAZEN: "Integraph Solution of Differential Equations," *J. F. I.*, Vol. 204, No. 5, pp. 575–615, Nov. 1927; also as M. I. T. E. E. Dept. Bul. No. 59.

31. BUSH, V.: "Operational Circuit Analysis," John Wiley and Sons, Inc., 1929.

32. BUSH, V.: "The Differential Analyzer," *J. F. I.*, Vol. 212, No. 4, pp. 447–488, Oct. 1931.

Reprinted from *Inst. Mech. Eng. Proc.* **159**:25–34 (1948)

Basic Principles of Automatic Control Systems

By Professor A. Porter, M.Sc., Ph.D.*

The importance of automatic control systems in a wide field of industrial and military applications has been accentuated during the past few years. The object of this paper is to review some of the basic principles of the subject, with special reference to automatic regulating systems.

A major difficulty at present is the lack of standardization of the terminology associated with automatic control systems; in this paper an attempt has been made to co-ordinate, albeit on a small scale, generally recognized industrial usage with the recommendations of the Ministry of Supply Servo Nomenclature Panel. It is convenient, for example, to consider an automatic regulator system as a special type of servo-system. In order to present the nomenclature in a manner which may be readily assimilated, the operation of a simple automatic speed regulator is described in detail.

The performance of an automatic control system is usually assessed by (a) the speed of response of the system subsequent to a sudden disturbance, (b) the nature of the response, and (c) the magnitude of the steady-state errors. In the case of complex control systems, such as fire-control systems, it is sometimes desirable to study, in addition, the frequency response characteristics of each main element, and to determine the overall performance of the system by the application of vector methods. The latter have been used widely in the solution of acoustical problems, and in the design of electronic feed-back amplifiers, and their adaptation to the analogous problems of servo-system design has considerably facilitated progress.

The stabilization of automatic control systems, and the elimination of steady-state errors can often be achieved by the incorporation of subsidiary feed-back loops. For example, a comparison of the basic operation of a typical position control servo-system with that of a typical automatic regulator shows that certain lags in the operation of the latter can be short-circuited by introducing "disturbance feed-back". This approach to the problem of improving the performance of these systems does not appear to have been treated extensively in the literature. Its value is demonstrated in the paper by comparing the responses of certain idealized automatic thermal regulating systems, some incorporating "disturbance feed-back", and others with straightforward controllers.

INTRODUCTION

Modern industrial processes and machines frequently require a degree of accurate and reliable operation which is difficult to achieve manually. This also applies to the operation of certain military equipments. In the design of such processes, machines, and equipments, it is often necessary to incorporate automatic controls.

Perhaps the most spectacular advances in the design of automatic control systems during the past few years have been made in the development of special equipment for the Armed Forces. Examples of important Service applications are: automatic computers and predictors; automatic remote power control of heavy gun mountings; gyro-stabilization of naval and airborne equipment; automatic remote indicating systems; completely automatic radar equipment, etc. Each of these systems incorporates at least one high-speed automatic position control system (i.e. a system which controls the angular position or linear displacement of a dynamic load) capable of a degree of performance far exceeding the standards of ten years ago.

One reason for the recent rapid advances in the subject has been the utilization of a wide variety of precise components and instruments such as (for example) accurate angular position measuring elements, high-amplification electronic amplifiers, high-efficiency hydraulic pumps and motors, metadyne generators, etc. Another equally, if not more important, reason for the substantial progress which has been made has been the close collaboration between the engineer and physicist. This association has resulted in new and more powerful techniques being evolved. In particular, the methods of harmonic synthesis and

The MS. of this paper was received at the Institution on 3rd July 1946.

* Professor of Instrument Technology, Military College of Science, Shrivenham; lately of the Department of the Director-General of Scientific Research (Defence), Ministry of Supply, and the National Physical Laboratory.

the study of the harmonic responses of control systems—for many years the research tools of specialists in communications, mechanical vibrations, and acoustics—have recently been adapted to the study of automatic controls with outstanding success (Hall 1943; Prinz 1944).†

Applications of automatic controls in the industrial field, although somewhat less spectacular than Service applications, have increased considerably in number and efficiency, and in some elaborate processes are almost indispensable. The main reasons for incorporating them are: (1) to increase efficiency and reliability of plant operation; (2) to improve the quality of the finished product; (3) in cases where manual control is impracticable on account of inaccessibility of equipment; and (4) to speed up processes. The manufacture of synthetic rubber and high-octane content aircraft fuel, the control of atomic energy processes, the cold rolling of steel strip, and the pasteurization of milk are typical examples of elaborate processes in which efficient plant operation can only be accomplished by introducing a multiplicity of automatic controls. In some processes, the manual control of temperature, pressure, liquid flow, viscosity, etc., is often impracticable because not only must the human operators deal with the effects of uncontrollable disturbances, but, in addition, operations react upon each other in an unpredictable manner, and still further complicate the already difficult tasks of the operators.

The paper has two main objects: first, to present some fundamental aspects of the design of automatic control systems; and second, to present an introduction to a modified approach to the problem of automatic regulator design which the author is studying at present.

Unfortunately, it will not be possible to present a detailed theoretical treatment of the subject in a single paper; this is not considered essential, since several papers on the theory of automatic control have already been published (Ivanoff 1934;

† An alphabetical list of references is given in the Appendix, p. 12

Hazen 1934; Callender, Hartree, Porter 1936; Prinz 1944; Hayes 1945; Whiteley 1946).

TERMINOLOGY

The terminology used in the study of automatic control systems presents a vexed question. No real attempt to standardize the nomenclature has been made until comparatively recently, and as a result many inconsistencies exist in the literature. Moreover, Service and industrial developments have progressed along independent paths, and it is difficult to form a glossary of terms which is acceptable to designers in both fields. The terminology used in this paper is based partly upon the recommendations of the Ministry of Supply Servo Nomenclature Panel, and partly upon generally recognized industrial usage.

A *control system* is defined as an assembly of elements (amplifiers, mechanical shafts and gears, steam valves, human operators, etc.) interconnected in such a way that the operation of each element is dependent upon the operation of at least one other element, and the purpose of which is to control some process or machine. The automatic control of water-level in a domestic cistern by a ball-tap arrangement, the tuning of a radio receiving set, the manual operation of a lathe, and the thermostatic control of modern central heating systems are examples of control systems in everyday use.

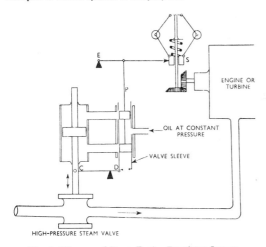

Fig. 1. Diagram of Steam Engine Regulator System

A *servo-system* or *servo-mechanism* is a control system which is power-amplifying, automatic, and error-actuated. An example is the automatic remote control of an anti-aircraft gun which is required to follow accurately the motion of the output shaft of a predictor. Power amplification is clearly necessary between the lightly loaded predictor shaft and the high inertia and friction load of the gun. The *error* is the angular misalignment (in azimuth for example) between the direction in which the gun is pointing and the correct direction indicated by the angular position of the predictor shaft. The operation of the servo-system is determined solely by the behaviour of the error (i.e. the system is error-actuated) which is usually measured by electrical methods. One major problem is to ensure that the error-measuring device does not react upon the input shaft and thereby cause spurious errors.

The majority of servo-systems in industry are required to maintain some physical or chemical condition of a process or plant at a specified value regardless of the effects of external disturbances; these systems are called *automatic regulators*. The automatic speed regulator system shown schematically in Fig. 1 is a typical and appropriate example, because the centrifugal speed governor, invented by James Watt towards

the end of the eighteenth century, was probably the first servo-system of widespread application in industry.

Referring to Fig. 1, it is clear that the speed-regulating system is fully automatic, error-actuated, and, in view of the incorporation of the hydraulic amplifier, it is power-amplifying. Under normal operating conditions, the position of the sleeve S of the governor, and the corresponding displacement of the high-pressure steam valve remain fixed until, for some reason, the speed of the engine changes. Suppose the speed of the engine increases above the desired speed, as a result, for example, of a reduction in the *load* or *demand*, or an increase in the steam pressure (i.e. increased *power supply*). The fly-balls rise and thereby lift the sleeve which in turn lifts the pilot valve P. This causes the upper port of the hydraulic amplifier to be opened, and oil pressure is applied in the main valve, the piston of which moves in a direction tending to close the high-pressure steam valve. In addition, the *feed-back* member CD lifts the pilot valve sleeve and causes the flow of oil to the main valve to be throttled down and eventually to be cut off. As a result, in the steady-state condition, the displacement of the steam valve from the position corresponding to zero error in speed is closely proportional to the displacement of the pilot valve. If the displacement of the governor sleeve is proportional to the error in speed, and the flow characteristic of the steam valve is linear, the regulator is classified as a *proportional control system*.

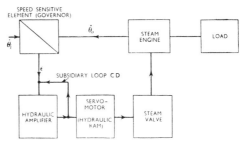

Fig. 2. Flow Diagram of Automatic Speed Regulator

$\dot{\theta}_i$ = Required operating speed.
$\dot{\theta}_o$ = Actual operating speed.
$\epsilon = \dot{\theta}_i - \dot{\theta}_o$ = Error.

On the other hand, if the pivoted member CD is omitted, the speed of opening or closing the steam valve is roughly proportional to the displacement of the pilot valve P. This mode of control is one in which the restoring action of the controller depends upon the time integral of the error; it is frequently called *proportional floating control*, or, if combined with proportional control in order to eliminate steady-state errors, it is called *reset* control. It is stressed that, in general, the proportional floating mode of control is unsatisfactory on account of the inherent tendency of such controllers to give self-oscillating operation. When combined with proportional and first derivative of error control, however, floating control is very desirable because, in addition to minimizing steady-state errors, it effectively increases the speed of response of servo-systems (Callender and others 1936, p. 437).

An appreciation of the operation of a servo-system is enhanced by the introduction of the concept of control signal. In a control system, a *control signal* is defined as the physical quantity passed on from one element to another. For example, in the speed regulator considered previously, the speed-sensitive element (i.e. the centrifugal governor) transmits a control signal in the form of a mechanical displacement to the pilot valve. The signal is passed from the pilot valve to the main valve as an oil pressure, then to the high-pressure steam valve as a displacement, and finally to the engine as a steam pressure.

A convenient way of representing the basic operation of a control system is by means of a *flow diagram*; the main elements are represented by "blocks" and the control signals by the interconnecting lines. The flow diagram of the steam engine speed regulator is shown in Fig. 2. The system is clearly a closed loop system; this characteristic is common to all servo-systems, automatic regulators, and to the majority of manually operated systems. In addition to the main control loop, a subsidiary loop, representing the feed-back link (CD in Fig. 1), is also shown in Fig. 2. Such so-called subsidiary loops are common in the design of servo-systems; their main uses are :—

(i) To provide a convenient means of obtaining a close approximation to derivative of error control.
(ii) To provide a means of obtaining integrated error control (e.g. automatic reset control).
(iii) To modify the characteristics of the control system. For example, the feed-back link CD of the speed regulator modifies the mode of operation of the control system from floating to proportional.

The most difficult problem in the design of a servo-system is to obtain high speed of response of the system without causing undesirable oscillations. Moreover, the existence of *exponential time-lags* and *finite time-lags* in the operation of certain elements in the control loop magnifies this difficulty. In extreme circumstances, the corrective action of the controller may be "lagging" to such an extent that the tendency is to increase rather than to decrease the error; it is difficult to avoid instability under these conditions.

Consider the operation of the speed regulator; lags are inherent in the operation of the centrifugal governor, the hydraulic amplifier and steam valve, and a finite time-lag is introduced on account of the steam valve being located some

of the error. In this connexion, controller action refers to the displacement of a control valve or rheostat from some arbitrary zero or, in the case of a position control system, to the torque applied to the load. The idealized operating characteristics of floating and proportional controls are illustrated in Figs. 3a and 3b respectively. The corresponding equations of control may be written :—

Floating control:—

$$V = K_o \int_o^\infty \epsilon \, dt \quad \ldots \ldots \quad (1)$$

or

$$\frac{dV}{dt} = K_o \epsilon \quad \ldots \ldots \quad (2)$$

in which V denotes the valve (or rheostat) position, K_o the rate of valve opening (or closing) per unit error, and ϵ the error, i.e. the difference between the input signal θ_i and the output signal θ_o.

Thus

$$\epsilon = \theta_i - \theta_o \quad \ldots \ldots \quad (3)$$

Proportional Control:—

$$V = K_1 \epsilon + A \quad \ldots \ldots \quad (4)$$

in which K_1 is the displacement of the valve from normal zero position per unit of error,[*] and A the normal valve position when $\epsilon = 0$.

The value of the coefficient K_1 usually determines the speed of response of a simple servo-system; a high value of K_1 gives high speed of response. In the case of a speed regulator, K_1 may be varied by varying the position of the fulcrum along the link CD, or by changing the position of the pivot E (see Fig. 1).

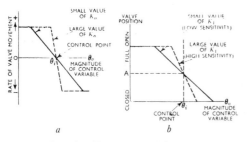
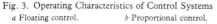

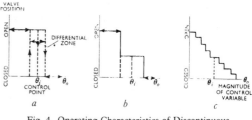

Fig. 3. Operating Characteristics of Control Systems

a Floating control.   *b* Proportional control.

Fig. 4. Operating Characteristics of Discontinuous Controllers

a Two-position controller.   *b* Three-position controller.
c Multi-position controller.

distance from the engine (see Fig. 1). These lags retard the corrective action of the controller, and, if the system is subjected to such extraneous disturbances as changes in temperature and pressure of steam or demand changes, it is usually necessary to take special precautions to remove oscillatory tendencies of the system. The process of ensuring that the response of a control system, subsequent to the application of a disturbance, is adequately damped is called *stabilization*. One method of stabilizing the speed regulator is to increase the coefficient of viscous friction at the sleeve of the governor. An alternative method will be considered in a later section of the paper.

CHARACTERISTICS OF CONTINUOUS AND DISCONTINUOUS SERVO-SYSTEMS

Servo-systems may be divided into two main classes according to the fundamental mode of operation. These are "continuous" and "discontinuous" systems respectively. Examples of the former class have already been considered in the previous section; for example, proportional control and floating control (integrated error control) are necessarily continuous because the action of the controller is a continuous function of the error (or integral of error) in both cases.

It is convenient to define the operating characteristics of a control system as the action of the controller plotted as a function

A discontinuous servo-system is one in which the action of the controlling element is not a continuous function of the error. For example, if the sleeve of the speed-regulating governor operated a two-position switch which controlled the direction of operation of the servo-motor positioning the steam valve, the regulator would be a discontinuous servo-system. Another example is the thermostatically operated switch which controls an electric heater. In addition to the simple "on–off" or two-position controller, the three-position and multi-position types of discontinuous control are also utilized in industry. Idealized characteristics of each of these types of control are illustrated in Figs. 4a, 4b, and 4c.

Although the majority of industrial servo-systems are discontinuous, the basic theory of these systems has been neglected. This is due to the non-linear nature of the problem; even the most elementary on–off control systems are difficult to handle mathematically. Nevertheless, designers of control systems have shown that on–off controllers give very satisfactory results in the control of processes in which lags and extraneous disturbances are small.

This paper will deal almost exclusively with continuous and linear control systems.

[*] The coefficient K_1 partially determines the "sensitivity" of a servo-system. In a position control system, it is called the "stiffness coefficient" of the system.

TRANSIENT AND STEADY-STATE BEHAVIOUR OF SERVO-SYSTEMS

The performance of a servo-system is usually based upon two considerations :—

(1) The behaviour of the system when subjected to a sudden disturbance ; this is called the transient response of the system. A convenient method of determining the transient response of a speed-regulated engine, for example, would be to move suddenly the steam valve from its steady-state position, and to record the subsequent motion of the sleeve of the governor.

(2) The magnitude of the steady-state error arising, for example, when the output of the servo-system is required to follow a steadily changing input signal, or, in the case of an automatic regulator, when a change in the demand or load occurs.

It is impossible to specify, in general terms, the most desirable transient behaviour of all servo-systems. Each system should be treated as a separate design problem. In some systems it may be worthwhile to tolerate an appreciable "overshoot" in order to achieve high speed of response. In other systems the major requirement may be exceptionally heavy damping ; this condition inevitably leads to appreciable deterioration in the speed of response. Some typical transient response curves for a servo-system with various degrees of damping are shown in Fig. 5 ; the input disturbance in each case is a conventional "unit function".

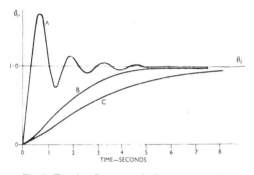

Fig. 5. Transient Response of a Servo-system with Various Degrees of Damping

Referring to Fig. 5, curve A shows the transient behaviour of an "under-damped" servo-system ; it has the desirable feature of rapid recovery, but the response is inherently oscillatory. Curve B shows the behaviour of a critically damped system, which, in many of the more elementary servo-systems, is the most desirable condition ; in more elaborate systems, excessive "overshoot" may occur even though the system is critically damped. Curve C shows the over-damped condition ; the speed of recovery of the system subsequent to a disturbance is unavoidably low.

The second important criterion is the magnitude of the steady-state error. In the operation of a position control system, this error may be due to the effect of :—

(1) Coulomb friction torque at the load shaft.
(2) Constant wind torque acting on the load.
(3) Viscous friction and inertia associated with the output shaft and load.

If the effective wind torque (acting, for example, on a large radar aerial system which may be regarded as the load) is assumed to be independent of the speed of rotation of the aerial, the effects of (1) and (2) above may be combined. The resulting steady-state error is proportional to the combined friction and windage torques, and inversely proportional to the stiffness coefficient of the control system.

Viscous friction, the damping effect of which may be simulated by electrical methods such as eddy-current brakes, causes a steady-state error frequently called velocity-lag, the magnitude of which is proportional to the coefficient of viscous friction, to the angular or linear speed of the load, and inversely proportional to the stiffness coefficient. Similarly, inertia or acceleration lag in a position control servo-system is the steady-state error proportional to the moment of inertia of the load and servo-motor referred, for example, to the output shaft, to the acceleration of the output shaft, and inversely proportional to the stiffness coefficient.

These various components of the total steady-state error of a position control system can be corrected by introducing a term dependent upon the time-integral of the error into the operation of the controller.

Consider now the operation of an automatic regulator. A steady-state error may arise in the operation of the control system when, for example, extraneous disturbances affect the system, or when changes in the normal demand or power supply occur. In industrial processes this error is called "droop". It is proportional to the change in demand or power supply (e.g. an increased demand or decreased power supply may give a positive droop), and inversely proportional to the sensitivity K_1 of the system. Most modern automatic regulators incorporate means for automatically correcting droop by introducing integrated error control ; these systems are called automatic reset controllers. It is necessary, however, to correct droop manually, by resetting the main control valve to a position corresponding to the modified conditions, when the controller is a simple on–off system or a proportional controller.

It is sometimes impossible to specify fully the criteria for good performance of a servo-system in terms of the transient response of the system and the magnitude of the steady-state errors. In these cases it is necessary to specify the harmonic response characteristics of the control. The harmonic response of a control system is the steady-state behaviour of the output of the system relative to the input expressed as an amplitude and phase. A range of simple harmonic input signals, covering a wide frequency band, is normally used in this determination. The results are plotted either as two Cartesian graphs showing (a) the ratio of output to input amplitudes, and (b) the phase angle between the output and input signals, as functions of frequency, or as a composite vector diagram, sometimes called a Nyquist Diagram (Nyquist 1932), which is a parametric polar graph showing the locus of the output vector with respect to a unit input vector with frequency as the parameter.

A designer, skilled in the art of interpreting harmonic response diagrams, can utilize this technique in the synthesis of a servo-system with a desired performance, provided a reasonable degree of linearity is anticipated. For example, having determined the harmonic response diagrams of the main components of the system, the harmonic response of the complete servo-system may be synthesized by applying the standard procedure for obtaining vector products. On the other hand, if the transient response of each component is known, the work involved in synthesizing the response of the complete system would be prohibitive.

Harmonic response investigations are also valuable in the determination of the resonance frequencies of a control system, and the degree of the "resonance". Normally, an amplification factor of 1·40 to 1·60 at a particular frequency is the maximum which can be tolerated.

In certain classes of servo-systems it is desirable for the system to discriminate between real and spurious error signals ; in these cases the servo-system must behave as a band-pass filter. This situation arises in auto-following radar equipment where the true error signal is combined with a considerable amount of "noise". A knowledge of the harmonic response characteristics is essential in the design of these systems.

THE SIGNIFICANCE OF FEED-BACK

A fundamental property of all error-actuated control systems is the existence of at least one feed-back link. This leads to the conception of the closed loop control system shown, for example, as a flow diagram in Fig. 2. In addition to this main feed-back link, which has been called *monitoring feed-back* by the Servo Nomenclature Panel, most servo-systems incorporate one or more subsidiary links. The main reasons for the introduction of these have already been given in the section on Terminology, p. 4. The object of this section is to consider the problem in more detail.

It is convenient to represent a typical subsidiary loop of a servo-system by the simple flow diagram shown in Fig. 6. The input signal is the normal servo-error signal ϵ, and the output signal ϵ' of the subsidiary loop actuates the subsequent stages in the main control loop (Bode 1945, p. 31).

Let the "block" X represent a straightforward electric, hydraulic or pneumatic amplifier having a "gain factor" K in the frequency band under consideration, and let the feed-back element, depicted by the "block" Y, carry out some operation upon the signal ϵ', represented by the operational expression $F(p)\epsilon'$. If, for example, the element Y exerts a simple exponential lagging action upon the signal ϵ', $F(p)\epsilon'$ can be written :—

$$F(p)\epsilon' = \frac{\epsilon'}{1+pT} = \lambda \quad . \quad . \quad . \quad . \quad (5)$$

in which T denotes the time constant of the system, and p is equivalent to d/dt. Equation (5) written in non-operational form is :—

$$\lambda + T\frac{d\lambda}{dt} = \epsilon' \quad . \quad . \quad . \quad . \quad . \quad (6)$$

The lagging operation may be realized in practice by means of a RC electrical network, or the equivalent spring-dashpot mechanical network.

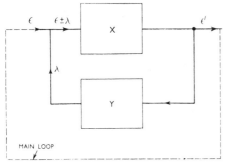

Fig. 6. Subsidiary Loop of Servo-system

If the feed-back is degenerative (i.e. if the feed-back signal is subtracted from the main signal) the equation of the system is :—

$$\epsilon' = K(\epsilon - \lambda) \quad . \quad . \quad . \quad . \quad . \quad (7)$$

or, substituting for λ,

$$\epsilon' = \frac{K}{1+KF(p)}\epsilon \quad . \quad . \quad . \quad . \quad . \quad (8)$$

Suppose the operation of the feed-back element is defined by equation (5); substitution for $F(p)\epsilon'$ in equation (8) gives

$$\epsilon' = \frac{K}{K+1} \times \frac{1+Tp}{1+\dfrac{Tp}{1+K}}\epsilon \quad . \quad . \quad . \quad (9)$$

If the gain factor of the amplifier is high, e.g. if $K\gg1$, equation (9) may be written approximately as

$$\epsilon' = (1+Tp)\epsilon$$

or

$$\epsilon' = \epsilon + T\frac{d\epsilon}{dt} \quad . \quad . \quad . \quad . \quad . \quad (10)$$

This type of feed-back loop provides a signal dependent upon the servo-error ϵ and upon a good approximation to the first time derivative of the error signal. It is introduced into the main servo-loop to ensure adequate damping of the system. The complete control system is described as a "proportional plus first derivative" controller.

Alternatively, suppose the feed-back link considered above is regenerative (i.e. the feed-back signal is added to the main signal), the equation of system is

$$\epsilon' = K(\epsilon + \lambda) \quad . \quad . \quad . \quad . \quad . \quad (11)$$

The equation corresponding to (9) above is, in this case

$$\epsilon' = \frac{K(1+Tp)}{1-K+Tp}\epsilon \quad . \quad . \quad . \quad . \quad (12)$$

If the gain factor is nearly equal to unity, i.e. $(1-K)\to 0$, equation (12) may be written approximately as

$$\epsilon' = \frac{K}{Tp}\epsilon + K\epsilon$$

or

$$\epsilon' = K\epsilon + \frac{K}{T}\int \epsilon\, dt \quad . \quad . \quad . \quad . \quad (13)$$

The output signal ϵ' of the subsidiary loop system, which actuates the main control system, is the sum of two components, one proportional to the input signal ϵ, and the other proportional to the integrated input signal. If ϵ is the main servo-system error, the effect of applying a controller defined by equation (13) is to eliminate steady-state errors. The system is called a "proportional plus integral" controller.

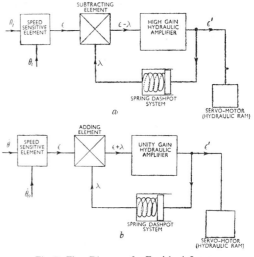

Fig. 7. Flow Diagrams for Feed-back Loops

a Degenerative feed-back loop for stabilizing automatic speed regulator.
b Regenerative feed-back loop for correcting "droop" in automatic speed regulator.

Either of the feed-back loops considered above may be incorporated into the automatic speed regulator. For example, if the system tends to "hunt", the degenerative feed-back loop is utilized, or, if appreciable steady-state errors arise as a result of large load or supply changes, the regenerative loop is used. The flow diagrams shown in Figs. 7a and 7b correspond to the two cases.

AUTOMATIC REGULATORS: INCORPORATION OF SUBSIDIARY FEED-BACK LOOPS *

It is generally agreed that all automatic control systems have a common basic theory, although several important differences

* The author's attention has been drawn recently to two excellent papers by Mr. Paul G. Kaufmann, A.M.I.Mech.E., published in War Emergency Proceedings No. 9 (Proc. I.Mech.E., 1945, vol. 153, pp. 237–257). These papers cover many of the ideas introduced in this section, and constitute a major contribution to the subject.

in design procedures exist between automatic regulators and position control servo-systems. A comparison of the fundamental characteristics of these two classes of control systems has suggested a modified approach to the problem of improving the performance of some automatic regulators; this approach will be considered in detail.

The relatively large difference in the magnitude of the time scale is usually regarded as the most important basic difference between the operation of position and process control systems. For example, the lags inherent in a typical position control system rarely have time constants exceeding 1 second, whilst in process control systems time constants of the order of minutes or even hours are common. In classifying automatic control systems, some control engineers find it convenient to link together position control systems with automatic regulators such as automatic voltage and speed regulators, etc., which have high speeds of response. Alternatively, process control systems controlling such physical states as the temperature, liquid level, humidity, etc., of a process are classified as slow-speed response systems. This distinction is sometimes misleading, however, and the author prefers to classify servo-systems as (a) systems in which the input signal varies as a

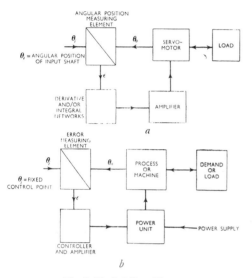

ANGULAR POSITION
MEASURING
ELEMENT

θ_i = ANGULAR POSITION OF INPUT SHAFT

DERIVATIVE
AND/OR
INTEGRAL
NETWORKS

ERROR
MEASURING
ELEMENT

θ_i = FIXED
CONTROL POINT

CONTROLLER
AND AMPLIFIER

a

b

Fig. 8. Typical Flow Diagrams

a Position-control servo-system.
b Automatic regulator.

continuous function of time, and (b) systems in which the input signal is constant (e.g. automatic regulators).

Typical flow diagrams of a simple position control servo and an automatic regulator are shown in Figs. 8a and 8b respectively. A common fundamental property of both systems is the closed main control loop. Apart from differences in the magnitudes of the lags, the two systems only differ in respect of the nature of the "external" (i.e. outside the main loop) disturbances or influences which affect their behaviour.

Suppose the variations in the input signal θ_i are considered as an external disturbance in the case of the position control servo and variations in the normal demand D_N and power supply S_N are considered as external disturbances in the case of the regulator. If these disturbances are negligible, the main control loops of both the servo and regulator are, if correctly adjusted, very stable systems. The aim of the designer is to ensure that the effect of all external influences upon the inherent stability of the main control loops is as small as possible.

It is not only the nature and magnitude of the external

disturbances but also their location in the main control loop which, in addition to the lags inherent in the system, determine the performance of a controller. In a position control servo-system the major external disturbance acting on the main loop is the variation of θ_i. Additional disturbances such as windage torques acting upon the load, variations in oil or air pressure, the effect of temperature changes on mechanical and electrical components, etc., are generally negligible compared with the effect on the control loop of continuous variations in the input signal. This characteristic, common to the majority of position control servo-systems in which the input signal changes (e.g. gun and radar control systems) is most desirable from the point of view of the designer. The location of the major disturbing influence, as shown in Fig. 8a, is ideal because, being associated with the error-measuring element, its effect is detected and measured almost instantaneously. The resulting error signal can then initiate the correcting action of the control system with minimum delay.

The design of an automatic regulator presents a basically different problem. The input signal θ_i is constant in this case, and subsequent to the "starting-up" or transient period, it does not constitute a disturbance on the normal operation of the system. There are, however, other important external disturbances which affect the behaviour of the control system, and which are the primary cause for the necessity of incorporating automatic regulation. Moreover, as shown in Fig. 8b, these disturbances are usually associated either with the process under control or with the source of the power supply. Some of them are:—

(1) Changes in the demand from the normal operating value. This type of disturbance, in addition to causing an error in the system, modifies the fundamental characteristics of the control (for continuous changes in demand the control parameters of the system are continuous functions of time); the effect may be continuous or discontinuous. For example, in a voltage regulator system the demand change may be a continuous function of time, whereas in a temperature control system it may be discontinuous.

(2) Changes in the power supply from normal. These may include, for example, changes in mains voltage, steam pressure, gas temperature, etc. This type of disturbance affects the sensitivity of the control directly.

(3) Changes in external conditions which affect plant operation; e.g. changes in ambient temperature.

The effect of each of the above disturbances is not communicated to the error-measuring element immediately, and, especially in the case of disturbances arising from changes in power supply, some appreciable time interval may elapse before the controller is fully "aware" of the magnitude of the disturbance. In general, the more "remote" (in terms of the flow diagram) the disturbance from the error detecting element, the larger the additional lag introduced into the system; for example, referring to Fig. 8b, the power supply disturbance can be regarded as more remote than the demand disturbance. Consider the operation of a temperature-control system as an example of a simple regulator. The effect of a change in the demand or power supply is not transmitted instantaneously to the temperature-sensitive element because a certain time interval is required for thermal potentials to be established, and for heat transfers in the plant to be effected (this time delay is determined by the magnitude of the thermal capacities and resistances inherent in the system).

The basis of the modified approach to the problem of speeding-up the response of an automatic regulator, without introducing undesirable oscillations, is the introduction of feed-back loops which "short-circuit" the additional lags referred to previously. The object is to ensure that the occurrence and magnitude of a major external disturbance are transmitted without delay to a suitable point in the main control loop, regardless of the location of the disturbance.

If signals proportional to the magnitude of the disturbances ΔD and ΔS are fed back and combined with the normal error signal ϵ, as indicated in the dotted portion of the flow diagram shown in Fig. 9, the operation of the controller anticipates the effect of the disturbances, and the correcting action is accelerated. This procedure has been called "disturbance feed-back". Moreover, suppose the disturbance signals are modified by elements F_1 and F_2 (if the disturbances are given as voltages, for example, F_1 and F_2 are electrical networks), as shown in Fig. 9, which provide signals γ_1 and γ_2 given approximately by

$$\gamma_1 = \Delta D + T_1 \frac{d(\Delta D)}{dt} \quad . \quad . \quad . \quad . \quad (14)$$

and

$$\gamma_2 = \Delta S + T_2 \frac{d(\Delta S)}{dt} \quad . \quad . \quad . \quad . \quad (15)$$

the response of the controller should be still further accelerated. The introduction of the derivative terms in (14) and (15) should be particularly effective in regulator systems in which demand and power supply changes are continuous. This technique is to a certain extent analogous to the feed-back stabilization of a position control servo-system in which a signal proportional to the velocity of the output shaft is fed back and, after modification by a network, combined with the error signal. It should be possible, therefore, to determine the most effective type of disturbance feed-back network by methods already well established in the design of position control systems.

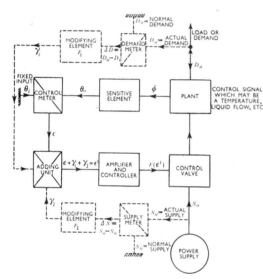

Fig. 9. Flow Diagram of Typical Process Control System with Disturbance Feed-back

DISTURBANCE FEED-BACK APPLIED TO SIMPLE THERMAL CONTROL SYSTEMS

The following elementary analysis is restricted to the study of a temperature-regulating system in which the only important disturbance is due to changes in the demand. It is also convenient to make the following assumptions:—

(1) The "distributed" parameters of the system can be "lumped".
(2) The system is linear.
(3) Lags in the process and controller, apart from the lags associated with the thermal capacity of the tank and its contents, are neglected.
(4) The controller is assumed to be a proportional controller.

Straightforward disturbance feed-back will be considered rather than modified disturbance feed-back; this simplifies the mathematics appreciably.

Fig. 10 shows a schematic diagram of a temperature-controlled

tank through which liquid is flowing. The rate of flow of the liquid is measured by the flowmeter at X, and if the value deviates from a normal predetermined rate, a signal, proportional to the deviation, can be fed back and added to the normal error signal at P.*

Let θ_i = Desired temperature of outflowing liquid.
$\quad \theta_o$ = Actual temperature of outflowing liquid.
$\quad \epsilon$ = $\theta_i - \theta_o$ = Error.
$\quad H_i$ = Heat entering the main tank per second (calories per sec.).
$\quad H_o$ = Heat leaving the main tank per second.
$\quad C_s$ = Thermal capacity of heating jacket J; e.g. the "supply-side" capacity (calories per degree).
$\quad C_d$ = Thermal capacity of main tank and liquid; e.g. the "demand-side" capacity.
$\quad k$ = Thermal conductivity of tank wall, measured in calories per second per degree.
$\quad D_n$ = Normal rate of flow of liquid in grams per second.
$\quad D_a$ = Actual rate of flow of liquid.
$\quad \Delta D$ = $D_a - D_n$ = Disturbing function.
$\quad s$ = Specific heat of liquid.

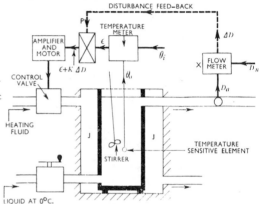

Fig. 10. Diagram of Simple Temperature-control System

The simplified equations of the control system are :—
Heat transfer equation :

$$\frac{C_s}{k} \times \frac{C_d}{D_a s} \times \frac{d^2 H_o}{dt^2} + \left(\frac{C_s}{k} + \frac{C_s}{D_a s} + \frac{C_d}{D_a s}\right)\frac{dH_o}{dt} + H_o = H_i \quad (16)$$

in which

$$H_o = D_a s \theta_o \quad . \quad . \quad . \quad . \quad . \quad (17)$$

The equation of the proportional controller is

$$H_i = K_1 \epsilon + D_n s \theta_i \quad . \quad . \quad . \quad . \quad (18)$$

The constant term $D_n s \theta_i$ is introduced into equation (18) to satisfy the requirements when $\epsilon = 0$. In the steady-state condition with $\epsilon = 0$, the rate of heat flow into the tank must be just sufficient to satisfy the normal demand; in these circumstances equation (18) can be written

$$H_i = H_o = D_n s \theta_i$$

Suppose the system, assumed to be initially in a steady state, is suddenly subjected to an increase in the rate of flow of the liquid from the normal value D_n to a new value D_a. This disturbance can be expressed as follows :—

At $t \leqslant 0$, rate of liquid flow = D_n grams per second.
At $t \geqslant 0$, rate of liquid flow = $D_n + \Delta D = D_a$ grams per second.

* This does not correspond exactly to disturbance feed-back; the real disturbance corresponds to the change in demand (measured in calories per second) from normal demand. Nevertheless, the deviation in rate of liquid flow is roughly proportional to the change in demand.

It is desirable to consider the transient behaviour of several related control systems for purposes of comparison; these systems will be designated system A, system B, etc.

System A. This is an idealized system in which the supply-side capacity C_s can be neglected. The system has a proportional controller, but no disturbance feed-back. Substituting for H_i and H_o in equation (16), and writing $C_s = 0$, gives the control equation

$$C_d\frac{d\theta_o}{dt}+(K_1+D_as)\theta_o = (K_1+D_ns)\theta_i \quad . \quad . (19)$$

The initial conditions associated with the disturbing function defined previously are :—

At $\qquad t = 0,\ \theta_o = \theta_i,$ and $\dfrac{d\theta_o}{dt} = \dfrac{\varDelta Ds\theta_i}{C_d} \quad . \quad . \quad . (20)$

The solution of equation (19) with initial conditions given in equation (20) is

$$(\theta_o-\theta_i) = -\frac{\varDelta Ds}{K_1+D_as}\Big(1-e^{-\frac{(K_1+D_as)t}{C_d}}\Big)\theta_i \ . \ (21)$$

RATE OF FLOW OF LIQUID   RATE OF FLOW OF LIQUID

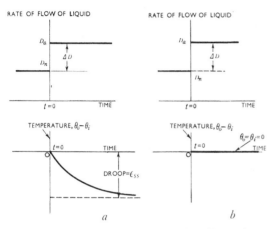

Fig. 11. Behaviour of Systems with Different Forms of Control

a Single-capacity thermal control system with proportional control subjected to sudden change in demand.
b System with proportional control plus disturbance feed-back.

This function is plotted in Fig. 11a. The steady-state error ϵ_{ss} (i.e. droop) is given by

$$= \frac{D_as\theta_o-D_ns\theta_i}{K_1} = \frac{\text{Change in demand}}{\text{Sensitivity}} \quad . \quad . (22)$$

System B. Consider the effect of introducing disturbance feed-back into the ideal system examined in the previous case. Equation (18) is modified as follows :—

$$H_i = K_1\epsilon+Q\varDelta D+D_ns\theta_i \quad . \quad . \quad . \quad . (23)$$

in which $Q\varDelta D$ is the disturbance feed-back term. The coefficient Q, which has the dimensions of calories per gram, determines the magnitude of the feed-back signal.

Combining equations (16), (17), and (23), and substituting $C_s = 0$ gives

$$C_d\frac{d\theta_o}{dt}+(K_1+D_as)\theta_o = (K_1+D_as)\theta_i+\varDelta Ds\theta_i-Q\varDelta D \ . (24)$$

In order to eliminate the steady-state error, the amount of feed-back is adjusted so that

$$Q = s\theta_i \quad . \quad . \quad . \quad . \quad . \quad . (25)$$

The equation of the control system can now be written

$$C_d\frac{d\theta_o}{dt}+(K_1+D_as)\theta_o = (K_1+D_as)\theta_i \quad . \quad . (26)$$

and the initial conditions are: at $t = 0,\ \theta_o = \theta_i,$ and $d\theta_o/dt = 0$. The solution of equation (26) is $\theta_o = \theta_i$; the corresponding diagram is Fig. 11b.

It would appear, therefore, that the behaviour of the ideal system B is unaffected by a sudden change in the demand if the correct amount of disturbance feed-back is introduced. A comparison of Figs. 11a and 11b shows the obvious advantage of incorporating disturbance feed-back. Admittedly, the performance of the normal system could be improved considerably by introducing integrated error control (for eliminating droop), and first derivative of error control (to ensure adequate damping), but it is quite impracticable to obtain by these means the perfect control indicated in Fig. 11b.

System C. In this system the effect of the supply-side capacity will be included. Assuming a proportional controller, the equation of the system may be obtained by combining

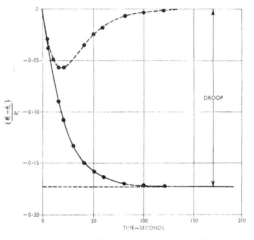

Fig. 12. Response of Critically Damped Thermal System when subjected to a Sudden 50 per cent Increase in Demand
————————— Proportional controller.
— — — — — Proportional controller with disturbance feed-back.

equations (16), (17), and (18). The combined control equation is

$$\frac{C_sC_d}{k}\frac{d^2\theta_o}{dt^2}+\Big(\frac{C_sD_as}{k}+C_s+C_d\Big)\frac{d\theta_o}{dt}+(K_1+D_as)\theta_o$$
$$= (K_1+D_ns)\theta_i \quad . \quad (27)$$

The disturbing function is the same as for the previous cases, e.g. :—

At $\qquad t < 0,\ D_a = D_n$
At $\qquad t > 0,\ D_a = D_n+\varDelta D$ $\qquad . \quad . \quad . \quad . (28)$

Moreover, the steady-state error is the same as for system A and is given by equation (22).

An examination of equation (27) indicates that the coefficients of θ_o and $d\theta_o/dt$ each contains a term depending upon the value of D_a. If the demand changes continuously, the basic characteristics of the control system must also change continuously. The solution of equation (27) is not difficult to handle if the disturbing function is the straightforward step function defined in equations (28). On the other hand, if the disturbance $\varDelta D$ changes continuously, the solution can only be obtained conveniently by numerical or mechanical (e.g. differential analyser)

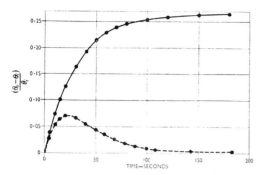

Fig. 13. Response of Critically Damped Thermal System when subjected to a Sudden 50 per cent Decrease in Demand

——————————— Proportional controller.
— — — — — Proportional controller with disturbance feed-back.

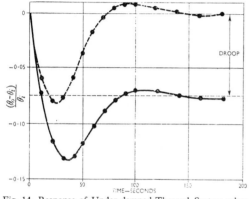

Fig. 14. Response of Under-damped Thermal System when subjected to a Sudden 50 per cent Increase in Demand

— · —————————— Proportional controller.
— · — · ——————— Proportional controller with disturbance feed-back.

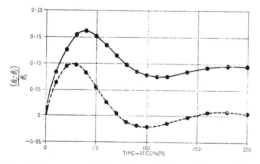

Fig. 15. Response of Under-damped Thermal System when subjected to a Sudden 50 per cent Decrease in Demand

— · ——————— · · Proportional controller.
— · · — · — · — Proportional controller with disturbance feed-back.

methods. In this respect at least the theoretical treatment of automatic regulating systems is more difficult than the treatment of position control systems, in which the control coefficients can normally be regarded as constant.

Solutions of equations (27) and (28) have been obtained with the following conditions :—

(1) Critically damped system; rate of flow of liquid increased by 50 per cent.
(2) Critically damped system; rate of flow of liquid decreased by 50 per cent.
(3) Under-damped system; rate of flow of liquid increased by 50 per cent.
(4) Under-damped system; rate of flow of liquid decreased by 50 per cent.

Typical values of the control constants (C_s, C_d, and k) and the control sensitivity parameter K_1 were chosen to satisfy the above conditions, and the corresponding solutions of the control equations were obtained. The results are presented as graphs showing the transient behaviour of the systems; plots of $(\theta_o - \theta_i)/\theta_i$ as a function of t are shown in the full curves of Figs. 12, 13, 14, and 15. Conditions (1) and (3) above give a positive droop, and conditions (2) and (4) give a "negative" droop.

System D. This system is identical with system C except for the introduction of disturbance feed-back. The equation of the controller, assuming the amount of the feed-back is adjusted to eliminate steady-state errors, is

$$H_i = K_1\epsilon + (D_n + \Delta D)s\theta_i = K_1\epsilon + D_a s\theta_i \quad . \quad . (29)$$

and the equation of the complete control system is

$$\frac{C_s C_d}{k}\frac{d^2\theta_o}{dt^2} + \left(\frac{C_s D_a s}{k} + C_s + C_d\right)\frac{d\theta_o}{dt} + (K_1 + D_a s)\theta_o$$
$$= (K_1 + D_a s)\theta_i \quad . \quad . (30)$$

with initial conditions given by equations (28).

Solutions of equation (29) have been obtained for each of the four sets of conditions applied to system C, and corresponding results are plotted together on the same graph; e.g. the responses of the system with disturbance feed-back are shown as dotted curves in Figs. 12, 13, 14, and 15.

The desirability of introducing disturbance feed-back, from the point of view of rapid response coupled with adequate damping and zero steady-state error, is clearly demonstrated in these results. An examination of the graphs reveals that the advantages to be gained by introducing disturbance feed-back are accentuated as the inherent damping of the system is increased. For example, the improvement achieved when the system is critically damped is more marked than when the system is under-damped. This is an important point, in view of the fact that the designer usually strives to obtain a critically damped system.

System E. The effectiveness of the normal proportional controller when the system is subjected to changes in demand is considerably impaired on account of the existence of steady-state errors; these are unavoidable in the circumstances. Moreover, this factor certainly biases the previous results still more in favour of proportional control with disturbance feed-back. In order to determine the extent to which the results obtained with the normal controller can be improved by eliminating these errors, a term dependent upon the time integral of the error is introduced into the operation of the controller in system E. In this case the equation of the controller, corresponding to equation (18) is

$$H_i = K_o\!\int\!\epsilon\,dt + K_1\epsilon + D_n s\theta_i \quad . \quad . \quad . \quad . (31)$$

and the complete equation of the system is

$$\frac{C_s C_d}{k}\frac{d^3\theta_o}{dt^3} + \left(\frac{C_s D_a s}{k} + C_s + C_d\right)\frac{d^2\theta_o}{dt^2} + (D_a s + K_1)\frac{d\theta_o}{dt} + K_o\theta_o$$
$$= K_o\theta_i \quad . \quad . (32)$$

The disturbing function ΔD is again defined by equations (28).

Two typical solutions of equation (32) have been obtained for systems having a small value and a large value of K_o respectively.

The same values of the control constants and of K_1 as were used in the critically damped case considered previously are used in system E. The results are shown in Fig. 16. and the corresponding curve obtained with proportional control combined with disturbance feed-back is shown in the same

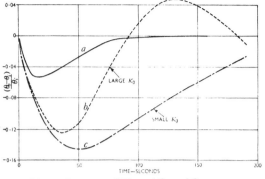
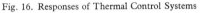

Fig. 16. Responses of Thermal Control Systems

a Proportional controller with disturbance feed-back.
b and *c* Proportional plus floating controllers when systems are subjected to sudden 50 per cent increase in demand.

diagram. A small value of K_o causes the system to be sluggish, whilst a large value of K_o causes oscillations. The improved performance obtainable with disturbance feed-back is clearly demonstrated. Admittedly, the results with the proportional plus integral controller could be appreciably improved by

introducing derivative of error control, but this also applies to the system with disturbance feed-back. Moreover, the introduction of a term dependent upon the rate of change of the disturbance into the latter system might be expected to improve still further the already high performance.

Acknowledgements. The author wishes to acknowledge the appreciable help he has received from Mr. A. M. Walker, of Sydney Sussex College, Cambridge, who carried out all the quantitative work involved in the study of disturbance feed-back. Acknowledgements are also due to the Director-General of Scientific Research (Defence), and to the Director, National Physical Laboratory, for permission to present this paper.

APPENDIX

BIBLIOGRAPHY

BODE, H. W. 1945 "Network Analysis and Feed-back Amplifier Design", Van Nostrand, New York.

CALLENDER, A., HARTREE, D. R., PORTER, A. 1936 Phil. Trans. Roy. Soc. A, 1936, vol. 235, p. 415.

HALL, A. C. 1943 "The Analysis and Synthesis of Linear Servomechanisms", Electrical Engineering Doctorate Thesis, Massachusetts Institute of Technology.

HAYES, K. A. 1945 Proc. of a Joint Conference of the Inst. Physics and the Inst. Chemical Eng. on "Automatic Controllers".

HAZEN, H. L. 1934 Jl. Franklin Inst. (U.S.A.), vol. 218, No. 5, p. 543.

IVANOFF, A. 1934 Jl. Inst. Fuel, vol. 7, No. 33, p. 117.

NYQUIST, H. 1932 Bell System Tech. Jl., vol. 11, p. 126.

PRINZ, D. G. 1944 *Jl. Scientific Instruments*, vol. 21, No. 4, p. 53.

WHITELEY, A. L. 1946 Jl. I.E.E., vol. 93, No. 34, p. 353.

Part III

ENERGY CONTROL IN LIVING
SYSTEMS: HOMEOSTASIS

Editor's Comments
on Paper 19

19   CANNON
Organization for Physiological Homeostasis

The importance of control of energy in living organisms has already been mentioned in the Introduction, together with the name *homeostasis*, to describe the regulatory mechanisms associated with this control. We reproduce as Paper 19 the fundamental article by Walter B. Cannon (1871–1945) "Organization for Physiological Homeostasis" (1929), which provides an excellent summary (with historical notes) of this aspect of energy control.

It should be recalled that the idea of the importance of a regulatory mechanism to maintain the stability of an organism was first emphasized by the great French physiologist Claude Bernard (1813–1878). This resulted from his philosophical interest in the general problems of physiology, in which he was led to place emphasis on what he called the "internal environment" *(milieu interieur)*, the stability of which he regarded as the precondition of an independent life. This idea was the precursor of Cannon's considerations. The latter coined the term *homeostasis* and discussed its application to all aspects of the endeavor of the living organism to adjust itself to and so survive in a variable external environment.

Cannon's article, which he followed up in 1932 with his well-known book *The Wisdom of the Body* (W. W. Norton, New York, second, revised edition, 1939), provides an excellent review of the subject as it stood in his time. It evidently had a strong influence on future work, particularly in connection with the cybernetics of N. Wiener, as is made clear in later articles in this volume.

Cannon was professor of physiology at Harvard University from 1906–1942.

19

Reprinted from *Physiol. Rev.* **9**(3):399–431 (1929)

ORGANIZATION FOR PHYSIOLOGICAL HOMEOSTASIS

WALTER B. CANNON

The Laboratories of Physiology in the Harvard Medical School

Biologists have long been impressed by the ability of living beings to maintain their own stability. The idea that disease is cured by natural powers, by a *vis medicatrix naturae*, an idea which was held by Hippocrates, implies the existence of agencies ready to operate correctively when the normal state of the organism is upset. More precise modern references to self-regulatory arrangements are found in the writings of prominent physiologists. Pflüger (1877) recognized the natural adjustments leading toward the maintenance of a steady state of organisms when he laid down the dictum, "The cause of every need of a living being is also the cause of the satisfaction of the need." Similarly Fredericq (1885) declared, "The living being is an agency of such sort that each disturbing influence induces by itself the calling forth of compensatory activity to neutralize or repair the disturbance. The higher in the scale of living beings, the more numerous, the more perfect and the more complicated do these regulatory agencies become. They tend to free the organism completely from the unfavorable influences and changes occurring in the environment." Further, Richet (1900) emphasized the general phenomenon,—"The living being is stable. It must be in order not to be destroyed, dissolved or disintegrated by the colossal forces, often adverse, which surround it. By an apparent contradiction it maintains its stability only if it is excitable and capable of modifying itself according to external stimuli and adjusting its response to the stimulation. In a sense it is stable because it is modifiable—the slight instability is the necessary condition for the true stability of the organism."

To Claude Bernard (1878) belongs the credit of first giving to these general ideas a more precise analysis. He pointed out that in animals with complex organization the living parts exist in the fluids which bathe them, i.e., in the blood and lymph, which constitute the "milieu interne" or "intérieur"—the internal environment, or what we may call the *fluid matrix* of the body. This fluid matrix is made and controlled by the organism itself. And as organisms become more independent, more free from changes in the outer world, they do so by preserving uniform their own inner world in spite of shifts of outer circumstances. "It is the fixity of the 'milieu intérieur' which is the condition of free and independent life," wrote Bernard (1878, i, pp. 113 and 121), "all the vital mechanisms, however varied they may be, have only one object, that of preserving constant the conditions of life in the internal environment." "No more pregnant sentence," in Haldane's (1922) opinion, "was ever framed by a physiologist."

DEFINITION OF HOMEOSTASIS. The general concept suggested in the foregoing quotations may be summarized as follows. The highly developed living being is an open system having many relations to its surroundings—in the respiratory and alimentary tracts and through surface receptors, neuromuscular organs and bony levers. Changes in the surroundings excite reactions in this system, or affect it directly, so that internal disturbances of the system are produced. Such disturbances are normally kept within narrow limits, because automatic adjustments within the system are brought into action, and thereby wide oscillations are prevented and the internal conditions are held fairly constant. The term "equilibrium" might be used to designate these constant conditions. That term, however, has come to have exact meaning as applied to relatively simple physico-chemical states in closed systems where known forces are balanced. In an exhaustive monograph L. J. Henderson (1928) has recently treated the blood from this point of view, i.e., he has defined, in relation to circumstances which affect the blood, the nice arrangements within the blood itself, which operate to keep its respiratory functions stable. Besides these arrangements, however, is the integrated coöperation of a wide range of organs—brain and nerves, heart, lungs, kidneys, spleen—which are promptly brought into action when conditions arise which might alter the blood in its respiratory services. The present discussion is concerned with the physiological rather than the physical arrangements for attaining constancy. The coördinated physiological reactions which maintain most of the steady states in the body are so complex, and are so peculiar to the living organism, that it has been suggested (Cannon, 1926) that a specific designation for these states be employed—*homeostasis*.

Objection might be offered to the use of the term *stasis*, as implying something set and immobile, a stagnation. Stasis means, however, not only that, but also a condition; it is in this sense that the term is employed. *Homeo*, the abbreviated form of *homoio*, is prefixed instead of *homo*, because the former indicates "like" or "similar" and admits some variation, whereas the latter, meaning the "same," indicates a fixed and rigid constancy. As in the branch of mechanics called "statics," the central concept is that of a steady state produced by the action of forces; *homeostatics* might therefore be regarded as preferable to homeostasis. The factors which operate in the body to maintain uniformity are often so peculiarly physiological that any hint of immediate explanation in terms of relatively simple mechanics seems misleading. For these various reasons the term homeostasis was selected. Of course, the adjectival form, *homeostatic*, would apply to the physiological reactions or agencies or to the circumstances which relate to steady states in the organism.

CLASSIFICATION OF HOMEOSTATIC CONDITIONS. According to Bernard (1878, ii, p. 7), the conditions which must be maintained constant in the fluid matrix of the body in order to favor freedom from external limitations are water, oxygen, temperature and nutriment (including salts, fat and sugar).

Naturally during the past fifty years new insight has been acquired and therefore a more ample classification than that just given should be possible. Any classification offered now, however, will probably be found to be incomplete; other materials and environmental states, whose homeostasis is essentially important for optimal activity of the organisms, are likely to be discovered in the future. Moreover, in any classification there will be cross-relations among the homeostatic states; a uniform osmotic pressure in the body fluids, for example, is dependent on constancy within them of the proportions of water, salts and protein. The classification suggested below, therefore, should not be regarded as more than a serviceable grouping of homeostatic categories; it may claim only the merit of having served as a basis for studying the means by which the organism achieves stability:

A. Material supplies for cellular needs.
    1. Material serving for the exhibition of energy, and for growth and repair—glucose, protein, fat.
    2. Water.
    3. Sodium chloride and other inorganic constituents except calcium.
    4. Calcium.
    5. Oxygen.
    6. Internal secretions having general and continuous effects.

B. Environmental factors affecting cellular activity.
   1. Osmotic pressure.
   2. Temperature.
   3. Hydrogen-ion concentration.

Each item in the foregoing list exists in a relatively uniform condition of the fluid matrix in which the living cells of the organism exist. There are variations of these conditions, but normally the variations are within narrow limits. If these limits are exceeded serious consequences may result or there may be losses from the body. A few examples will make clear these relations:

A reduction of the glucose in the blood to about 70 mgm. per cent (e.g., by insulin) induces the "hypoglycemic reaction" (Fletcher and Campbell, 1922), and a reduction below 45 mgm. per cent brings on convulsions and possibly coma and death; an increase of the percentage above 170 to 180 mgm. results in loss via the kidneys. Too much water in the body fluids results in "water intoxication," characterized by headache, nausea, dizziness, asthenia, incoördination (Rowntree, 1922); on the other hand, too little water results in lessened blood volume, greater viscosity, and the appearance of fever (Keith, 1922; Crandall, 1899). Sodium (with the attendant chloride ion) is especially important in maintaining constant the osmotic properties of the plasma; if the percentage concentration rises from 0.3 to 0.6 per cent, water is drawn from the lymph and cells, and fever may result (Freund, 1913; Cushny, 1926, p. 19); on the other hand, if the concentration is reduced, toxic symptoms appear—marked reflex irritability, followed by weakness, shivering, paresis and death (see Grünwald, 1909). The normal level of calcium in the blood is about 10 mgm. per cent; if it falls to half that concentration, twitchings and convulsions are likely to occur (Mac-Callum and Voegtlin, 1909); if it rises to twice that concentration, profound changes take place in the blood, which may cause death (Collip, 1926). The normal daily variations of body temperature in man range between 36.3°C. and 37.3°C.; though it may fall to 24°C. and not be fatal (Reincke, 1875), that level is much lower than is compatible with activity; and if the temperature persists at 42–43°C., it is dangerous because of the coagulation of certain proteins in nerve cells (Halliburton, 1904). The hydrogen-ion concentration of the blood may vary between approximately pH 6.95 and pH 7.7; at a pH of about 6.95 the blood becomes so acid that coma and death result (Hasselbalch and Lundsgaard, 1912); above pH 7.7 it becomes so alkaline that tetany appears (Grant and Goldman, 1920). The heart rate (of the dog) has been seen to decrease from 75 beats per minute to 50 when the pH fell from 7.4 to 7.0; and to increase from 30 per minute to about 85 when the pH rose from 7.0 to 7.8 (Andrus and Carter, 1924). The foregoing instances illustrate the importance of homeostasis in the body fluids. Ordinarily the shifts away from the mean position do not reach extremes which impair the activities of the organism or endanger its existence. Before those extremes are reached agencies are automatically called into service which act to bring back towards the mean position the disturbed state. The interest now turns to an enquiry into the character of these agencies.

An inductive unfolding of the devices employed in maintaining homeostatic conditions—an examination of each of the conditions with the object of learning how it is kept constant—would require more space than is permitted here. It will be possible, however, to define in broad terms the agencies of homeostasis and to illustrate the operation of some of those agencies by reference to the specific cases. Thus the account may be much abbreviated.

Two general types of homeostatic regulation can be distinguished dependent on whether the steady state involves *supplies* or *processes*.

HOMEOSTASIS BY REGULATING SUPPLIES. The characteristic feature of the homeostasis of supplies is provision for A, storage as a means of adjustment between occasional abundance and later privation and need, and for B, overflow or discharge from the body when there is intolerable excess. Two types of storage can be distinguished: a temporary flooding of interstices of areolar tissue by the plenteously ingested material, which may be designated *storage by inundation*; and an inclusion of the material in cells or in other relatively fixed and permanent structures—*storage by segregation*. We shall consider illustrations of these two types.

STORAGE BY INUNDATION. The analogy implied in this phrase is that of a bog or swamp into which water soaks when the supply is bountiful and from which the water seeps back into the distributing system when the supply is meager. There appears to be such an arrangement in the loose areolar connective tissue found under the skin and around and between muscles and muscle bundles, and also in other parts of the body. Connective tissue is distinguished from other kinds in being richest in extracellular colloid, in having a close relation to blood vessels—indeed, it serves as a support for the blood vessels—and in exposing an enormous surface area. In such structures chiefly do the agencies rule which hold not only mobile water but also substances dissolved in it, i.e., electrolytes and glucose. Here there are few cells, but instead "a spongy cobweb of delicate filaments," each of which is composed of minute fibrils bound together by a small amount of "cement substance" (Lewis and Bremer, 1927). Within the fine mesh of these collagenous fibres occur mucoid and small amounts of albumin and globulin. In this mesh and bound by it in some manner water and its dissolved substances appear to be held. Probably the proportions of stored water, electrolytes and glucose do not vary beyond a fairly limited range. Because there is evidence, however, that water and electrolytes, at least, may be affected somewhat independently with regard to their retention and elimination, they will be considered separately.

Water. The evidence for water storage is best demonstrated in experiments which withdraw water from its reservoirs and which permit an examination of the amount held in them. After hemorrhage all tissues lose water. By comparing one side of the body with the other in the same animal (the cat) Skelton (1927) found that most of the water which leaves the tissues after bleeding comes from the muscles and the skin—i.e., where loose areolar tissue is most abundant; the amount per 100 grams of tissue, however, is much less from the muscles than from the skin. The observations by Engels (1904) on dogs are in harmony with those of Skelton. Engels found that though 48 per cent of the total body water is in muscles, as might be expected from the great bulk of muscle tissue, about 12 per cent is in the skin, nearly half again as much as is in the fluid blood. And after injecting 0.6–0.9 per cent sodium chloride solution into a vein for an hour, he discovered that 690 grams had been retained and that the muscles and the skin had taken up the solution to almost the same per cent.

That the water stored in the tissues passes out from them as it is needed is shown by the studies of Wettendorf (1901) on the state of the blood during water deprivation. His dogs were, of course, continually losing water through respiratory surfaces and kidneys. Yet one of his animals thirsted for 3 days with no change in the freezing point of the blood, and another for 4 days with a depression of only 0.01°C. Clearly this constancy must be due to the seepage of water from the reservoirs to the blood as fast as it is lost from the body.

Just how the water is brought to the reservoirs, how it is held there, and how it is released as required for preserving the osmotic homeostasis of the blood, is not yet satisfactorily explained. Doubtless a change in the balance between filtration pressure through the capillary walls and osmotic pressure of the proteins, as expounded by Starling (1909), plays an important rôle. And naturally, conditions affecting the capillary wall (e.g., increasing its permeability), raising or lowering intracapillary blood pressure, or altering the concentration of the plasma proteins would affect the water content of the tissues. Diffusion pressure would likewise take part in the complex of active factors. Furthermore, as Adolph (1921) and Baird and Haldane (1922) have shown, the taking of sodium chloride can markedly influence the retention of water in the body. Probably other electrolytes likewise play a rôle.. There is evidence also that the H- and OH-ion concentration may be important—a shift towards an alkaline reaction causing imbibition of water by connective tissue and an opposite shift resulting in release (Schade, 1925). That the thyroid gland is a determinative agent is indicated by the great increase of protein in the plasma and of albumin in the tissues in myxedema, and the disappearance of these conditions, together with a large release of water and sodium chloride, when thyroxin is administered (see Thompson, 1926). How these various factors coöperate when water and sodium chloride are needed in the circulation— after hemorrhage, for example—is not clear, and urgently calls for investigation. Krogh (1922) has written concerning the arrangement of water mobilization, "The nature of such a mechanism is entirely unknown and I should not like to venture even a guess regarding it"— and yet it is of primary significance for the organism.

Sodium chloride. There is good evidence that the sodium and chloride ions in the plasma may vary independently, and that of the two the base is much the more constant element (see Gamble and Ross, 1925; Gamble and McIver, 1925, 1928). In a study of steady conditions in the fluid matrix, therefore, the emphasis might properly be laid on the homeostasis of the fixed base. Since most of the facts now available, however, have come from experiments in which the behavior of sodium chloride has been examined, the present treatment of the subject must be concerned with that salt. The evidence for storage of sodium chloride in the body is found in retention under different conditions. With a fairly constant chloride intake abundant sweating and attendant loss of chloride through the skin are accompanied by a great reduction of the chloride output in the urine—a condition which continues although thereafter a diet rich in salt is taken; by this method of study a compensatory retention of 10 to 14 grams of sodium chloride has been observed (Cohnheim, Kreglinger, and Kreglinger, 1909). Further, the taking of concentrated sodium chloride by mouth results in the appear-

ance in the urine of only a part of the amount ingested—most of it is
retained in storage in the body; and even if thereupon enough water is
drunk to produce a diuresis the urine has a low salt content, i.e., the
salt is not given up readily from its storage place (Baird and Haldane,
1922).

When a search is made for the sodium chloride reserve in the body
the highest percentage of chloride is found in the skin and the lowest in
the muscles—indeed, on a chloride-rich diet one-third of the chloride
of the body may be in the skin, and after an intravenous infusion of a
sodium-chloride solution the skin may hold the stored chloride to an
amount varying in different experiments between 28 and 77 per cent
of the amount injected. This evidence is supported by observations
on animals fed a chloride-poor diet. Under these circumstances between
one and two-tenths of the chloride content of the body is lost, and of this
amount between 60 and 90 per cent comes from the skin, though the skin
is only 16 per cent of the total body weight (Padtberg, 1910). It is
noteworthy that the blood gives up relatively little of its chloride
content; again the circulating fluid is kept constant by supplies from
tissue storage.

It is well to recognize that the way in which sodium chloride is held
in the skin, whether by adsorption on surfaces in areolar tissue or by
solution in the interstitial fluid of the areolar spaces, is not known.
Probably it is osmotically inactive. That sodium chloride and water
are closely related in storage, however, seems to be well established·
(see Adolph, 1921).

Glucose. The first, temporary depository for excessive blood sugar,
as for excessive sodium chloride, is the skin. When sugar or other
readily digestible carbohydrate is a large constituent of the diet the
glycemic concentration rises commonly from about 100 to 170 mgm.
per cent (Hansen, 1923). During this period of high percentage of
sugar in the blood there is also a high percentage in the skin (Folin,
Trimble and Newman, 1927). This appears to be again an example
of storage by inundation. No chemical change occurs in the sugar.
No special device is required either to deposit it in the temporary
reservoir or to remove it therefrom. As the circulating sugar is utilized
or placed in more permanent storage in the liver and in muscle cells,
the glycemic level falls. Thereupon the more concentrated glucose,
which has overflowed into the spaces of the skin and possibly into other
regions where alveolar tissue is abundant, gradually runs back into the
blood again and then follows the usual courses of the blood glucose into
use or into the fixed reserves.

STORAGE BY SEGREGATION. As previously stated, this mode of storage, commonly within cells, is stable and lasting. It is seen, for example, in carbohydrate reserves as glycogen, in protein reserves as irregular masses in liver cells, in fat reserves as adipose tissue, and in calcium reserves as the trabeculae of the long bones. It differs from storage by inundation in being subject to much more complicated control. Storage by inundation may be regarded as a process of outflow from the blood stream and backflow into it according to the degree of abundance—a relatively simple process. Storage by segregation commonly involves changes of physical state or of molecular configuration and appears to be subject to nervous or neuro-endocrine government. This rather tentative statement is used because of the large gaps in our knowledge, which further consideration will reveal. We shall consider the segregated storage of carbohydrate, protein, fat and calcium.

Carbohydrate. The best example of homeostasis by means of segregation is offered by the arrangements for storage and release of carbohydrate. As is well known, when carbohydrate food is plentiful the glycogen reserves in the liver are large; in prolonged muscular work these reserves may be almost wholly discharged (Kulz, 1880); and yet, while they are being discharged, the blood sugar is maintained at concentrations which neither result in the possibility of sugar loss through the kidneys, nor in the possibility of disturbance from hypoglycemia (Campos, Cannon, Lundin and Walker, 1929). A mechanism must exist, therefore, to release sugar from the liver as it is needed.

An insight into the action of factors which prevent the fall of the blood sugar to a seriously low level may be obtained by a study of the effects of insulin. As stated above, the reduction of the glycemic concentration to about 70 mgm. per 100 cc. by insulin induces the "hypoglycemic reaction," characterized by pallor, rapid pulse, dilated pupils and profuse sweating. These are signs of sympathetic innervation. That this is part of a general display of activity by the sympathetic division of the autonomic system is shown by the involvement of the adrenal medulla. Using the denervated heart as an indicator, Cannon, McIver and Bliss (1923, 1924) found that as the blood sugar fell a critical point was reached at about 70 mgm. per cent, when the heart began to beat faster—a phenomenon which failed to appear if the adrenal glands had been inactivated. If the blood sugar continued to fall the heart rate became faster, thus indicating a greater output of adrenin; and if the blood sugar rose, either because of intravenous injection of glucose or because of a physiological reaction, the heart beat returned to its original slow rate, thus indicating a subsidence of the extra discharge of adrenin. Since medulliadrenal secretion is controlled by splanchnic impulses, and since such impulses in coöperation with secreted adrenin are highly effective in causing an increase of blood sugar (Bulatao and Cannon, 1925; Britton, 1928), it is clear that the reduction of the glycemic percentage below a critical level calls forth an agency—the sympathico-adrenal system—to correct the condition. These observations have been confirmed by Abe (1924) who used the denervated iris to signal a greater output of adrenin, and by Houssay, Lewis and Molinelli (1924) who used for that purpose an adrenal-jugular anastomosis between two dogs. If, in spite of the increasingly active service of this agency as the blood sugar falls, the fall is not checked, convulsions occur (at about 45 mgm. per cent) and each convulsion is associated with a maximal display of sympathico-adrenal activity. If the liver is well supplied with glycogen such activity can restore the blood sugar to the normal level and thus abolish the conditions which brought on the convulsive attacks (McCormick, Macleod, Noble and O'Brien, 1923).

The importance of this agency has been demonstrated by experiments on healthy non-anesthetized animals in which the adrenal glands had been inactivated. The fall of blood sugar after insulin was less retarded at the critical level in cats thus altered, and the convulsive seizures were induced sooner and with smaller doses than in animals with active glands (Cannon, McIver and Bliss, 1924). This increased sensitiveness to insulin after medulliadrenal inactivation has also been proved true of rats (Lewis, 1923), of rabbits (Sundberg, 1923) and of dogs (Lewis and Magenta, 1925; Hallion and Gayet, 1925). The evidence that a small dose of insulin, mildly effective in a normal animal, causes in an animal treated with ergotamine profound hypoglycemia, with convulsions and collapse (Burn, 1923), is in harmony with this testimony, for the drug, though without influence on the action of insulin, paralyzes the protective sympathico-adrenal mechanism. Section of the splanchnic nerves, according to Lewis and Magenta, renders animals more sensitive than does removal of one adrenal and denervation of the other by splanchnic section; under the latter circumstances, it should be noted, one splanchnic is still innervating the liver.

Operation of agencies opposed to those just considered occurs when the blood sugar tends to rise. The efficacy of these agencies is revealed when an excess of glucose is ingested. The blood sugar rises to a level close to that at which it escapes through the kidneys, but normally it does not often surpass that level (Hansen, 1923). The excess sugar, apart from that set aside by inundation, is either stored in the liver or in muscles, or is converted to fat, or is promptly utilized. There is evidence that the process of storage by segregation in hepatic and muscle cells is dependent on secretion of insulin: 1. Removal of the pancreas results in prompt appearance of hyperglycemia and a great reduction of the hepatic glycogen reserves. 2 .The administration of insulin to sugar-fed *depancreatized* dogs reduces the blood sugar to the normal percentage and causes glycogen to accumulate again in large amounts in the liver (Banting, Best, Collip and Noble, 1922). 3. Insulin in small doses causes a deposit of glycogen in the liver of phlorizinized rabbits and cats, even though no sugar is provided (Cori, 1925). 4. Insulin injected into decapitated, eviscerated cats causes a decided increase in the glycogen deposit in the muscles, especially when extra blood sugar is provided (Best, Hoet and Marks, 1926). 5. Islet cells in a remnant of the pancreas degenerate, showing signs of overwork (Allen, 1920; Homans, 1914), when carbohydrate is fed; from this evidence it appears that hyperglycemia stimulates the islet cells to secrete. This stimulation may be direct, as proved by absence of diabetes if a portion of the pancreas is transplanted under the skin and the rest of the gland is then removed and by the appearance of the disease when the engrafted piece is extirpated (Minkowski, 1908), by prompt reduction of hyperglycemia from injected glucose although the vagi have been severed (Banting and Gairns, 1924), or by reduction of diabetic hyperglycemia when a pancreas is connected with blood vessels in the neck (Gayet and Guillaumie, 1928). There is evidence also of a nerve control of insulin secretion, for stimulation of the right vagus reduces blood sugar, but not if the vessels of the pancreas are tied (Britton, 1925); and, according to Zunz and La Barre (1927), after union of the pancreatic vein of dog A to the jugular of dog B injection of glucose into A causes the blood sugar to fall in B—an effect that does not occur if the vagi of A have been cut or atropine given. That this nervous control is not necessary does not prove that it is useless when present—for example, the heart after denervation will continue beating and will maintain the circulation! It may be that the vagus provides a fine adjustment for insulin secretion as the sympathetic does for secretion of adrenin.

The general scheme which has been presented above is represented diagrammatically in figure 1. As Hansen (1923) has pointed out, there are normal oscillations in blood sugar occurring within a relatively

narrow range.   Possibly these ups and downs result from action of the
opposing factors, depressing or elevating the glycemic level.   If known
elevating agencies (normally and primarily the sympathico-adrenal
apparatus) are unable to bring forth sugar from storage in the liver, the
glycemic level falls from about 70 to about 45 mgm. per cent, whereupon
serious symptoms (convulsions and coma) may supervene.   The range
between 70 and 45 mgm. per cent may be regarded as the *margin of
safety*.   On the other hand, if the depressing agency (the insular or vago-

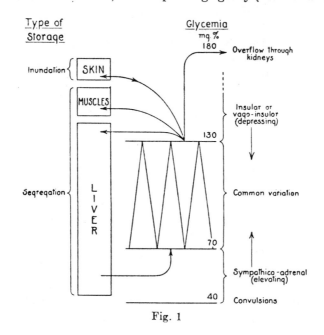

Fig. 1

insular apparatus) is ineffective, the glycemic level rises to about 180
mgm. per cent and then sugar begins to be lost through the kidneys.   The
range from 100 or 120 to 180 mgm. per cent may be regarded as the *margin of economy*—beyond that, homeostasis is dependent on wasting the
energy contained in the sugar and the energy possibly employed by the
body to bring it as glucose into the blood.[1]

[1] Evidence opposed to the foregoing views has been brought forward recently
by Cori and Cori (1928).   They state that "the most prominent effect of epi-
nephrin is observed in the peripheral tissues and consists in a mobilization of
muscle glycogen and in a decreased utilization of blood sugar"; and that insulin
causes a rapid disappearance of the hepatic stores, due to increased use of blood

Protein. The homeostasis of protein is perhaps widely manifested in the constancy of body structure. That would include the blood, however, and since we are concerned with the conditions which keep uniform the fluid matrix of the body, we shall pay particular attention only to that.

The importance of constancy of the plasma proteins need not be

---

sugar in peripheral tissues and to "compensatory mobilization of liver glycogen." These declarations, so contrary to evidence long accepted, call for comment. First, they gave doses of adrenalin (0.2 mgm. per k.) and of insulin (7.5 units per k.) far beyond physiological limits (equivalent to 14 cc. of adrenin and 525 units of insulin in a man of 70 k.!). Pronounced physiological effects have been obtained in white rats (which they used) with a dose of adrenin *one-twentieth* of their dose. Although they argue that their adrenin doses were slowly and fairly evenly absorbed, they present no actual evidence; and the fact that the highest blood sugar in their experiments came early and was associated with the lowest glycosuria indicates both that their argument is ill-based and that the huge doses disturbed the circulation. Further, "mobilization of muscle glycogen" consists, they explain, in a change of the glycogen to lactic acid, and from this circulating lactic acid a reconstruction of glycogen by the liver. But the glycogen in muscle is there for use; to "mobilize" it without use is like withdrawing forces from the firing line and settling them in barracks! Again, in declaring that adrenin causes hyperglycemia "because the utilization of blood sugar is diminished" they neglect the evidence: 1, that intravenous injection of adrenin raises blood sugar with almost no latent period (Tatum, 1921); 2, that emotional excitement can raise blood sugar 30 per cent or more in a few minutes, but not after adrenalectomy (Britton, 1928), and that the same phenomena are seen when the splanchnics are stimulated (Macleod, 1913), and 3, that after an adrenin injection the blood sugar rises quickly in the liver veins and only later is equaled in the portal or femoral vein (Vosburgh and Richards, 1903)—all evidence against their views, because the hyperglycemia comes too soon and is too clearly of hepatic origin to be ascribed to failure of use of glucose by peripheral tissues. Moreover, their belief that adrenin causes "decreased utilization of blood sugar" is contradicted by the observation that when glucose and adrenin are supplied to the heart-lung preparation sugar consumption rises to about four times the former amount (Patterson and Starling, 1913), and that dogs exhausted by running can be made to continue (i.e., using sugar in their muscles) and will put forth from 17 to 44 per cent additional energy if they are given subcutaneously small doses of adrenin (0.02–0.04 mgm. per k., sometimes repeated) but not if a large dose (0.17 mgm. per k.—n.b., *less* than that used by the Coris) is given (Campos, Cannon, Lundin and Walker, 1929). Finally, although they mention a "compensatory mobilization of liver glycogen" as the cause of depleted hepatic stores after their enormous doses of insulin, they do not hint at the nature of the compensatory process, though they report low blood-sugar levels which would set in action the sympathico-adrenal apparatus. For these various reasons the views advanced by Cori and Cori seem not to warrant a surrender of the well established conceptions of the action of adrenin and insulin.

emphasized. Because they exert osmotic pressure and do not ordinarily escape through capillary walls, they prevent the salts dissolved in the blood from passing freely into perivascular spaces and out from the body through the renal glomeruli (Starling, 1909). When Barcroft and Straub (1910) removed much of the blood from a rabbit, separated the corpuscles, and reinjected them suspended in an equal volume of Ringer's solution instead of plasma, so that the difference was merely a reduction of the colloid, urine secretion was increased *forty times*. The development of "shock"—engorgement of liver, spleen, kidneys, intestinal mucosa and lungs, with accumulation of fluid in the intestine—noted by Whipple, Smith and Belt (1920) when the plasma proteins were reduced to 1 per cent, tells the same story. But not only does homeostasis of the plasma proteins provide for homeostasis of the blood volume; at least one of them (fibrinogen) is, in case of hemorrhage, essential for preservation of the blood itself. The very existence of the fluid matrix of the body is dependent, therefore, on constancy of the proteins in the plasma—and usually they are remarkably constant in various conditions of health and disease.

Blood is the one tissue in the body from which protein can be quantitatively removed and its restoration then studied. When by plasmapharesis the plasma proteins are reduced from about 6 to about 2 per cent there is a prompt rise in their concentration within fifteen minutes, a more gradual restoration thereafter during the first twenty-four hours to about 40 per cent, and full recovery in two to seven days. It may be that the prompt rise is relative, due to escape of salt solution from the blood vessels, although arguments have been advanced that it results from emergency discharge of the proteins from storage (Kerr, Hurwitz and Whipple, 1918). The slower recovery seems certainly dependent on the liver, for the following reasons: 1. If the liver has been injured by phosphorus or chloroform, restoration of the plasma proteins is delayed. 2. Dogs with an Eck fistula may have no restoration for the first three days after plasmapharesis. And 3, fibrinogen, which usually is completely restored within twenty-four hours, is not thus restored if the liver is unable to act (Foster and Whipple, 1922; Meek, 1912).

The evidence that the liver is important for homeostasis of proteins in the blood plasma raises the question whether protein is stored there. Results obtained by histological and biochemical methods have agreed in supporting the conclusion that hepatic cells can carry reserve protein as well as reserve carbohydrate. The early observations of Afanassiew

(1883) that the liver cells of dogs given an abundance of "albuminates" increase in size and contain protein granules between the structural strands, have been confirmed by Berg (1914, 1922), by Cahn-Bronner (1914), by Stübel (1920) and by Noël (1923). In sum, these recent histological studies show that when animals are fed protein there appear in the hepatic cells fine droplets or masses, which react to Millon's reagent, which yield the ninhydrin reaction of a simple protein, which disappear on fasting, and which reappear on feeding protein or amino acids. The biochemical analyses by Seitz (1906), who found that in fed animals the total nitrogen of the liver in relation to that in the rest of the body is from two to three times as great as in fasting animals, have been supported by the results obtained by Tichmeneff (1914). He starved mice for two days, then killed half of them and after giving the others an abundance of cooked meat killed them and compared the livers of the two groups. Expressed in percentage of body weight, the livers of the meat-fed animals increased about 20 per cent, with the hepatic nitrogen content augmented between 53 and 78 per cent.

Although the experiments on homeostasis of plasma proteins indicate that the liver is an important source of these materials in case of need, and although the testimony cited above would justify consideration of the liver as a storage place for protein, the modes of storage and release are almost wholly unknown. Stübel (1920) did, indeed, observe that the small protein droplets or masses in the hepatic cells could be greatly reduced by injecting adrenin subcutaneously. If these masses help to supply essential protein elements for blood clotting, as the dependence of fibrinogen on the liver would imply, their liberation by adrenin and by conditions which would excite the sympathico-adrenal apparatus might account for certain phenomena of faster clotting. Coagulation is more rapid after adrenin injections (see Cannon and Gray, 1914; La Barre, 1925; Hirayama, 1925), after splanchnic stimulation (Cannon and Mendenhall, 1914), or after large hemorrhage which calls the sympathetic into action (Gray and Lunt, 1914), but only if the blood is allowed to flow through the liver and intestines. In this category also is the very rapid clotting of blood taken at the height of the hypoglycemic reaction (Macleod, 1924), when sympathico-adrenal activity is maximal.

It is quite possible that protein is stored in other places than the liver, and also that the thyroid gland is an important agency for controlling both storage and release. Boothby, Sandiford and Slosse (1925) have reported that with a uniform nitrogen intake a negative nitrogen balance exists while thyroxin is establishing a new higher metabolic level. After its establishment there is a smaller deposit of nitrogen in the body. Now if thyroid dosage is stopped (while the uniform nitrogen intake continues), a positive balance obtains until a new lower metabolic level is reached, i.e., more nitrogen is deposited. These effects are much more marked in a person afflicted with myxedema than in a normal person. Indeed, as Boothby has suggested, the "edema" of myxedema may be an abnormal amount of deposit protein in and beneath the skin. The efficacy of thyroid therapy, previously noted, in reducing the increased proteins of the plasma and the albumin of the tissues, in cases of myxedema, supports the view that the thyroid gland is somehow associated with protein regulation and metabolism.

Although the foregoing review has brought out the primary importance of homeostasis of the proteins of the plasma for maintaining the volume and character of both the intravascular and extravascular fluid matrix of the organism, and for protecting the organism against loss of the essential part of the matrix—the blood—, it has revealed also how much still needs to be learned. Here, as with other useful material, constancy is attained by storage, which stands between plenty and need, and in this respect the liver plays an important rôle. The sympathico-adrenal apparatus seems to influence release from storage, and also varying activity of the thyroid gland may be determinative. Are special agencies required to manage the laying by of the reserves? We do not know.

Fat. According to Bloor (1922) the concentration of fat, cholesterol and lecithin in the blood is fairly constant in the same species of animals, but may differ greatly in different species. As is well known, ingestion of fat produces an "alimentary lipemia" which may cause the fat content of the blood to rise as high as 3 per cent in the dog and 2 per cent in man. A relatively large increase in the fat content of the blood appears to be without serious consequence. In pathological states —e.g., in diabetes—the lipemic percentage may rise to 10, 15 and even to 20 per cent without producing obvious symptoms. On the other hand the normal blood fat is remarkably persistent. Carbohydrate and protein alone may be fed for considerable periods without reduction of the lipemic level, and fasting for short periods may actually be accompanied by a rise (Schulz, 1896), although after two weeks of fasting the level may undergo a slow fall. Whether total absence of fat from the blood could be produced, and if so, whether that condition would be attended by disturbances, are questions yet to be answered.

The constancy of the lipemic level for many days in spite of relative or complete starvation implies that there is a governing agency which brings the fat from storage into the blood stream. As Lusk (1928, p. 107) has remarked, "The length of life under the condition of starvation generally depends upon the quantity of fat present in the organism at the start." Fat is stored in the liver, if carbohydrate is not fed (Rosenfeld, 1903); it is also stored under the skin, beneath serous coats (e.g., around the kidneys), in the omentum, and between and in the muscle fibres. What leads to fat storage in some individuals to a greater extent than in others is unknown. In hypothyroidism there may be a generally diffused obesity, an obesity which rapidly disappears under thyroid therapy. A slight scratch in the surface of the brain stem between the infundibular process and a mammillary body produces adiposity (Bailey and Bremer, 1921), as does a tumor or other lesion of this region. Grafe (1927) cites instances of unilateral hypertrophy or atrophy of fatty tissues and suggests that the disposal of fat is under a sympathetic control managed from the hypothalamic region. It is pertinent to note that kittens allowed to live after unilateral sympathectomy until they have doubled their weight, have no demonstrable differences in the amount or distribution of the fat on the two sides of the body (Cannon, Newton, Bright, Menkin and Moore, 1929).

If the regulation of fat storage is obscure, the regulation of its release is even more so. When fat is needed for maintaining the energies of the body it is removed from adipose tissue until the fat cells are practically empty. Yet, even when death from starvation occurs, the fat content of other tissues may not be very different from normal (Terroine, 1914). What causes the fat to move from the adipose stores is not known. Lusk's remark, that "the fasting organs attract fat from the fat deposits of the body, and it is brought to them in the circulating blood," was probably not intended to be explanatory, and it is not. Possibly the reversible reaction mediated by lipase, as described by Kastle and Loevenhart (see Loevenhart, 1902), may be an important factor in maintaining homeostasis of the lipemic level—the enzyme favoring storage when the level is raised and favoring release when the level falls. But on these points more information is desirable.

Calcium. The special and diverse uses of calcium—for the growth of the skeleton and teeth, for the repair of broken bone, for the maintenance of proper conditions of irritability of nervous and muscular tissues, for the coagulation of blood, and for the production of serviceable milk—render it a highly important element in the bodily economy. Like sugar and protein and fat, calcium may be in great demand on exceptional occasions. Under such circumstances, however, the amount in the blood must not be much reduced, for serious consequences ensue. As previously noted, there is normally a homeostasis of calcium in the blood at approximately 10 mgm. per cent. If the blood calcium is lowered to less than 7 mgm. per cent, as may be done by removal of the parathyroid glands (without change in the percentage of sodium and potassium), or by injection of sodium citrate, twitchings and tetanic convulsions occur, with a severity measured by the degree of deficit of available calcium; and these symptoms are quickly relieved by injecting a soluble calcium salt sufficient to restore the normal percentage (MacCallum and Voegtlin, 1909; MacCallum and Vogel, 1913; Trendelenburg and Goebel, 1921). On the other hand, if the blood calcium is raised above approximately 20 mgm. per cent, by injection of parathyroid extract, profound changes are produced in the blood—the viscosity is greatly increased, the osmotic pressure rises, the blood phosphates are doubled, and there are four times the normal amount of non-protein and urea nitrogen—conditions associated with vomiting, coma, and a failing circulation (Collip, 1926). Obviously homeostasis of blood calcium is of capital importance.

As in the homeostasis of other materials, that of calcium is made possible by storage, built up in times of abundance and utilized in time of need. The recent studies of Bauer, Aub and Albright (1929) have demonstrated that the trabeculae of the long bones are easily made to disappear by a persistent diet deficient in calcium and by growth, and that they are readily restored by feeding a calcium-rich diet. The trabeculae serve, therefore, as a storehouse of conveniently available calcium.

How the homeostasis of calcium is regulated has not been determined. The following evidence associates the parathyroid glands with the regulation: 1. Partial or complete removal of these glands results in a lowering of the calcium content of the blood, as mentioned above, and in a defective deposit of dentine in growing teeth and in a defective development of the callus about a bone fracture (Erdheim, 1911). 2. A diet poor in calcium induces parathyroid hyperplasia (Marine, 1913; Luce, 1923); pregnancy and lactation do likewise, without, however, reduction of the calcium percentage in the blood. 3. Diseases characterized by defects in calcification of bone—e.g., rickets and osteomalacia—are attended by hypertrophy of the parathyroid glands (see Strada, 1909; Weichselbaum, 1914). And 4, implantation of parathyroids in a parathyroidectomized rat restores the power to deposit dentine having a normal calcium content (Erdheim, 1911). But *how* the parathyroids control calcium homeostasis—whether they act directly or are stimulated by nerves, whether they act alone and by increased or decreased activity effect storage or release, or whether they coöperate with other agencies perhaps antagonistic—all this still needs investigation.

The pharmacodynamic action of thyroxin seems to involve the thyroid as well as the parathyroid glands in calcium metabolism (Aub, Bauer, Heath and Ropes, 1929). Administration of thyroxin greatly increases the calcium losses in urine and feces whether persons are normal or myxedematous. There is evidence also that in hyperthyroidism the bones show osteoporosis and that calcium excretion is much augmented. Possibly the parathyroids serve for deposit and the thyroids for release of calcium—thyroxin raises the blood calcium in a low-calcium tetany. Such hints, however, must be regarded with suspicion until put to test, for, as shown in the experiments on the action of adrenin and insulin, mentioned above, powerful pharmacodynamic agents given in doses exceeding the physiological range can produce complicated and indirect effects.

The homeostatic functions of hunger and thirst. In the foregoing discussions storage has been emphasized as a regulatory mediation between supply and demand. Back of storage, however, and assuring provisions which can be stored, are powerful motivating agencies—appetites and hunger and thirst. Because of pleasurable previous experiences with food and drink appetites invite to renewal of these experiences; thereby material for the reserves is taken in. If the reserves are not thus provided for, hunger and thirst appear as imperious stimuli. Hunger is characterized by highly disagreeable pangs which result from strong contractions of the empty stomach—pangs which disappear when food is taken (Cannon and Washburn, 1912; Carlson, 1916). Thirst is an uncomfortable sensation of dryness and stickiness in the mouth, which can be explained as due to failure of the salivary glands (which need water to make saliva) to keep the mouth moist; when water is swallowed and absorbed they, as well as the rest of the body, are provided with it and since they can consequently moisten the mouth, the thirst disappears (Cannon, 1918). By these automatic mechanisms the necessary materials for storage of food and water are assured.

OVERFLOW. Previously, in relation to the homeostasis of blood sugar, the use of overflow as a means of checking an upward variation of constituents of the blood has been mentioned. Not only excessive sugar, but excessive water, excessive sodium and potassium and chloride ions are discharged by the kidneys. In accordance with the modern theory of urine formation (Cushny, 1926), these are all "threshold substances." They are resorbed by the kidney tubules only in such relations to one another as to preserve the normal status in the blood. All in excess of that is allowed to escape from the body.

It is interesting to note that these substances are primarily stored by flooding or inundation. When these reserve supplies are adequate, however, the ability of the overflow factor to maintain homeostasis is marvelous. The feat reported by Haldane and Priestley (1915) of drinking 5.5 liters of water in six hours—an amount exceeding by one-third the estimated volume of the blood—which was passed through the kidneys with such nicety that at no time was there appreciable reduction of the hemoglobin percentage, is a revelation not only of the efficacy of the kidney as a spillway but also of the provision in the body for maintaining a constancy of its fluid matrix.

The lungs as well as the kidneys serve for overflow. As is well known, a slight excess of carbonic acid in the arterial blood is followed by greatly increased pulmonary ventilation. Thus the extra carbon dioxide is so promptly and effectively eliminated that the alveolar air is kept nearly constant (Haldane, 1922). By this means provision is made for extra carbon dioxide to flow out from the blood over a dam set at a fixed level. In consequence, in usual circumstances, the hydrogen-ion concentration of the blood is fairly evenly maintained, and the harmful effects of an excessive shift in the alkaline or acid direction is avoided.

HOMEOSTASIS BY REGULATING PROCESSES. There are steady states in the body which do, indeed, involve the utilization of materials, but which are so much more notably dependent on altering the rate of a continuous process that they can reasonably be placed in a separate category. We shall consider two of them, the maintenance of neutrality, and the maintenance of a uniform temperature (in homeothermic animals). The physiological adjustments involved in these processes are so commonly known that a mere outline of them, without detailed description or many references, will be sufficient to illustrate the mode of regulation.

Maintenance of neutrality. The importance of confining the changes in the hydrogen-ion concentration of the blood to a narrow range has already been emphasized. This concentration is determined by the ratio, $H_2CO_3 : NaHCO_3$, in the blood. On going to a high altitude the tension of carbonic acid is lessened, the ratio is lowered and the pH rises.

Under these circumstances the blood alkali also is lessened until the pH is restored. And on returning to sea level the opposite process occurs and continues until there is a normal adjustment again, due, according to Y. Henderson (1925), to "calling an increased amount of alkali into the blood," a result probably attained, however, by the passage of acid elements from the blood into the tissues or the urine.

Back of these adjustments between the blood and the tissues, however, are the preventive measures which protect the blood from danger by anticipatory action. Acid metabolites are continuously being produced in the living cells and if allowed to accumulate in them these substances interfere with or prevent further action. Elaborate arrangements are ready in the organism to forestall that contingency. To be sure, the facilities for controlling non-volatile acid are limited. But it can be dealt with in a variety of ways. The lactic acid, for example, which is developed in muscular contraction, is in part promptly neutralized—the phosphocreatine recently discovered by Fiske and Subbarow (1929) appears to be capable of functioning in an extraordinarily effective manner in neutralizing lactic acid within the muscle cells. Another part of the acid is soon oxidized; and the rest is rebuilt into neutral glycogen. For continued effectiveness of all three of these methods of disposal there must be provided an adequate supply of oxygen. Although muscles, and probably other tissues as well, go into "oxygen debt" by acting in spite of accumulating lactic acid, that state is characterized by a diminished capacity to do work, great according to the debt, by a prescription on the amount of debt allowable, and by the definite requirement of ultimate payment. When non-volatile acid is burned to volatile carbonic acid, however, it is in a form which can be carried away and disposed of to an almost unlimited amount, with only slight change in the reaction of the blood. During vigorous muscular work as much oxygen as possible must be delivered. There is practically no storage of oxygen in the body. Air-breathing animals are surrounded by an ocean of oxygen—the problem is solely that of conveyance from the boundless external supply to the exigent tissues. For that purpose circulatory and respiratory processes must be greatly accelerated. Fortunately these adjustments, required to get rid of the volatile acid, are precisely those required to bring to the tissues the oxygen which serves to make the acid volatile and readily discharged.

In vigorous muscular effort the pulmonary ventilation may be increased from 6 liters per minute to 60 or 80 liters or more, due to the effects of acid in the respiratory center. Under these circumstances the sympathico-adrenal system is active (Cannon and Britton, 1927) and it is altogether probable that thereby the bronchioles are dilated at a time when wider passageways would facilitate the to-and-fro movement of larger volumes of air.   There is an ampler return of blood to the heart per minute, because of contraction of the splanchnic vessels, because of the pressures excited by the active muscles on capillaries and veins within them, and because of the pumping action of the diaphragm. Thus the heart receives a greater charge of blood and puts forth a greater amount per beat.   And because of lessened vagal tone, increased tension on the venous side, and participation of the sympathetic acceler-ators, the heart, well charged, may beat twice as fast as it does at rest. With a much larger minute output from the heart and a constricted splanchnic area the arterial pressure is markedly raised.   In the active muscles the arterioles are dilated and the closed capillaries are opened; and through these more numerous channels the high head of arterial pressure drives an abundant blood stream.   Evidence indicates that the total circulation rate may be augmented as much as four times. 'But not only are the corpuscles utilized more effectively by being made to move faster, the *number* of corpuscles is increased by discharge from storage in the spleen (see Barcroft, 1926)—an effect which, like splanch-nic constriction and cardiac acceleration, the sympathetic system helps to produce (see Izquierdo and Cannon, 1928).   In the laboring muscles, where acid is being produced and where especially oxygen is needed, the excess carbon dioxide itself facilitates the unloading of oxygen from the corpuscles and also its own carriage away to the lungs.   In these ways the local flow may be increased as much as 9 times and the oxygen delivery may be increased as much as 18 times what it is during rest (Bainbridge, 1923).   Thus, in spite of the fact that in a short time more lactic acid by far can be produced by muscular work than could be neutralized by the buffers in the blood—a condition which must inevi-tably cause death—the reaction of the blood is altered to only a minor degree.

No more admirable example of homeostasis can be mentioned than that of the pH of the fluid matrix of the body.   It is managed by accelerating and retarding the continuous processes of pulmonary ventilation and the flow of blood.   The physico-chemical changes within the blood itself, which will not be considered in this article, greatly diminish the effects of slight variations in these physiological processes. The degree of respiration is largely influenced by the hydrogen-ion con-centration in the cells of the respiratory center, but they in turn are influenced by an increase of the concentration of carbonic acid in the blood. Again the disturbance brings its own cure, and as the concentration is lowered by the heavier breathing, the heavier breathing ceases. The adjustment of the circulation may be similarly managed. The faster heart, the vascular constriction (except in active areas), and the contracted spleen point to functioning of the sympathetic system. Even slight voluntary activity calls the system into service (Cannon and Britton, 1927), and asphyxia is a highly effective stimulus for it (Cannon and Carrasco-Formiguera, 1922). The centers for sympathetic control may be influenced like the respiratory center—acidity may develop in them as a consequence of oxygen-want or carbonic acid excess and the primary result may be stimulation. This suggestion is supported by the experiments of Mathison (1911), showing that asphyxia and also extra carbon dioxide in the respired air raise arterial blood pressure, and by the observation of Cannon, Linton and Linton (1924) that muscle metabolites bring into action the sympathico-adrenal system. In vigorous muscular work the remarkably close correlation between the adjustment of the respiratory and the circulatory apparatus to the needs of the organism might thus be explained,—though both systems are started into faster service by impulses incidental to a voluntary act, they might be maintained in the performance of their extra task by the increased hydrogen-ion concentration in the blood and later they would gradually return to their quiet routine functions because their extra activity had resulted in reducing the hydrogen-ion concentration to the resting level.

Maintenance of uniform temperature. The importance of uniform temperature in providing conditions favorable for a constant rate of the chemical changes in the body requires no emphasis. And the danger of a rise of temperature a relatively few degrees above the normal, as well as the depressant effect of a fall much below the normal, likewise is well recognized. For general considerations, to be discussed presently, it is pertinent, however, to mention briefly the changes which take place when the body temperature tends to rise or fall. If the change is in the direction of a rise, relaxation of peripheral vessels occurs, thus exposing warm blood to the surface where heat may escape to colder surroundings; or when that is ineffective, sweating takes place, the skin is cooled by evaporation, and the abundant blood flowing through the skin loses heat thereby. Polypnea plays a part similar to sweating, and is especially serviceable in animals not well provided with sweat glands. If, on the other hand, the change is in the direction of a fall, there is a constriction of peripheral vessels and an erection of hairs and feathers

which enmesh near the skin a layer of poorly conducting air; when these
means of conserving heat do not check the fall of temperature, adrenin,
capable of increasing heat production, is set free in the blood stream
(Cannon, Querido, Britton and Bright, 1927); and when the heat thus
produced does not suffice, shivering is resorted to as the final automatic
protection against a temperature drop.   This highly efficient arrangement for maintaining homeostasis of body temperature involves only an
acceleration or retardation of the processes of heat production and heat
loss which are constantly going on.   The delicate thermostat which
operates the regulation appears to be located in the subthalamus
(Isenschmid, 1926; Rogers, 1920), and to be influenced directly by
changes in the temperature of the blood (Kahn, 1904; O'Connor, 1919;
Sherrington, 1924), and also reflexly (see Hill, 1921).   The noteworthy
features of the total arrangement, apart from its efficiency, are the
varieties of the devices for homeostasis, their appearance in a sequence
of defences against change, and the close involvement of the sympathetic system in the conservation, production and dissipation of heat.

THE RÔLE OF THE AUTONOMIC NERVOUS SYSTEM IN HOMEOSTASIS.
The homeostatic regulators act automatically.   Although skeletal
muscles and the diaphragm are, of course, under control of the cerebral
cortex, their functions in the regulation of temperature (shivering) and
neutrality (faster breathing) are managed low in the brain stem.   And
for the most part the regulators are not under voluntary government.
Commonly the autonomic system, or that system in coöperation with
endocrine organs, is called into action.   Illustrations of these facts are
seen in the vago-insular and the sympathico-adrenal influences on the
glycemic level, the vagal and sympathetic effects on the heart rate and
the sympathetic effects on blood vessels during vigorous muscular effort,
and the sympathico-adrenal function in accelerating heat production
when the body temperature tends to fall.

The facts just mentioned emphasize a distinction long recognized
between the "voluntary" and the "involuntary" or "vegetative" functions of the nervous system.   It is desirable to remove from physiology
terms having psychological and botanical implications.   The two relations of the nervous system—towards the external and towards the
internal environment—naturally suggest that distinctions should be
based on these opposite functions.   The "voluntary" or cerebro-
spinal system, elaborately outfitted with exteroceptors and with muscles
which operate bony levers, is arranged for altering the external environment or the position of the organism in that environment by laboring,

running or fighting. These may appropriately be regarded as *exterofec-tive* activities, and the "voluntary" system therefore as the exterofective system. The exterofective activities, however, must produce coincident changes in the internal environment—e.g., utilizing blood sugar and discharging into the blood acid waste and extra heat. Under these circumstances the "involuntary" nervous system plays its part by acting on the heart, smooth muscles and glands in such ways as will preserve the "fitness" of the internal environment for continued exterofective action. This interofective function of the "involuntary" nervous system justifies calling it the *interofective* system. Inactivity of the exterofective system establishes a basal state for the organism, because minimal functioning of that system is accompanied by minimal function-ing also of the interofective system. Exterofective action is reflected in interofective action, which rises as the internal disturbance rises and subsides as the disturbance subsides.

The interofective system has been referred to thus far as if it were single instead of consisting of three divisions, with distinctive general functions. These functions were summarized by Cannon (1914) as follows—the sacral, a group of reflexes for emptying hollow organs which become filled; the cranial, a series of reflexes protective and conservative and upbuilding in their function; and the mid- or sympathetic division, a mobilizer of bodily forces for struggle. Similar views have been expressed by Hess (1926) who has recognized the "histotropic" function of the cranial division in promoting the welfare of the tissues, and the "ergotropic" function of the sympathetic division in operating to in-crease the facilities for doing work. The ideas developed in the fore-going discussion modify only slightly the views expressed in 1914. The emphasis is somewhat differently placed, to be sure, if the maintenance of conditions favorable for exterofective activity is regarded as the chief function of the autonomic system. Thus, although the sacral autonomic division has as one of its functions the perpetuation of the race, it is also serviceable in emptying the bladder and rectum of loads which might interfere with extreme physical effort—an interpretation which is con-sistent with the well-known effects of strong emotion (e.g., fear) in voiding these viscera as a preparation for struggle. Likewise, though the cranial division has other conservative functions it notably exhibits its conservative uses in providing and preserving the measures which are required to keep the fluid matrix constant when profound disturbance might occur: it gives the heart opportunity for rest and recuperation by checking its rate in quiet times, it promotes the gastro-intestinal move-

ments and excites secretion of the digestive juices and thereby assures
the reserves of energy-yielding material, and it appears to be further
useful through the vago-insular system in bringing into storage some
of these reserves. These divisions, the sacral and the cranial, however,
operate indirectly and somewhat remotely to protect homeostasis. It
is the mid- or sympathetic division which acts directly and promptly
to prevent changes in the internal environment; by mobilizing reserves
and by altering the rate of continuous processes; as repeatedly noted in
the foregoing pages, this division works to keep constant the fluid
matrix of the body and therefore may properly be regarded as the
special and immediate agent of homeostasis. The idea that the sympathetic division—or the sympathico-adrenal apparatus, for the nerve
impulses and adrenin coöperate—serves to assure homeostasis, that for
this purpose it functions reciprocally when the exteroceptive system
functions, is not a fundamental modification of the "emergency theory"
(Cannon, 1914a). Recent studies have shown that if emergencies do
not arise, if marked changes in the outer world or vigorous reactions to
it do not occur, the sympathico-adrenal apparatus is not a necessity
and can be wholly removed without consequent disorder (Cannon,
Newton,.Bright, Menkin and Moore, 1929). Limitations appear when
circumstances alter the internal environment; it is then that the importance of the sympathico-adrenal apparatus becomes evident. As has
been shown above, however, this apparatus plays its rôle not only in
preserving homeostasis during grave crises which demand supreme
effort, but also in the minor exterofective adjustments which might
change the fluid matrix of the body.

SOME POSTULATES REGARDING HOMEOSTATIC REGULATION. About
four years ago Cannon (1925) advanced six tentative propositions concerned with physiological factors which maintain steady states in the
body. It will be pertinent to consider them again now with reference
to the foregoing discussion of homeostasis.

1. "In an open system such as our bodies represent, compounded of
unstable material and subjected continually to disturbing conditions,
constancy is in itself evidence that agencies are acting, or ready to act,
to maintain this constancy." This is a confident inference—an inference
based on some insight into the ways by which certain steady states
(e.g., glycemia, body temperature, and neutrality of the blood) are
regulated and a confidence that other steady states are similarly regulated. The instances cited in the previous pages have illustrated various
agencies employed in the organism to that end. Although we do

not know how constancy of plasma proteins, lipemia and blood calcium, for example, is brought about, probably it results from as nice devices as those operating in the better known cases of homeostasis. Of course this realm of interest is full of problems—highly significant problems—inviting attempts at solution, and as they are solved the confidence expressed in the first postulate may be justified.

2. "If a state remains steady it does so because any tendency towards change is automatically met by increased effectiveness of the factor or factors which resist the change." Thirst, the hypoglycemic reaction, the respiratory and circulatory response to a blood shift towards acidity, the thermogenic functions, all become more intense as the disturbance of homeostasis is more pronounced, and they all subside promptly when the disturbance is relieved. Similar conditions probably prevail in other steady states. Of course, the state may not remain steady, as in pathological weakness or defect, and for that reason the postulate was made conditional. As Lotka (1925) has pointed out, this conditional statement, required for living beings and due to their lack of permanent stability, sharply distinguishes the proposal from the strict principle of Le Chatelier true for simple physical or chemical systems. Indeed, as Y. Henderson (1925) has remarked, the physiological and the chemical conceptions of equilibrium are quite different. "The one invokes energy to maintain itself, or if disturbed to recover . . . . the other in seeking balance only goes down hill dynamically."

3. "Any factor which operates to maintain a steady state by action in one direction does not also act at the same point in the opposite direction." This proposal, which should have been limited to *physiological* action, is related to the questions discussed in the footnote on p. 410. Does adrenin in physiological doses both discharge glycogen from the liver and increase glycogen storage there? Does insulin likewise act oppositely in relation to the hepatic glycogen reserves? In the footnote mentioned, reasons were given for not crediting the evidence for opposed action by a single one of these agents. An agent may exist which has an influence of a tonic type—a moderate activity—which can be varied up or down, and which can act at a given point in "high" concentration but not in a "low" concentration. The adrenal medulla, which is subject to control of opposed nervous influences (Cannon and Rapport, 1921), may be cited as an agent of that type.

4. "Homeostatic agents, antagonistic in one region of the body, may be coöperative in another region." The sympathico-adrenal and the vago-insular influences are opposed in action on the liver, but they

appear to be collaborators in their action on muscles, e.g., leading to effective use of sugar by muscle cells (Burn and Dale, 1924). Too little is known about the effects of these agents to permit this postulate to be of much significance at present.

5. "The regulating system which determines a homeostatic state may comprise a number of coöperating factors brought into action at the same time or successively." This statement is well illustrated in the arrangements for protection against a fall of temperature in which series of defences are used one after another, and also in the elaborate and complex arrangements for maintaining uniform reaction of the blood.

6. "When a factor is known which can shift a homeostatic state in one direction it is reasonable to look for automatic control of that factor or for a factor or factors having an opposing effect." This postulate is implied in earlier postulates. It is expressed as a reiteration of the confidence that homeostasis is not accidental but is a result of organized government, and that search for the governing agencies will result in their discovery.

The reader has had occasion to be impressed by the large gaps in our knowledge not only of homeostatic conditions but also of the arrangements which establish and maintain them. Repeatedly the phrase "is not known" has had to be employed. It is remarkable that features so characteristic of living beings as the steady states should have received so little attention. Innumerable questions remain to be answered. Little is known, for example, about the effective stimuli for such homeostatic reactions as are well recognized. Are there receptors which are affected in blood-sugar regulation or are the regulatory factors worked by direct action on cerebral centers? Again, there are homeostatic agencies which were not considered above, such as the extra erythrocytes produced in organisms living at high altitudes, the thicker hair growth during prolonged cold weather, and also steady states which have not been mentioned, such as the stabilization of phosphorus in relation to calcium, the evidence from constancy of basal metabolism that there is constancy in the thyroxin content of the blood, indeed the evidence from other steady states (as weight, and sex character) that other endocrine products are uniformly circulating—the questions presented by these and many other reactions which are serviceable in preserving uniformity in the fluid matrix offer a fascinating field for research.

In the two preceding sections the functions of the divisions of the autonomic system in relation to homeostasis were defined and some postulates regarding homeostasis were presented, not with the idea that the statements should be taken as conclusive but rather that they might prove suggestive for further investigation. Indeed, that point of view should be recognized as prevailing throughout this review. It is the writer's belief that the study of the particular activities of the various organs of the body has progressed to a degree which will permit to a greater extent than is generally recognized an examination of the interplay of these organs in the organism as a whole. Their relations to their internal environment seemed to offer a suggestive approach to a survey of their possible integrative functions. In such a venture errors are sure to creep in which must later be corrected, and crude ideas are sure to be projected which must later be refined. Though the present account of agencies which regulate steady states in the body is likely to prove inadequate and provisional, there is no question of the great importance of the facts of homeostasis with which it deals. This account may at least serve to rouse interest in them and their importance. The facts are significant as outstanding features of biological organization and activity. They are significant also in understanding the complex disorders of the body, for in a state normally kept regular by a group of coöperating parts, full insight into irregularity is obtained only by learning their mode of coöperation. Again, effective methods of attaining homeostasis are significant in comparison with the methods in systems where steady states are not yet well developed; the regulation of homeostasis in higher animals is probably the result of innumerable evolutionary trials, and knowledge of the stability which has finally been achieved is suggestive in relation to the less efficient arrangements operating in lower animals and also in relation to attempts at securing stability in social and economic organizations. Finally, continued analysis of biological processes in physical and chemical terms must await a full understanding of the ways in which these processes are roused to perform their service and are then returned to inactivity. Indeed, regulation in the organism is the central problem of physiology. For all these reasons further research into the operation of agencies for maintaining biological homeostasis is desirable.

BIBLIOGRAPHY

ABE, Y. 1924. Arch. f. exper. Path. u. Pharm., ciii, 73.
ADOLPH, E. F. 1921. Journ. Physiol., lv, 114.
AFANASSIEW, M. 1883. Pflüger's Arch., xxx, 385.
ALLEN, F. M. 1920. Journ. Exper. Med., xxxi, 363, 381.
ANDRUS, E. C. AND E. P. CARTER. 1924. Heart, xi, 106.

Aub, J. C., W. Bauer, C. Heath and M. Ropes. 1929. Journ. Clin. Invest., vii, 97.

Bailey, P. and F. Bremer. 1921. Arch. Int. Med., xxviii, 773.

Bainbridge, F. A. 1923. The physiology of muscular exercise. London, p. 90.

Baird, M. M. and J. B. S. Haldane. 1922. Journ. Physiol., lvi, 259.

Banting, F. G., C. H. Best, J. B. Collip and E. C. Noble. 1922. Trans. Roy. Soc. Canada, xvi, 13.

Banting, F. G. and S. Gairns. 1924. Amer. Journ. Physiol., lxviii, 24.

Barcroft, J. and H. Staub. 1910. Journ. Physiol., xli, 145.

Barcroft, J. 1926. Ergebn. d. Physiol., xxv, 818.

Bauer, W., J. C. Aub and F. Albright. 1929. Journ. Exper. Med., xlix, 145.

Berg, W. 1914. Biochem. Zeitschr., lxi, 428; München. med. Wochenschr., lxi, 434.
   1922. Pflüger's Arch., cxciv, 102; cxcv, 543.

Bernard, C. 1878. Les Phénomènes de la Vie. Paris, two vols.

Best, C. H., J. P. Hoet and H. P. Marks. 1926. Proc. Roy. Soc. (London), B, c, 32.

Bloor, W. R. 1922. Endocrinology and metabolism. New York, iii, 204.

Boothby, W. M., I. Sandiford and J. Slosse. 1925. Ergebn. d. Physiol., xxiv, 733.

Britton, S. W. 1925. Amer. Journ. Physiol., lxxiv, 291.
   1928. Amer. Journ. Physiol., lxxxvi, 340.

Bulatao, E. and W. B. Cannon. 1925. Amer. Journ. Physiol., lxxii, 295.

Burn, J. H. 1923. Journ. Physiol., lvii, 318.

Burn, J. H. and H. H. Dale. 1924. Journ. Physiol., lix, 164.

Cahn-Bronner, C. E. 1914. Biochem. Zeitschr., lxvi, 289.

Campos, F. A. de M., W. B. Cannon, H. Lundin and T. T. Walker. 1929. Amer. Journ. Physiol., lxxxvii, 680.

Cannon, W. B. and A. L. Washburn. 1912. Amer. Journ. Physiol., xxix, 441.

Cannon, W. B. 1914a. Amer. Journ. Physiol., xxxiii, 356.
   1914b. Amer. Journ. Psychol., xxv, 256.

Cannon, W. B. and H. Gray. 1914. Amer. Journ. Physiol., xxxiv, 232.

Cannon, W. B. and W. L. Mendenhall. 1914. Amer. Journ. Physiol., xxxiv, 245, 251.

Cannon, W. B. 1918. Proc. Roy Soc. (London), B, xc, 283.

Cannon, W. B. and D. Rapport. 1921. Amer. Journ. Physiol., lviii, 308.

Cannon, W. B. and R. Carrasco-Formiguera. 1922. Amer. Journ. Physiol., lxi, 215.

Cannon, W. B., M. A. McIver and S. W. Bliss. 1923. Boston Med. and Surg. Journ., clxxxix, 141.
   1924. Amer. Journ. Physiol., lxix, 46.

Cannon, W. B., J. R. Linton and R. R. Linton. 1924. Amer. Journ. Physiol., lxxi, 153.

Cannon, W. B. 1925. Trans. Cong. Amer. Physicians and Surgeons, xii 31; 1926, Jubilee Volume for Charles Richet, p. 91.

Cannon, W. B. and S. W. Britton. 1927. Amer. Journ. Physiol., lxxix, 433.

Cannon, W. B., A. Querido, S. W. Britton and E. M. Bright. 1927. Amer. Journ. Physiol., lxxix, 466.

Cannon, W. B., H. F. Newton, E. M. Bright, V. Menkin and R. M. Moore. 1929. Amer. Journ. Physiol., lxxxix, 84.

CARLSON, A. J. 1916. The control of hunger in health and disease. Chicago.

COHNHEIM, O., KREGLINGER AND KREGLINGER, JR. 1909. Zeitschr. f. physiol. Chem., lxiii, 429.

COLLIP, J. B. 1926. Journ. Biol. Chem., lxiii, 395.

CORI, C. F. 1925. Journ. Pharm. Exper. Therap., xxv, 1.

CORI, C. F. AND G. T. CORI. 1928. Journ. Biol. Chem., lxxix, 309.

CRANDALL, F. M. 1899. Arch. Pediat., xvi, 174.

CUSHNY, A. R. 1926. The secretion of urine. 2nd ed., London.

ENGELS, W. 1904. Arch. f. exper. Path. u. Pharm., li, 346.

ERDHEIM, J. 1911. Zeitschr. f. Pathol., vii, 175, 238, 259.

FISKE, C. H. AND Y. SUBBAROW. 1929. Journ. Biol. Chem., lxxxi, 656.

FLETCHER, A. A. AND W. R. CAMPBELL. 1922. Journ. Metab. Res., ii, 637.

FOLIN, O., H. C. TRIMBLE AND L. H. NEWMAN. 1927. Journ. Biol. Chem., lxxv, 263.

FOSTER, D. P. AND G. H. WHIPPLE. 1922. Amer. Journ. Physiol., lviii, 393, 407.

FREDERICQ, L. 1885. Arch. de Zoöl Exper. et Gén., iii, p. xxxv.

FREUND, H. 1913. Arch. f. exper. Path. u. Pharm., lxxiv, 311.

GAMBLE, J. L. AND S. G. ROSS. 1925. Journ. Clin. Invest., i, 403.

GAMBLE, J. L. AND M. MCIVER. 1925. Ibid., i, 531.

        1928. Journ. Exper. Med., xlviii, 859.

GAYET, R. AND M. GUILLAUMIE. 1928. Compt. rend. Soc. de Biol., cxvii, 1613.

GRAFE, E. 1927. Oppenheimer's Handbuch der Biochemie, Jena, ix, 68.

GRANT, S. B. AND A. GOLDMAN. 1920. Amer. Journ. Physiol., lii, 209.

GRAY, H. AND L. K. LUNT. 1914. Amer. Journ. Physiol., xxxiv, 332.

GRÜNWALD, H. F. 1909. Arch. f. exper. Path. u. Pharm., lx, 360.

HALDANE, J. S. AND J. G. PRIESTLEY. 1915. Journ. Physiol., l, 296.

HALDANE, J. S. 1922. Respiration. New Haven, pp. 21–22; p. 383.

HALLIBURTON, W. D. 1904. Biochemistry of muscle and nerve. Philadelphia, p. 111.

HALLION, L. AND R. GAYET. 1925. Compt. rend. Soc. de Biol., xcii, 945.

HANSEN, K. M. 1923. Acta. Med. Scand., lviii, Suppl. iv.

HASSELBALCH, K. A. AND C. LUNDSGAARD. 1912. Skand. Arch. f. Physiol., xxvii, 13.

HENDERSON, L. J. 1928. Blood. New Haven.

HENDERSON, Y. 1925. Physiol. Rev., v, 131.

HESS, W. R. 1926. Klin. Wochenschr., v, 1353.

HILL, L. 1921. Journ. Physiol., liv, p. cxxxvi.

HIRAYAMA, S. 1925. Tohoku Journ. Exper. Med., vi, 160.

HOMANS, J. 1914. Journ. Med. Res., xxv, 63.

HOUSSAY, B. A., J. T. LEWIS AND E. A. MOLINELLI. 1924. Compt. rend. Soc. de Biol., xci, 1011.

ISENSCHMID, I. 1926. Handb. d. norm. u. path. Physiol., Berlin, xvii, 56.

IZQUIERDO, J. J. AND W. B. CANNON. 1928. Amer. Journ. Physiol., lxxxiv, 545.

KAHN, R. H. 1904. Arch. f. Physiol., Suppl. Bd., p. 81.

KEITH, N. M. 1923. Amer. Journ. Physiol., lxiii, 394.

KERR, W. J., S. H. HURWITZ AND G. H. WHIPPLE. 1918. Amer. Journ. Physiol., xlvii, 379.

KROGH, A. 1922. The anatomy and physiology of the capillaries. New Haven, p. 227.

Kulz, E. 1880. Pflüger's Arch., xxiv, 41.

LaBarre, J. 1925. Arch. Internat. de Physiol., xxv, 265.

Lewis, F. T. and J. L. Bremer. 1927. A textbook of histology. Philadelphia, p. 72.

Lewis, J. T. 1923. Compt. rend. Soc. de Biol., lxxxix, 1118.

Lewis, J. T. and M. Magenta. 1925. Compt. rend. Soc. de Biol., xcii, 821.

Loevenhart, A. S. 1902. Amer. Journ. Physiol., vi, 331.

Lotka, A. J. 1925. Elements of physical biology. Baltimore, p. 284.

Luce, E. M. 1923. Journ. Pathol. and Bact., xxvi, 200.

Lusk, G. 1928. The science of nutrition. 4th ed., Philadelphia.

MacCallum, W. G. and C. Voegtlin. 1909. Journ. Exper. Med., xi, 118.

MacCallum, W. G. and K. M. Vogel. 1913. Journ. Exper. Med., xviii, 618.

Macleod, J. J. R. 1913. Diabetes; its pathological physiology. London, p. 61.
1924. Physiol. Rev., iv, 51.

Marine, D. 1913. Proc. Soc. Exper. Biol. Med., xi, 117.

Mathison, G. C. 1911. Journ. Physiol., xlii, 283.

McCormick, N. A., J. J. R. Macleod, E. C. Noble and K. O'Brien. 1923. Journ. Physiol., lvii, 224.

Meek, W. J. 1912. Amer. Journ. Physiol., xxx, 161.

Minkowski, O. 1908. Arch. f. exper. Path. u. Pharm., Suppl. Bd., 399.

Noël, R. 1923. Presse méd., xxxi, 158.

O'Connor, J. M. 1919. Journ. Physiol., lii, 267.

Padtberg, J. H. 1910. Arch. f. exper. Path. u. Pharm., lxiii, 60.

Patterson, S. W. and E. H. Starling. 1913. Journ. Physiol., xlvii, 143.

Pflüger, E. F. W. 1877. Pflüger's Arch., xv, 57.

Reincke, J. J. 1875. Deutsch. Arch. f. Klin. Med., xvi, 12.

Richet, C. 1900. Dictionnaire de Physiologie, Paris, iv, 721.

Rogers, F. T. 1920. Arch. Neurol. und Psychiatr., iv, 148.

Rosenfeld, G. 1903. Ergebn. d. Physiol., ii, pt. 1, 86.

Rowntree, L. G. 1922. Physiol. Rev., ii, 158.

Schade, H. 1925. Wasserstoffwechsel. Oppenheimer's Handb. der Biochem., 2nd ed., Jena, p. 175.

Schulz, F. N. 1896. Pflüger's Arch., lxv, 299.

Seitz, W. 1906. Pflüger's Arch., cxi, 309.

Sherrington, C. S. 1924. Journ. Physiol., lxviii, 405.

Skelton, H. P. 1927. Arch. Int. Med., xl, 140.

Smith, H. P., A. E. Belt and G. H. Whipple. 1920. Amer. Journ. Physiol., lii, 54.

Starling, E. H. 1909. The fluids of the body. Chicago.

Strada, F. 1909. Pathologica, i, 423.

Stübel, H. 1920. Pflüger's Arch., clxxxv, 74.

Sundberg, C. G. 1923. Compt. rend. Soc. de Biol., lxxxix, 807.

Tatum, A. L. 1921. Journ. Pharm. Exper. Therap., xviii, 121.

Terroine, E. F. 1914. Journ. de Physiol. et de Pathol. Gén., xvi, 408.

Thompson, W. O. 1926. Journ. Clin. Invest., ii, 477.

Tichmeneff, N. 1914. Biochem. Zeitschr., lix, 326.

Trendelenburg, P. and W. Goebel. 1921. Arch. exper. Path. u. Pharm., lxxxix, 171.

Vosburgh, C. H. and A. N. Richards. 1903. Amer. Journ. Physiol., ix, 35.

Weichselbaum, A. 1914. Verhandl. d. deutsch. Naturf. u. Aerzte, 85.

Wettendorf, H. 1901. Trav. du Lab. de Physiol., Inst., Solvay, iv, 353.

Whipple, G. H., H. P. Smith and A. E. Belt. 1920. Amer. Journ. Physiol., lii, 72.

Zunz, E. and J. LaBarre. 1927. Compt. rend. Soc. de Biol., xcvi, 421, 708.

Part IV

FEEDBACK AMPLIFIERS

Editor's Comments
on Papers 20 and 21

20   BODE
Relations Between Attenuation and Phase in Feedback Amplifier Design

21   NYQUIST
Regeneration Theory

The concept of feedback energy control has naturally led to the development of amplifiers, of which the electrical type is the most important for its many applications in all branches of electronics. We therefore reproduce here two classic articles on this subject. The first, Paper 20, is by H. W. Bode (b. 1905): "Relations between Attenuation and Phase in Feedback Amplifier Design" (1940), and the second, Paper 21, is by H. Nyquist (1889–1976): "Regeneration Theory" (1932).

Written in clear and graceful style, with frequent use of attractive analogies, the Bode paper handles the problem of stability in feedback circuits. As has been emphasized previously in this volume this is one of the serious problems connected with feedback energy control devices. Unless the feedback conditions with respect to phase are carefully adjusted, strong oscillations in energy flow may ensue, thus jeopardizing the whole control idea. Bode takes up in considerable mathematical detail the incidence of this problem in feedback amplifiers. For most profitable reading the paper demands a sound background in electrical network theory.

Nyquist's paper is a highly mathematical analysis of feedback (here called regeneration) in electrical circuits, with special emphasis on stability. It is an earlier paper than that of Bode and hence a pioneer in the field. It had a profound influence on subsequent work on feedback networks. References to Nyquist diagrams have been frequent in the later literature and indeed occur in many places in Bode's paper.

H.W. Bode was a member of the Technical Staff of the Bell Telephone Laboratories from 1926–1967, and since the latter date has been a professor at Harvard University.

H. Nyquist was a member of the Technical Staff of the Bell Telephone Laboratories from 1934–1954.

Reprinted with permission from pp. 421–446 of *Bell Syst. Tech. J.* **19**(3):
421–454 (1940)

Relations Between Attenuation and Phase in Feedback Amplifier Design

By H. W. BODE

INTRODUCTION

THE engineer who embarks upon the design of a feedback amplifier must be a creature of mixed emotions. On the one hand, he can rejoice in the improvements in the characteristics of the structure which feedback promises to secure him.[1] On the other hand, he knows that unless he can finally adjust the phase and attenuation characteristics around the feedback loop so the amplifier will not spontaneously burst into uncontrollable singing, none of these advantages can actually be realized. The emotional situation is much like that of an impecunious young man who has impetuously invited the lady of his heart to see a play, unmindful, for the moment, of the limitations of the $2.65 in his pockets. The rapturous comments of the girl on the way to the theater would be very pleasant if they were not shadowed by his private speculation about the cost of the tickets.

In many designs, particularly those requiring only moderate amounts of feedback, the bogy of instability turns out not to be serious after all. In others, however, the situation is like that of the young man who has just arrived at the box office and finds that his worst fears are realized. But the young man at least knows where he stands. The engineer's experience is more tantalizing. In typical designs the loop characteristic is always satisfactory—except for one little point. When the engineer changes the circuit to correct that point, however, difficulties appear somewhere else, and so on ad infinitum. The solution is always just around the corner.

Although the engineer absorbed in chasing this rainbow may not realize it, such an experience is almost as strong an indication of the existence of some fundamental physical limitation as the census which the young man takes of his pockets. It reminds one of the experience of the inventor of a perpetual motion machine. The perpetual motion machine, likewise, always works—except for one little factor. Evidently, this sort of frustration and lost motion is inevitable in

---

[1] A general acquaintance with feedback circuits and the uses of feedback is assumed in this paper. As a broad reference, see H. S. Black, "Stabilized Feedback Amplifiers," *B. S. T. J.*, January, 1934.

feedback amplifier design as long as the problem is attacked blindly. To avoid it, we must have some way of determining in advance when we are either attempting something which is beyond our resources, like the young man on the way to the theater, or something which is literally impossible, like the perpetual motion enthusiast.

This paper is written to call attention to several simple relations between the gain around an amplifier loop, and the phase change around the loop, which impose limits to what can and cannot be done in a feedback design. The relations are mathematical laws, which in their sphere have the same inviolable character as the physical law which forbids the building of a perpetual motion machine. They show that the attempt to build amplifiers with certain types of loop characteristics *must* fail. They permit other types of characteristic, but only at the cost of certain consequences which can be calculated. In particular, they show that the loop gain cannot be reduced too abruptly outside the frequency range which is to be transmitted if we wish to secure an unconditionally stable amplifier. It is necessary to allow at least a certain minimum interval before the loop gain can be reduced to zero.

The question of the rate at which the loop gain is reduced is an important one, because it measures the actual magnitude of the problem confronting both the designer and the manufacturer of the feedback structure. Until the loop gain is zero, the amplifier will sing unless the loop phase shift is of a prescribed type. The cutoff interval as well as the useful transmission band is therefore a region in which the characteristics of the apparatus must be controlled. The interval represents, in engineering terms, the price of the ticket.

The price turns out to be surprisingly high. It can be minimized by accepting an amplifier which is only conditionally stable.[2] For the customary absolutely stable amplifier, with ordinary margins against singing, however, the price in terms of cutoff interval is roughly one octave for each ten db of feedback in the useful band. In practice, an additional allowance of an octave or so, which can perhaps be regarded as the tip to the hat check girl, must be made to insure that the amplifier, having once cut off, will stay put. Thus in an amplifier with 30 db feedback, the frequency interval over which effective control of the loop transmission characteristics is necessary is at least four octaves, or sixteen times, broader than the useful band. If we raise the feedback to 60 db, the effective range must be more than a hundred times the useful range. If the useful band is itself large these factors

[2] Definitions of conditionally and unconditionally stable amplifiers are given on page 432.

may lead to enormous effective ranges. For example, in a 4 megacycle amplifier they indicate an effective range of about 60 megacycles for 30 db feedback, or of more than 400 megacycles if the feedback is 60 db.

The general engineering implications of this result are obvious. It evidently places a burden upon the designer far in excess of that which one might anticipate from a consideration of the useful band alone. In fact, if the required total range exceeds the band over which effective control of the amplifier loop characteristics is physically possible, because of parasitic effects, he is helpless. Like the young man, he simply can't pay for his ticket. The manufacturer, who must construct and test the apparatus to realize a prescribed characteristic over such wide bands, has perhaps a still more difficult problem. Unfortunately, the situation appears to be an inevitable one. The mathematical laws are inexorable.

Aside from sounding this warning, the relations between loop gain and loop phase can also be used to establish a definite method of design. The method depends upon the development of overall loop characteristics which give the optimum result, in a certain sense, consistent with the general laws. This reduces actual design procedure to the simulation of these characteristics by processes which are essentially equivalent to routine equalizer design. The laws may also be used to show how the characteristics should be modified when the cutoff interval approaches the limiting band width established by the parasitic elements of the circuit, and to determine how the maximum realizable feedback in any given situation can be calculated. These methods are developed at some length in the writer's U. S. Patent No. 2,123,178 and are explained in somewhat briefer terms here.

RELATIONS BETWEEN ATTENUATION AND PHASE IN PHYSICAL NETWORKS [3]

The amplifier design theory advanced here depends upon a study of the transmission around the feedback loop in terms of a number of general laws relating the attenuation and phase characteristics of physical networks. In attacking this problem an immediate difficulty presents itself. It is apparent that no entirely definite and universal

[3] Network literature includes a long list of relations between attenuation and phase discovered by a variety of authors. They are derived typically from a Fourier analysis of the transient response of assumed structures and are frequently ambiguous, because of failure to recognize the minimum phase shift condition. No attempt is made to review this work here, although special mention should be made of Y. W. Lee's paper in the *Journal for Mathematics and Physics* for June, 1932. The proof of the relations given in the present paper depends upon a contour integration in the complex frequency plane and can be understood from the disclosure in the patent referred to previously.

relation between the attenuation and the phase shift of a physical structure can exist. For example, we can always change the phase shift of a circuit without affecting its loss by adding either an ideal transmission line or an all-pass section. Any attenuation characteristic can thus correspond to a vast variety of phase characteristics.

For the purposes of amplifier design this ambiguity is fortunately unimportant. While no unique relation between attenuation and phase can be stated for a general circuit, a unique relation does exist between any given loss characteristic and the *minimum* phase shift which must be associated with it. In other words, we can always add a line or all-pass network to the circuit but we can never subtract such a structure, unless, of course, it happens to be part of the circuit originally. If the circuit includes no surplus lines or all-pass sections, it will have at every frequency the least phase shift (algebraically) which can be obtained from any physical structure having the given attenuation characteristic. The least condition, since it is the most favorable one, is, of course, of particular interest in feedback amplifier design.

For the sake of precision it may be desirable to restate the situations in which this minimum condition fails to occur. The first situation is found when the circuit includes an all-pass network either as an individual structure or as a portion of a network which can be replaced by an all-pass section in combination with some other physical structure.[4] The second situation is found when the circuit includes a transmission line. The third situation occurs when the frequency is so high that the tubes, network elements and wiring cannot be considered to obey a lumped constant analysis. This situation may be found, for example, at frequencies for which the transit time of the tubes is important or for which the distance around the feedback loop is an appreciable part of a wave-length. The third situation is, in many respects, substantially the same as the second, but it is mentioned separately here as a matter of emphasis. Since the effective band of a feedback amplifier is much greater than its useful band, as the introduction pointed out, the considerations it reflects may be worth taking into account even when they would be trivial in the useful band alone.

It will be assumed here that none of these exceptional situations is found. For the minimum phase condition, then, it is possible to derive

[4] Analytically this condition can be stated as follows: Let it be supposed that the transmission takes place between mesh 1 and mesh 2. The circuit will include an all-pass network, explicit or concealed, if any of the roots of the minor Δ_{12} of the principal circuit determinant lie below the real axis in the complex frequency plane. This can happen in bridge configurations, but not in series-shunt configurations, so that all ladder networks are automatically of minimum phase type.

a large number of relations between the attenuation and phase characteristics of a physical network. One of the simplest is

$$\int_{-\infty}^{\infty} B\,du = \frac{\pi}{2}(A_\infty - A_0),\tag{1}$$

where u represents $\log f/f_0$, f_0 being an arbitrary reference frequency, B is the phase shift in radians, and A_0 and A_∞ are the attenuations in nepers at zero and infinite frequency, respectively. The theorem states, in effect, that the total area under the phase characteristic plotted on a logarithmic frequency scale depends only upon the difference between the attenuations at zero and infinite frequency, and not upon the course of the attenuation between these limits. Nor does it depend upon the physical configuration of the network unless a nonminimum phase structure is chosen, in which case the area is necessarily increased. The equality of phase areas for attenuation characteristics of different types is illustrated by the sketches of Fig. 1.

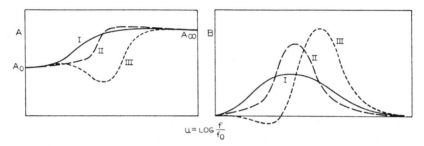

Fig. 1—Diagram to illustrate relation between phase area and change in attenuation.

The significance of the phase area relation for feedback amplifier design can be understood by supposing that the practical transmission range of the amplifier extends from zero to some given finite frequency. The quantity $A_0 - A_\infty$ can then be identified with the change in gain around the feedback loop required to secure a cut-off. Associated with it must be a certain definite phase area. If we suppose that the maximum phase shift at any frequency is limited to some rather low value the total area must be spread out over a proportionately broad interval on the frequency scale. This must correspond roughly to the cut-off region, although the possibility that some of the area may be found above or below the cut-off range prevents us from determining the necessary interval with precision.

A more detailed statement of the relationship between phase shift and change in attenuation can be obtained by turning to a second theorem. It reads as follows:

$$B(f_c) = \frac{1}{\pi} \int_{-\infty}^{\infty} \frac{dA}{du} \log \coth \frac{|u|}{2} du, \tag{2}$$

where $B(f_c)$ represents the phase shift at any arbitrarily chosen frequency f_c and $u = \log f/f_c$. This equation, like (1), holds only for the minimum phase shift case.

Although equation (2) is somewhat more complicated than its predecessor, it lends itself to an equally simple physical interpretation. It is clear, to begin with, that the equation implies broadly that the

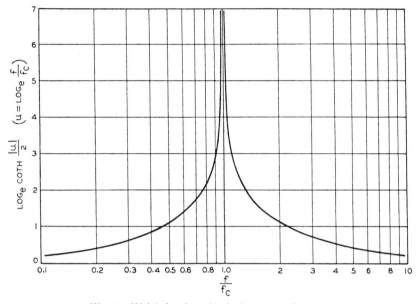

Fig. 2—Weighting function in loss-phase formula.

phase shift at any frequency is proportional to the derivative of the attenuation on a logarithmic frequency scale. For example, if dA/du is doubled B will also be doubled. The phase shift at any particular frequency, however, does not depend upon the derivative of attenuation at that frequency alone, but upon the derivative at all frequencies, since it involves a summing up, or integration, of contributions from the complete frequency spectrum. Finally, we notice that the contributions to the total phase shift from the various portions of the frequency spectrum do not add up equally, but rather in accordance with the function $\log \coth |u|/2$. This quantity, therefore, acts as a weighting function. It is plotted in Fig. 2. As we might expect physically it is much larger near the point $u = 0$ than it is in other regions. We can, therefore, conclude that while the derivative of attenuation at all frequencies enters into the phase shift at any particular frequency $f = f_c$ the derivative in the neighborhood of f_c is relatively much more important than the derivative in remote parts of the spectrum.

As an illustration of (2), let it be supposed that $A = ku$, which corresponds to an attenuation having a constant slope of $6 k$ db per octave. The associated phase shift is easily evaluated. It turns out, as we might expect, to be constant, and is equal numerically to $k\pi/2$ radians. This is illustrated by Fig. 3. As a second example, we may consider

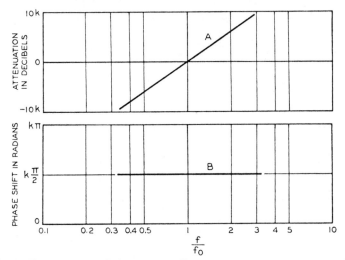

Fig. 3—Phase characteristic corresponding to a constant slope attenuation.

a discontinuous attenuation characteristic such as that shown in Fig. 4. The associated phase characteristic, also shown in Fig. 4, is proportional to the weighting function of Fig. 2.

The final example is shown by Fig. 5. It consists of an attenuation characteristic which is constant below a specified frequency f_b and has a constant slope of $6 k$ db per octave above f_b. The associated phase characteristic is symmetrical about the transition point between the two ranges. At sufficiently high frequencies, the phase shift approaches the limiting $k\pi/2$ radians which would be realized if the constant slope were maintained over the complete spectrum. At low frequencies the phase shift is substantially proportional to frequency· and is given by the equation

$$B = \frac{2k}{\pi} \frac{f}{f_b} .$$ (3)

Solutions developed in this way can be added together, since it is apparent from the general relation upon which they are based that the phase characteristic corresponding to the sum of two attenuation

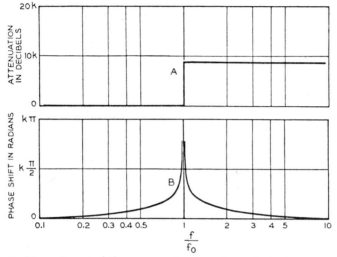

Fig. 4—Phase characteristic corresponding to a discontinuity in attenuation.

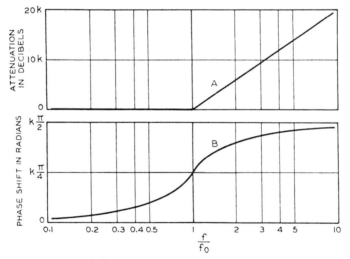

Fig. 5—Phase characteristic corresponding to an attenuation which is constant below a prescribed frequency and has a constant slope above it.

characteristics will be equal to the sum of the phase characteristics corresponding to the two attenuation characteristics separately. We can therefore combine elementary solutions to secure more complicated characteristics. An example is furnished by Fig. 6, which is built up from three solutions of the type shown by Fig. 5. By proceeding sufficiently far in this way, an approximate computation of the phase characteristic associated with almost any attenuation characteristic can be made, without the labor of actually performing the integration in (2).

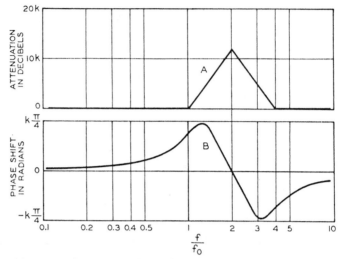

Fig. 6—Diagram to illustrate addition of elementary attenuation and phase characteristics to produce more elaborate solutions of the loss-phase formula.

Equations (1) and (2) are the most satisfactory expressions to use in studying the relation between loss and phase in a broad physical sense. The mechanics of constructing detailed loop cut-off characteristics, however, are simplified by the inclusion of one other, somewhat more complicated, formula. It appears as

$$\int_0^{f_0} \frac{A\,df}{\sqrt{f_0^2 - f^2}\,(f^2 - f_c^2)} + \int_{f_0}^{\infty} \frac{B\,df}{\sqrt{f^2 - f_0^2}\,(f^2 - f_c^2)}$$

$$= \frac{\pi}{2f_c} \frac{B(f_c)}{\sqrt{f_0^2 - f_c^2}}, \qquad f_c < f_0$$

$$= -\frac{\pi}{2f_c} \frac{A(f_c)}{\sqrt{f_c^2 - f_0^2}}, \qquad f_c > f_0, \quad (4)$$

where f_0 is some arbitrarily chosen frequency and the other symbols have their previous significance.

The meaning of (4) can be understood if it is recalled that (2) implies that the minimum phase shift at any frequency can be computed if the attenuation is prescribed at all frequencies. In the same way (4) shows how the complete attenuation and phase characteristics can be determined if we begin by prescribing the attenuation below f_0 and the phase shift above f_0. Since f_0 can be chosen arbitrarily large or small this is evidently a more general formula than either (1) or (2), while it can itself be generalized, by the introduction of additional irrational factors, to provide for more elaborate patterns of bands in which A and B are specified alternately.

As an example of this formula, let it be assumed that $A = K$ for $f < f_0$ and that $B = k\pi/2$ for $f > f_0$. These are shown by the solid lines in Fig. 7. Substitution in (4) gives the A and B characteristics in the rest of the spectrum as

$$B = k \sin^{-1}\frac{f}{f_0}, \qquad\qquad f < f_0$$

$$A = K + k \log\left[\sqrt{\frac{f^2}{f_0^2} - 1} + \frac{f}{f_0}\right], \qquad f > f_0. \qquad (5)$$

These are indicated by broken lines in Fig. 7. In this particularly

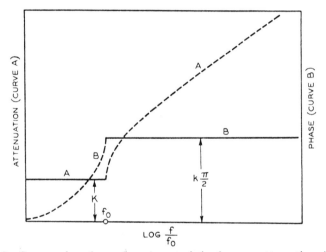

Fig. 7—Construction of complete characteristics from an attenuation characteristic specified below a certain frequency and a phase characteristic above it. The solid lines represent the specified attenuation and phase characteristics, and the broken lines their computed extensions to the rest of the spectrum.

simple case all four fragments can be combined into the single analytic formula

$$A + iB = K + k \log\left[\sqrt{1 - \frac{f^2}{f_0^2}} + i\frac{f}{f_0}\right]. \qquad (6)$$

This expression will be used as the fundamental formula for the loop cut-off characteristic in the next section.

OVERALL FEEDBACK LOOP CHARACTERISTICS

The survey just concluded shows what combinations of attenuation and phase characteristics are physically possible. We have next to determine which of the available combinations is to be regarded as representing the transmission around the overall feedback loop. The choice will naturally depend somewhat upon exactly what we assume that the amplifier ought to do, but with any given set of assumptions it is possible, at least in theory, to determine what combination is most appropriate.

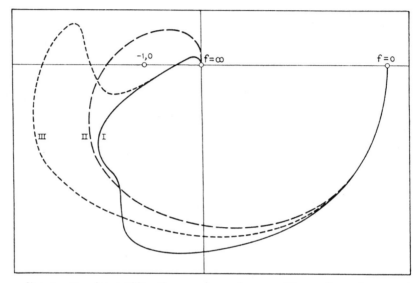

FIG. 8—Nyquist stability diagrams for various amplifiers. Curve I represents "absolute" stability, Curve II instability, and Curve III "conditional" stability. In accordance with the convention used in this paper the diagram is rotated through 180° from its normal position so that the critical point occurs at − 1, 0 rather than + 1, 0.

The situation is conveniently investigated by means of the Nyquist stability diagram [5] illustrated by Fig. 8. The diagram gives the path

[5] *Bell System Technical Journal*, July, 1932. See also Peterson, Kreer, and Ware, *Bell System Technical Journal*, October, 1934. The Nyquist diagrams in the present paper are rotated through 180° from the positions in which they are usually drawn, turning the diagrams in reality into plots of − $\mu\beta$. In a normal amplifier there is one net phase reversal due to the tubes in addition to any phase shifts chargeable directly to the passive networks in the circuit. The rotation of the diagram allows this phase reversal to be ignored, so that the phase shifts actually shown are the same as those which are directly of design interest.

traced by the vector representing the transmission around the feedback loop as the frequency is assigned all possible real values. In accordance with Nyquist's results a path such as II, which encircles the point − 1, 0, indicates an unstable circuit and must be avoided. A stable amplifier is obtained if the path resembles either I or III, neither of which encircles − 1, 0. The stability represented by Curve III, however, is only "Nyquist" or "conditional." The path will enclose the critical point if it is merely reduced in scale, which may correspond physically to a reduction in tube gain. Thus the circuit may sing when the tubes begin to lose their gain because of age, and it may also sing, instead of behaving as it should, when the tube gain increases from zero as power is first applied to the circuit. Because of these possibilities conditional stability is usually regarded as undesirable and the present discussion will consequently be restricted to "absolutely" or "unconditionally" stable amplifiers having Nyquist diagrams of the type resembling Curve I.

The condition that the amplifier be absolutely stable is evidently that the loop phase shift should not exceed 180° until the gain around the loop has been reduced to zero or less. A theoretical characteristic which just met this requirement, however, would be unsatisfactory, since it is inevitable that the limiting phase would be exceeded in fact by minor deviations introduced either in the detailed design of the amplifier or in its construction. It will therefore be assumed that the limiting phase is taken as 180° less some definite margin. This is illustrated by Fig. 9, the phase margin being indicated as $y\pi$ radians. At frequencies remote from the band it is physically impossible, in most circuits, to restrict the phase within these limits. As a supplement, therefore, it will be assumed that larger phase shifts are permissible if the loop gain is x db below zero. This is illustrated by the broken circular arc in Fig. 9. A theoretical loop characteristic meeting both requirements will be developed for an amplifier transmitting between zero and some prescribed limiting frequency with a constant feedback, and cutting off thereafter as rapidly as possible. This basic characteristic can be adapted to amplifiers with varying feedback in the useful range or with useful ranges lying in other parts of the spectrum by comparatively simple modifications which are described at a later point. It is, of course, contemplated that the gain and phase margins x and y will be chosen arbitrarily in advance. If we choose large values we can permit correspondingly large tolerances in the detailed design and construction of the apparatus without risk of instability. It turns out, however, that with a prescribed width of cutoff interval the amount of feedback which can be realized in the useful range is decreased as the assumed margins are increased, so that it is generally desirable to choose as small margins as is safe.

The essential feature in this situation is the requirement that the diminution of the loop gain in the cutoff region should not be accompanied by a phase shift exceeding some prescribed amount. In view of the close connection between phase shift and the slope of the attenuation characteristic evidenced by (2) this evidently demands that the amplifier should cut off, on the whole, at a well defined rate which is not too fast. As a first approximation, in fact, we can choose the cutoff

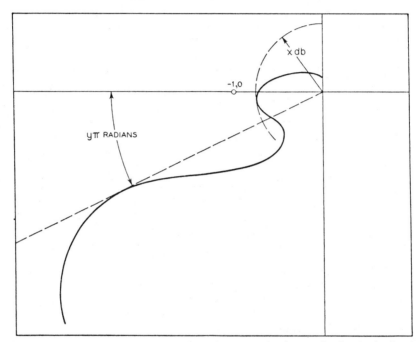

Fig. 9—Diagram to illustrate definitions of phase and gain margins for the feedback loop.

characteristic as an exactly constant slope from the edge of the useful band outward. Such a characteristic has already been illustrated by Fig. 5 and is shown, replotted,[6] by the broken lines in Fig. 10. If we choose the parameter corresponding to k in Fig. 5 as 2 the cutoff rate is 12 db per octave and the phase shift is substantially 180° at high frequencies. This choice thus leads to zero phase margin. By choosing a somewhat smaller k on the other hand, we can provide a definite

[6] To prevent confusion it should be noticed that the general attenuation-phase diagrams are plotted in terms of relative loss while loop cutoff characteristics, here and at later points, are plotted in terms of relative gain.

margin against singing, at the cost of a less rapid cutoff. For example, if we choose $k = 1.5$ the limiting phase shift in the $\mu\beta$ loop becomes 135°, which provides a margin of 45° against instability, while the rate of cutoff is reduced to 9 db per octave. The value $k = 1.67$, which corresponds to a cutoff rate of 10 db per octave and a phase margin of 30°, has been chosen for illustrative purposes in preparing Fig. 10. The loss margin depends upon considerations which will appear at a later point.

Although characteristics of the type shown by Fig. 5 are reasonably satisfactory as amplifier cutoffs they evidently provide a greater phase

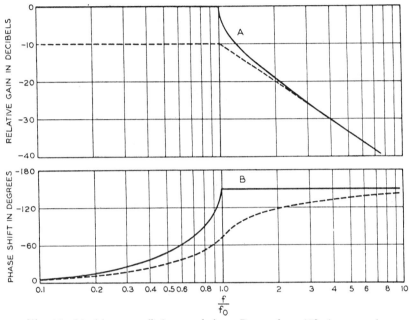

Fig. 10—Ideal loop cutoff characteristics. Drawn for a 30° phase margin.

margin against instability in the region just beyond the useful band than they do at high frequencies. In virtue of the phase area law this must be inefficient if, as is supposed here, the optimum characteristic is one which would provide a constant margin throughout the cutoff interval. The relation between the phase and the slope of the attenuation suggests that a constant phase margin can be obtained by increasing the slope of the cutoff characteristic near the edge of the band, leaving its slope at more remote frequencies unchanged, as shown by the solid lines in Fig. 10. The exact expression for the required curve can be found from (6), where the problem of determining such a characteristic appeared as an example of the use of the general formula (4).

At high frequencies the new phase and attenuation characteristics merge with those obtained from the preceding straight line cutoff, as Fig. 10 indicates. In this region the relation between phase margin and cutoff slope is fixed by the k in the equation (6) in the manner already described for the more elementary cutoff. At low frequencies, however, the increased slope near the edge of the band permits $6 k$ db more feedback.

It is worth while to pause here to consider what may be said, on the basis of these characteristics, concerning the breadth of cutoff interval required for a given feedback, or the "price of the ticket," as it was expressed in the introduction. If we adopt the straight line cutoff and assume the k used in Fig. 10 the interval between the edge of the useful band and the intersection of the characteristic with the zero gain axis is evidently exactly 1 octave for each 10 db of low frequency feedback. The increased efficiency of the solid line characteristic saves one octave of this total if the feedback is reasonably large to begin with. This apparently leads to a net interval one or two octaves narrower than the estimates made in the introduction. The additional interval is required to bridge the gap between a purely mathematical formula such as (6), which implies that the loop characteristics follow a prescribed law up to indefinitely high frequencies, and a physical amplifier, whose ultimate loop characteristics vary in some uncontrollable way. This will be discussed later. It is evident, of course, that the cutoff interval will depend slightly upon the margins assumed. For example, if the phase margin is allowed to vanish the cutoff rate can be increased from 10 to 12 db per octave. This, however, is not sufficient to affect the order of magnitude of the result. Since the diminished margin is accompanied by a corresponding increase in the precision with which the apparatus must be manufactured such an economy is, in fact, a Pyrrhic victory unless it is dictated by some such compelling consideration as that described in the next section.

Maximum Obtainable Feedback

A particularly interesting consequence of the relation between feedback and cutoff interval is the fact that it shows why we cannot obtain unconditionally stable amplifiers with as much feedback as we please. So far as the purely theoretical construction of curves such as those in Fig. 10 is concerned, there is clearly no limit to the feedback which can be postulated. As the feedback is increased, however, the cutoff interval extends to higher and higher frequencies. The process reaches a physical limit when the frequency becomes so high that parasitic effects in the circuit are controlling and do not permit the prescribed cutoff characteristic to be simulated with sufficient precision. For example, we are obviously in physical difficulties if the cutoff characteristic specifies a net gain around the loop at a frequency so high that the tubes themselves working into their own parasitic capacitances do not give a gain.

This limitation is studied most easily if the effects of the parasitic elements are lumped together by representing them in terms of the asymptotic characteristic of the loop as a whole at extremely high frequencies. An example is shown by Fig. 11. The structure is a

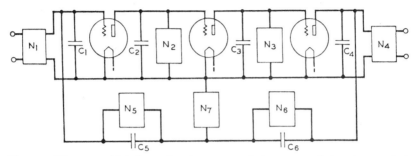

Fig. 11—Elements which determine the asymptotic loop transmission characteristic in a typical amplifier.

shunt feedback amplifier. The β circuit is represented by the T composed of networks N_5, N_6 and N_7. The input and output circuits are represented by N_1 and N_4 and the interstage impedances by N_2 and N_3. The C's are parasitic capacitances with the exception of C_5 and C_6, which may be regarded as design elements added deliberately to N_5 and N_6 to obtain an efficient high frequency transmission path from output to input. At sufficiently high frequencies the loop transmission will depend only upon these various capacitances, without regard to the N's. Thus, if the transconductances of the tubes are represented by G_1, G_2, and G_3 the asymptotic gains of the first two tubes are $G_1/\omega C_1$ and $G_2/\omega C_3$. The rest of the loop includes the third tube and the potentiometer formed by the capacitances C_1, C_4, C_5 and C_6. Its asymptotic transmission can be written as $G_3/\omega C$, where

$$C = C_1 + C_4 + \frac{C_1 C_4}{C_5 C_6} (C_5 + C_6).$$

Each of these terms diminishes at a rate of 6 db per octave. The complete asymptote is $G_1 G_2 G_3/\omega^3 C C_2 C_3$. It appears as a straight line with a slope of 18 db per octave when plotted on logarithmic frequency paper.

A similar analysis can evidently be made for any amplifier. In the particular circuit shown by Fig. 11 the slope of the asymptote, in units of 6 db per octave, is the same as the number of tubes in the circuit. The slope can evidently not be less than the number of tubes but it may be greater in some circuits. For example if C_5 and C_6 were omitted in Fig. 11 and N_5 and N_6 were regarded as degenerating into resistances the asymptote would have a slope of 24 db per octave and would lie below the present asymptote at any reasonably high frequency. In any event the asymptote will depend only upon the parasitic elements of the circuit and perhaps a few of the most significant design elements. It can thus be determined from a skeletonized version of the final structure. If waste of time in false starts is to be avoided such a determination should be made as early as possible, and certainly in advance of any detailed design.

The effect of the asymptote on the overall feedback characteristic is illustrated by Fig. 12. The curve $ABEF$ is a reproduction of the ideal

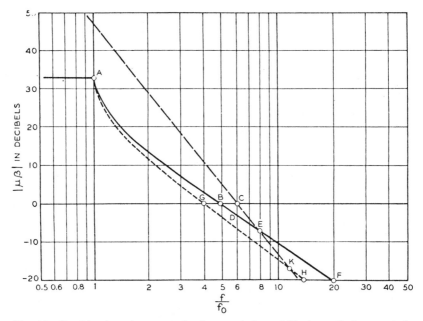

Fig. 12—Combination of asymptotic characteristic and ideal cutoff characteristic.

cutoff characteristic originally given by the solid lines in Fig. 10. It will be recalled that the curve was drawn for the choice $k = 5/3$, which corresponds to a phase margin of 30° and an almost constant slope, for the portion DEF of the characteristic, of about 10 db per octave. The straight line *CEK* represents an asymptote of the type just described, with a slope of 18 db per octave. Since the asymptote may be assumed to represent the practical upper limit of gain in the high-frequency region, the effect of the parasitic elements can be obtained by replacing the theoretical cutoff by the broken line characteristic *ABDEK*. In an actual circuit the corner at *E* would, of course, be rounded off, but this is of negligible quantitative importance. Since *EF* and *EK* diverge by 8 db per octave the effect can be studied by adding curves of the type shown by Fig. 5 to the original cutoff characteristic.

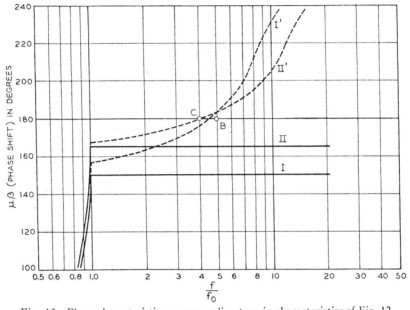

Fig. 13—Phase characteristics corresponding to gain characteristics of Fig. 12.

The phase shift in the ideal case is shown by Curve I of Fig. 13. The addition of the phase corresponding to the extra slope of 8 db per octave at high frequencies produces the total phase characteristic shown by Curve I'. At the point *B* where $|\mu\beta| = 1$, the additional phase shift amounts to 35 degrees. Since this is greater than the original phase margin of 30 degrees the amplifier is unstable when parasitic elements are considered. In the present instance stability can be regained by increasing the coefficient *k* to 1-5/6, which leads to the broken line characteristic *AGKII* in Fig. 12. This reduces the nominal phase margin to 15 degrees, but the frequency interval between *G* and *K* is so much greater than that between *B* and *E* that the added phase is reduced still more and is just less than 15° at the new cross over point *G*. This is illustrated by **II** and **II′** in Fig. 13. On the other hand, if the zero gain intercept of the asymptote *CEK* had occurred at a slightly lower frequency, no change in *k* alone would have been sufficient. It would have been necessary to reduce the amount of feedback in the transmitted range in order to secure stability.

The final characteristic in Fig. 13 reaches the limiting phase shift of 180° only at the crossover point. It is evident that a somewhat more efficient solution for the extreme case is obtained if the limiting 180° is approximated throughout the cutoff interval. This result is attained by the cutoff characteristic shown in Fig. 14. The characteristic con-

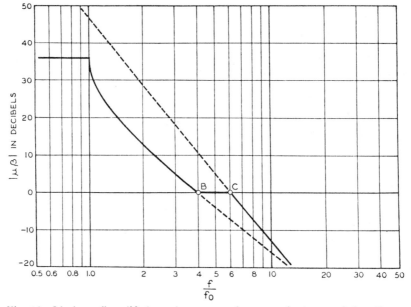

Fig. 14—Ideal cutoff modified to take account of asymptotic characteristic. Drawn for zero gain and phase margins.

sists of the original theoretical characteristic, drawn for *k* = 2, from the edge of the useful band to its intercept with the zero gain axis, the zero gain axis from this frequency to the intercept with the high-frequency asymptote, and the asymptote thereafter. It can be regarded as a combination of the ideal cutoff characteristic and two characteristics of the type shown by Fig. 5. One of the added characteristics starts at *B* and has a positive slope of 12 db per octave, since the ideal cutoff was drawn for the limiting value of *k*. The other starts at *C* and has the negative slope, − 18 db per octave, of the asymptote itself. As (3) shows, the added slopes correspond at lower frequencies to ap-

proximately linear phase characteristics of opposite sign.   If the
frequencies B and C at which the slopes begin are in the same ratio,
12 : 18, as the slopes themselves the contributions of the added slopes
will substantially cancel each other and the net phase shift throughout
the cutoff interval will be almost the same as that of the ideal curve
alone.   The exact phase characteristic is shown by Fig. 15.   It dips

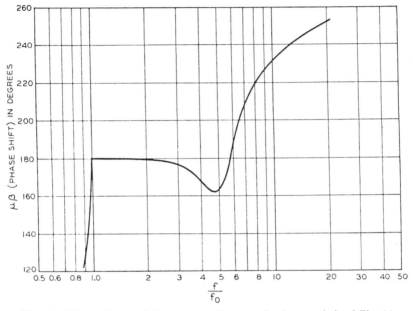

Fig. 15—Phase characteristic corresponding to gain characteristic of Fig. 14.

slightly below 180° at the point at which the characteristic reaches the
zero gain axis, so that the circuit is in fact stable.

The same analysis can evidently be applied to asymptotes of any
other slope.   This makes it easy to compute the maximum feedback
obtainable under any asymptotic conditions.   If f_0 and f_a are respectively the edge of the useful band and the intercept (C in Figs. 12 and
14) of the asymptote with the zero gain axis, and n is the asymptotic
slope, in units of 6 db per octave, the result appears as

$$A_m = 40 \log_{10} \frac{4f_a}{f_0}, \tag{7}$$

where A_m is the maximum feedback in db.[7]

[7] The formulæ for maximum feedback given here and in the later equation (8)
are slightly conservative.   It follows from the phase area law that more feedback
should be obtained if the phase shift were exactly 180° below the crossover and rose

For the sake of generality it is convenient to extend this formula to include also situations in which there exists some further linear phase characteristic in addition to those already taken into account. In exceptional circuits, the final asymptotic characteristic may not be completely established by the time the curve reaches the zero gain axis and the additional phase characteristic may be used to represent the effect of subsequent changes in the asymptotic slope. Such a situation might occur in the circuit of Fig. 11, for example, if C_5 or C_6 were made extremely small. The additional term may also be used to represent departures from a lumped constant analysis in high-frequency amplifiers, as discussed earlier, If we specify the added phase characteristic, from whatever source, by means of the frequency f_d at which it would equal $2n/\pi$ radians, if extrapolated, the general formula corresponding to (7) becomes

$$A_m = 40 \log_{10} \frac{4}{n f_0 f_a} \cdot \frac{f_a f_d}{f_a + f_d} . \tag{8}$$

It is interesting to notice that equations (7) and (8) take no explicit account of the final external gain of the amplifier. Naturally, if the external gain is too high the available μ circuit gain may not be sufficient to provide it and also the feedback which these formulæ promise. This, however, is an elementary question which requires no further discussion. In other circumstances, the external gain may enter the situation indirectly, by affecting the asymptotic characteristics of the β path, but in a well chosen β circuit this is usually a minor consideration. The external gain does, however, affect the parts of the circuit upon which reliance must be placed in controlling the overall loop characteristic. For example, if the external gain is high the μ circuit will ordinarily be sharply tuned and will drop off rapidly in gain beyond the useful band. The β circuit must therefore provide a decreasing loss to bring the overall cutoff rate within the required limit. Since the β circuit must have initially a high loss to correspond to the high final gain of the complete amplifier, this is possible. Conversely, if the gain of the amplifier is low the μ circuit will be relatively flexible and the β circuit relatively inflexible.

rapidly to its ultimate value thereafter. These possibilities can be exploited approximately by various slight changes in the slope of the cutoff characteristic in the neighborhood of the crossover region, or a theoretical solution can be obtained by introducing a prescribed phase shift of this type in the general formula (4). The theoretical solution gives a Nyquist path which, after dropping below the critical point with a phase shift slightly less than 180°, rises again with a phase shift slightly greater than 180° and continues for some time with a large amplitude and increasing phase before it finally approaches the origin. These possibilities are not considered seriously here because they lead to only a few db increase in feedback, at least for moderate n's, and the degree of design control which they envisage is scarcely feasible in a frequency region where, by definition, parasitic effects are almost controlling.

In setting up (7) and (8) it has been assumed that the amplifier will, if necessary, be built with zero margins against singing. Any surplus which the equations indicate over the actual feedback required can, of course, be used to provide a cutoff characteristic having definite phase and gain margins. For example, if we begin with a lower feedback in the useful band the derivative of the attenuation between this region and the crossover can be proportionately reduced, with a corresponding decrease in phase shift. We can also carry the flat portion of the characteristic below the zero gain axis, thus providing a gain margin when the phase characteristic crosses 180°. In reproportioning the characteristic to suit these conditions, use may be made of the approximate formula

$$A_m - A = (A_m + 17.4)y + \frac{n-2}{n}x + \frac{2}{n}xy, \tag{9}$$

where A_m is the maximum obtainable feedback (in db), A is the actual feedback, and x and y are the gain and phase margins in the notation of Fig. 9. Once the available margin has been divided between the x and

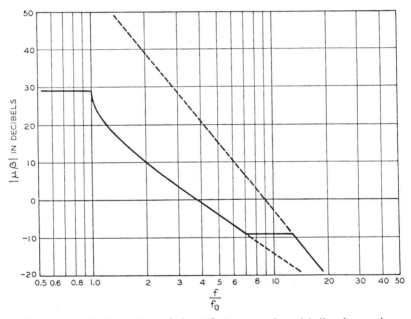

Fig. 16—Modified cutoff permitting 30° phase margin and 9 db gain margin.

y components by means of this formula the cutoff characteristic is, of course, readily drawn in. An example is furnished by Figs. 16 and 17, where it is assumed that $A_m = 43$ db, $A = 29$ db, $x = 9$ db, $n = 3$ and $y = 1/6$. The Nyquist diagram for the structure is shown by Fig. 18. It evidently coincides almost exactly with the diagram postulated originally in Fig. 9.

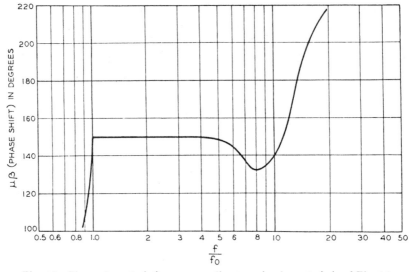

Fig. 17—Phase characteristic corresponding to gain characteristic of Fig. 16.

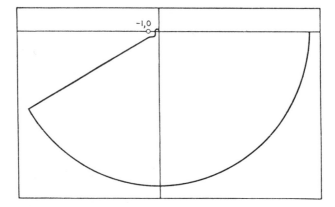

Fig. 18—Nyquist diagram corresponding to gain and phase characteristics of Figs. 16 and 17. As in Fig. 8 the diagram is rotated to place the critical point at − 1, 0 rather than + 1, 0.

With the characteristic of Fig. 16 at hand, we can return once more to the calculation of the total design range corresponding to any given feedback. From the useful band to the intersection of the cutoff characteristic with the zero gain axis the calculation is the same as that made previously in connection with Fig. 10. From the zero gain intercept to the junction with the asymptote, where we can say that design control is finally relaxed, there is, however, an additional interval of nearly two octaves. Although Fig. 16 is fairly typical, the exact breadth of the additional interval will depend somewhat on circumstances. It is increased by an increase in the asymptotic slope and reduced by decreasing the gain margin.

Relative Importance of Tubes and Circuit in Limiting Feedback [8]

The discussion just finished leads to the general conclusion that the feedback which can be obtained in any given amplifier depends ultimately upon the high-frequency asymptote of the feedback loop. It is a matter of some importance, then, to determine what fixes the asymptote and how it can be improved. Evidently, the asymptote is finally restricted by the gains of the tubes alone. We can scarcely improve upon the result secured by connecting the output plate directly to the input grid. Within this limit, however, the actual asymptotic characteristic will depend upon the configuration and type of feedback employed, since a given distribution of parasitic elements may evidently affect one arrangement more than another. The salient circuit problem is therefore that of choosing a general configuration for the feedback circuit which will allow the maximum efficiency of transmission at high frequencies.

The relative importance of tube limitations and circuit limitations is most easily studied if we replace (7) by

$$A_m = 40 \log_{10} \frac{4f_t}{nf_0} - \frac{2A_t}{n}, \tag{10}$$

where f_t is the frequency at which the tubes themselves working into their own parasitic capacitances have zero gain [9] and A_t is the asymptotic loss of the complete feedback loop in db at $f = f_t$. The first term

---

[8] The material of this section was largely inspired by comments due to Messrs. G. H. Stevenson and J. M. West.

[9] I.e., $f_t = \frac{G_m}{2\pi C}$, where G_m and C are respectively the transconductance and capacitance of a representative tube. The ratio $\frac{G_m}{C}$ is the so-called "figure of merit" of the tube. The analysis assumes that the interstage network is a simple shunt impedance, so that the parasitic capacitance does correctly represent its asymptotic behavior. More complicated four-terminal interstage networks, such as transformer coupling circuits and the like, are generally inadmissible in a feedback amplifier because of the high asymptotic losses and consequent high-phase shifts which they introduce.

of (10) shows how the feedback depends upon the intrinsic band width of the available tubes. In low-power tubes especially designed for the purpose f_t may be 50 mc or more, but if f_0 is small the first term will be substantial even if tubes with much lower values of f_t are selected. The second term gives the loss in feedback which can be ascribed to the rest of the circuit. It is evidently not possible to provide input and output circuits and a β-path without making some contribution to the asymptotic loss, so that A_t cannot be zero. In an amplifier designed with particular attention to this question, however, it is frequently possible to assign A_t a comparatively low value, of the order of 20 to 30 db or less. Without such special attention, on the other hand, A_t is likely to be very much larger, with a consequent diminution in available feedback.

In addition to f_t and A_t, (10) includes the quantity n, which represents the final asymptotic slope in multiples of 6 db per octave. Since the tubes make no contribution to the asymptotic loss at $f = f_t$ we can vary n without affecting A_t by changing the number of tubes in the circuit. This makes it possible to compute the optimum number of tubes which should be used in any given situation in order to provide the maximum possible feedback. If A_t is small the first term of (10) will be the dominant one and it is evidently desirable to have a small number of stages. The limit may be taken as $n = 2$ since with only one stage the feedback is restricted by the available forward gain, which is not taken into account in this analysis. On the other hand since the second term varies more rapidly than the first with n, the optimum number of stages will increase as A_t is increased. It is given generally by

$$n = \frac{A_t}{8.68} \tag{11}$$

or in other words the optimum n is equal to the asymptotic loss at the tube crossover in nepers.

This relation is of particular interest for high-power circuits, such as radio transmitters, where circuit limitations are usually severe but the cost of additional tubes, at least in low-power stages, is relatively unimportant. As an extreme example, we may consider the problem of providing envelope feedback around a transmitter. With the relatively sharp tuning ordinarily used in the high-frequency circuits of a transmitter the asymptotic characteristics of the feedback path will be comparatively unfavorable. For illustrative purposes we may assume that $f_a = 40$ kc. and $n = 6$. In accordance with (7) this would provide a maximum available feedback over a 10 kc. voice band of 17 db. It will also be assumed that the additional tubes for the low-power portions of the circuit have an f_t of 10 mc.[10]   The corresponding A_t is 33 nepers [11] so that equation (11) would say that the feedback would be increased by the addition of as many as 27 tubes to the circuit.   Naturally in such an extreme case this result can be looked upon only as a qualitative indication of the direction in which to proceed.   If we add only 4 tubes, however, the available feedback becomes 46 db while if we add 10 tubes it reaches 60 db.   It is to be observed that only a small part of the available gain of the added tubes is used in directly increasing the feedback.   The remainder is consumed in compensating for the unfortunate phase shifts introduced by the rest of the circuit.

[*Editor's Note:* Material has been omitted at this point.]

---

[10] In tubes operating at a high-power level f_t may, of course, be quite low.   It is evident, however, that only the tubes added to the circuit are significant in interpreting (11).   The additional tubes may be inserted directly in the feedback path if they are made substantially linear in the voice range by subsidiary feedback of their own. This will not affect the essential result of the present analysis.

[11] It is, of course, not to be expected that the actual asymptotic slope will be constant from 40 kc. to 10 mc.   Since only the region extending a few octaves above 40 kc. is of interest in the final design, however, the apparent A_t can be obtained by extrapolating the slope in this region.

21

Reprinted with permission from *Bell Syst. Tech. J.* **11**:126–147 (1932)

Regeneration Theory

By H. NYQUIST

Regeneration or feed-back is of considerable importance in many applications of vacuum tubes. The most obvious example is that of vacuum tube
oscillators, where the feed-back is carried beyond the singing point. Another
application is the 21-circuit test of balance, in which the current due to the
unbalance between two impedances is fed back, the gain being increased
until singing occurs. Still other applications are cases where portions of
the output current of amplifiers are fed back to the input either unin-
tentionally or by design. For the purpose of investigating the stability of
such devices they may be looked on as amplifiers whose output is connected
to the input through a transducer. This paper deals with the theory of
stability of such systems.

PRELIMINARY DISCUSSION

WHEN the output of an amplifier is connected to the input through
a transducer the resulting combination may be either stable or
unstable. The circuit will be said to be stable when an impressed small
disturbance, which itself dies out, results in a response which dies out.
It will be said to be unstable when such a disturbance results in a
response which goes on indefinitely, either staying at a relatively small
value or increasing until it is limited by the non-linearity of the
amplifier. When thus limited, the disturbance does not grow further.
The net gain of the round trip circuit is then zero. Otherwise stated,
the more the response increases the more does the non-linearity decrease
the gain until at the point of operation the gain of the amplifier is just
equal to the loss in the feed-back admittance. An oscillator under
these conditions would ordinarily be called stable but it will simplify
the present paper to use the definitions above and call it unstable.
Now, this fact as to equality of gain and loss appears to be an accident
connected with the non-linearity of the circuit and far from throwing
light on the conditions for stability actually diverts attention from the
essential facts. In the present discussion this difficulty will be avoided
by the use of a strictly linear amplifier, which implies an amplifier of
unlimited power carrying capacity. The attention will then be
centered on whether an initial impulse dies out or results in a runaway
condition. If a runaway condition takes place in such an amplifier, it
follows that a non-linear amplifier having the same gain for small
current and decreasing gain with increasing current will be unstable as
well.

Steady-State Theories and Experience

First, a discussion will be made of certain steady-state theories; and reasons why they are unsatisfactory will be pointed out. The most obvious method may be referred to as the series treatment. Let the complex quantity $AJ(i\omega)$ represent the ratio by which the amplifier and feed-back circuit modify the current in one round trip, that is, let the magnitude of AJ represent the ratio numerically and let the angle of AJ represent the phase shift. It will be convenient to refer to AJ as an admittance, although it does not have the dimensions of the quantity usually so called. Let the current

$$I_0 = \cos \omega t = \text{real part of } e^{i\omega t} \tag{a}$$

be impressed on the circuit. The first round trip is then represented by

$$I_1 = \text{real part of } AJe^{i\omega t} \tag{b}$$

and the nth by

$$I_m = \text{real part of } A^n J^n e^{i\omega t}. \tag{c}$$

The total current of the original impressed current and the first n round trips is

$$I_n = \text{real part of } (1 + AJ + A^2 J^2 + \cdots A^n J^n)e^{i\omega t}. \tag{d}$$

If the expression in parentheses converges as n increases indefinitely, the conclusion is that the total current equals the limit of (d) as n increases indefinitely. Now

$$1 + AJ + \cdots A^n J^n = \frac{1 - A^{n+1} J^{n+1}}{1 - AJ}. \tag{e}$$

If $|AJ| < 1$ this converges to $1/(1 - AJ)$ which leads to an answer which accords with experiment. When $|AJ| > 1$ an examination of the numerator in (e) shows that the expression does not converge but can be made as great as desired by taking n sufficiently large. The most obvious conclusion is that when $|AJ| > 1$ for some frequency there is a runaway condition. This disagrees with experiment, for instance, in the case where AJ is a negative quantity numerically greater than one. The next suggestion is to assume that somehow the expression $1/(1 - AJ)$ may be used instead of the limit of (e). This, however, in addition to being arbitrary, disagrees with experimental results in the case where AJ is positive and greater than 1, where the expression $1/(1 - AJ)$ leads to a finite current but where experiment indicates an unstable condition.

The fundamental difficulty with this method can be made apparent by considering the nature of the current expressed by (*a*) above. Does the expression cos ωt indicate a current which has been going on for all time or was the current zero up to a certain time and cos ωt thereafter? In the former case we introduce infinities into our expressions and make the equations invalid; in the latter case there will be transients or building-up processes whose importance may increase as *n* increases but which are tacitly neglected in equations (*b*) − (*e*). Briefly then, the difficulty with this method is that it neglects the building-up processes.

Another method is as follows: Let the voltage (or current) at any point be made up of two components

$$V = V_1 + V_2, \tag{f}$$

where *V* is the total voltage, V_1 is the part due directly to the impressed voltage, that is to say, without the feed-back, and V_2 is the component due to feed-back alone. We have

$$V_2 = AJV. \tag{g}$$

Eliminating V_2 between (*f*) and (*g*)

$$V = V_1/(1 - AJ). \tag{h}$$

This result agrees with experiment when $|AJ| < 1$ but does not generally agree when AJ is positive and greater than unity. The difficulty with this method is that it does not investigate whether or not a steady state exists. It simply assumes tacitly that a steady state exists and if so it gives the correct value. When a steady state does not exist this method yields no information, nor does it give any information as to whether or not a steady state exists, which is the important point.

The experimental facts do not appear to have been formulated precisely but appear to be well known to those working with these circuits. They may be stated loosely as follows: There is an unstable condition whenever there is at least one frequency for which AJ is positive and greater than unity. On the other hand, when AJ is negative it may be very much greater than unity and the condition is nevertheless stable. There are instances of $|AJ|$ being about 100 without the conditions being unstable. This, as will appear, accords closely with the rule deduced below.

Notation and Restrictions

The following notation will be used in connection with integrals:

$$\int_I \phi(z)dz = \lim_{M \to \infty} \int_{-iM}^{+iM} \phi(z)dz, \tag{1}$$

the path of integration being along the imaginary axis (see equation 9), i.e., the straight line joining $-iM$ and $+iM$;

$$\int_{s+} \phi(z)dz = \lim_{M \to \infty} \int_{-iM}^{iM} \phi(z)dz, \tag{2}$$

the path of integration being along a semicircle [1] having the origin for center and passing through the points $-iM$, M, iM;

$$\int_C \phi(z)dz = \lim_{M \to \infty} \int_{-iM}^{-iM} \phi(z)dz, \tag{3}$$

the path of integration being first along the semicircle referred to and then along a straight line from iM to $-iM$. Referring to Fig. 1 it

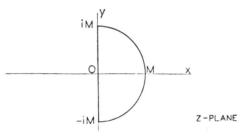

Fig. 1—Paths of integration in the z-plane.

will be seen that

$$\int_{s+} - \int_I = \int_C. \tag{4}$$

The total feed-back circuit is made up of an amplifier in tandem with a network. The amplifier is characterized by the amplifying ratio A which is independent of frequency. The network is characterized by the ratio $J(i\omega)$ which is a function of frequency but does not depend on the gain. The total effect of the amplifier and the network is to multiply the wave by the ratio $AJ(i\omega)$. An alternative way of characterizing the amplifier and network is to say that the amplifier is

[1] For physical interpretation of paths of integration for which $x > 0$ reference is made to a paper by J. R. Carson, "Notes on the Heaviside Operational Calculus," *B. S. T. J.*, Jan. 1930. For purposes of the present discussion the semicircle is preferable to the path there discussed.

characterized by the amplifying factor A which is independent of time, and the network by the real function $G(t)$ which is the response caused by a unit impulse applied at time $t = 0$. The combined effect of the amplifier and network is to convert a unit impulse to the function $AG(t)$. Both these characterizations will be used.

The restrictions which are imposed on the functions in order that the subsequent reasoning may be valid will now be stated. There is no restriction on A other than that it should be real and independent of time and frequency. In stating the restrictions on the network it is convenient to begin with the expression G. They are

$$G(t) \text{ has bounded variation, } -\infty < t < \infty. \tag{AI}$$

$$G(t) = 0, \qquad\qquad -\infty < t < 0. \tag{AII}$$

$$\int_{-\infty}^{\infty} |G(t)| dt \text{ exists.} \tag{AIII}$$

It may be shown [2] that under these conditions $G(t)$ may be expressed by the equation

$$G(t) = \frac{1}{2\pi i} \int_I J(i\omega) e^{i\omega t} d(i\omega), \tag{5}$$

where

$$J(i\omega) = \int_{-\infty}^{\infty} G(t) e^{-i\omega t} dt. \tag{6}$$

These expressions may be taken to define J. The function may, however, be obtained directly from computations or measurements; in the latter case the function is not defined for negative values of ω. It must be defined as follows to be consistent with the definition in (6):

$$J(-i\omega) = \text{complex conjugate of } J(i\omega). \tag{7}$$

While the final results will be expressed in terms of $AJ(i\omega)$ it will be convenient for the purpose of the intervening mathematics to define an auxiliary and closely related function

$$w(z) = \frac{1}{2\pi i} \int_I \frac{AJ(i\omega)}{i\omega - z} d(i\omega), \qquad 0 < x < \infty, \tag{8}$$

where

$$z = x + iy \tag{9}$$

and where x and y are real. Further, we shall define

$$w(iy) = \lim_{x \to 0} w(z). \tag{10}$$

[2] See Appendix II for fuller discussion.

The function will not be defined for $x < 0$ nor for $|z| = \infty$. As defined it is analytic [3] for $0 < x < \infty$ and at least continuous for $x = 0$.

The following restrictions on the network may be deduced:

$$\lim_{y \to \infty} y \, |J(iy)| \text{ exists.} \tag{BI}$$

$$J(iy) \text{ is continuous.} \tag{BII}$$

$$w(iy) = AJ(iy). \tag{BIII}$$

Equation (5) may now be written

$$AG(t) = \frac{1}{2\pi i} \int_I w(z)e^{zt}dz = \frac{1}{2\pi i} \int_{s+} w(z)e^{zt}dz. \tag{11}$$

From a physical standpoint these restrictions are not of consequence. Any network made up of positive resistances, conductances, inductances, and capacitances meets them. Restriction (AII) says that the response must not precede the cause and is obviously fulfilled physically. Restriction (AIII) is fulfilled if the response dies out at least exponentially, which is also assured. Restriction (AI) says that the transmission must fall off with frequency. Physically there are always enough distributed constants present to insure this. This effect will be illustrated in example 8 below. Every physical network falls off in transmission sooner or later and it is ample for our purposes if it begins to fall off, say, at optical frequencies. We may say then that the reasoning applies to all linear networks which occur in nature. It also applies to other linear networks which are not physically producible but which may be specified mathematically. See example 7 below.

A temporary wave $f_0(t)$ is to be introduced into the system and an investigation will be made of whether the resultant disturbance in the system dies out. It has associated with it a function $F(z)$ defined by

$$f_0(t) = \frac{1}{2\pi i} \int_I F(z)e^{zt}dz = \frac{1}{2\pi i} \int_{s+} F(z)e^{zt}dz. \tag{12}$$

$F(z)$ and $f_0(t)$ are to be made subject to the same restrictions as $w(z)$ and $G(t)$ respectively.

DERIVATION OF A SERIES FOR THE TOTAL CURRENT

Let the amplifier be linear and of infinite power-carrying capacity. Let the output be connected to the input in such a way that the

---

[3] W. F. Osgood, "Lehrbuch der Funktionentheorie," 5th ed., Kap. 7, § 1, Hauptsatz. For definition of "analytic" see Kap. 6, § 5.

amplification ratio for one round trip is equal to the complex quantity AJ, where A is a function of the gain only and J is a function of ω only, being defined for all values of frequency from 0 to ∞.

Let the disturbing wave $f_0(t)$ be applied anywhere in the circuit. We have

$$f_0(t) = \frac{1}{2\pi} \int_{-\infty}^{+\infty} F(i\omega) e^{i\omega t} d\omega \qquad (13)$$

or

$$f_0(t) = \frac{1}{2\pi i} \int_{s+} F(z) e^{zt} dz. \qquad (13')$$

The wave traverses the circuit and on completing the first trip it becomes

$$f_1(t) = \frac{1}{2\pi} \int_{-\infty}^{\infty} w(i\omega) F(i\omega) e^{i\omega t} d\omega \qquad (14)$$

$$= \frac{1}{2\pi i} \int_{s+} w(z) F(z) e^{zt} dz. \qquad (14')$$

After traversing the circuit a second time it becomes

$$f_2(t) = \frac{1}{2\pi i} \int_{s+} F w^2 e^{zt} dz, \qquad (15)$$

and after traversing the circuit n times

$$f_n(t) = \frac{1}{2\pi i} \int_{s+} F w^n e^{zt} dz. \qquad (16)$$

Adding the voltage of the original impulse and the first n round trips we have a total of

$$s_n(t) = \sum_{k=0}^{n} f_k(t) = \frac{1}{2\pi i} \int_{s+} F(1 + w + \cdots w^n) e^{zt} dz. \qquad (17)$$

The total voltage at the point in question at the time t is given by the limiting value which (17) approaches as n is increased indefinitely [4]

$$s(t) = \sum_{k=0}^{\infty} f_k(t) = \lim_{n \to \infty} \frac{1}{2\pi i} \int_{s+} S_n(z) e^{zt} dz, \qquad (18)$$

where

$$S_n = F + Fw + Fw^2 + \cdots Fw^n = \frac{F(1 - w^{n+1})}{1 - w}. \qquad (19)$$

[4] Mr. Carson has called my attention to the fact that this series can also be derived from Theorem IX, p. 49, of his Electric Circuit Theory. Whereas the present derivation is analogous to the theory expressed in equations (a)–(e) above, the alternative derivation would be analogous to that in equations (f)–(h).

Convergence of Series

We shall next prove that the limit $s(t)$ exists for all finite values of t. It may be stated as of incidental interest that the limit

$$\int_{s+} S_\infty(z)e^{izt}dz \tag{20}$$

does not necessarily exist although the limit $s(t)$ does. Choose M_0 and N such that

$$|f_0(\lambda)| \leq M_0. \qquad 0 \leq \lambda \leq t. \tag{21}$$

$$|G(t - \lambda)| \leq N. \quad 0 \leq \lambda \leq t. \tag{22}$$

We may write [5]

$$f_1(t) = \int_{-\infty}^{\infty} G(t - \lambda)f_0(\lambda)d\lambda. \tag{23}$$

$$|f_1(t)| \leq \int_0^t M_0 N d\lambda = M_0 N t. \tag{24}$$

$$f_2(t) = \int_{-\infty}^{\infty} G(t - \lambda)f_1(\lambda)d\lambda. \tag{25}$$

$$|f_2(t)| \leq \int_0^t M_0 N^2 t dt = M_0 N^2 t^2/2! \tag{26}$$

Similarly

$$|f_n(t)| \leq M_0 N^n t^n/n! \tag{27}$$

$$|s_n(t)| \leq M_0(1 + Nt + \cdots N^n t^n/n!). \tag{28}$$

It is shown in almost any text [6] dealing with the convergence of series that the series in parentheses converges to e^{Nt} as n increases indefinitely. Consequently, $s_n(t)$ converges absolutely as n increases indefinitely.

Relation Between $s(t)$ and w

Next consider what happens to $s(t)$ as t increases. As t increases indefinitely $s(t)$ may converge to zero, indicating a condition of stability, or it may go beyond any value however large, indicating a runaway condition. The question which presents itself is: *Referring to (18) and (19), what properties of $w(z)$ and further what properties of $AJ(i\omega)$ determine whether $s(t)$ converges to zero or diverges as t increases*

[5] G. A. Campbell, "Fourier Integral," *B. S. T. J.*, Oct. 1928, Pair 202.
[6] E.g., Whittaker and Watson, "Modern Analysis," 2d ed., p. 531.

indefinitely?   From (18) and (19)

$$s(t) = \lim_{n \to \infty} \frac{1}{2\pi i} \int_{s+} F\left(\frac{1}{1-w} - \frac{w^{n+1}}{1-w}\right) e^{zt} dz. \qquad (29)$$

We may write

$$s(t) = \frac{1}{2\pi i} \int_{s+} [F/(1-w)] e^{zt} dz - \lim_{n \to \infty} \frac{1}{2\pi i} \int_{s+} [Fw^{n+1}/(1-w)] e^{zt} dz \quad (30)$$

provided these functions exist.   Let them be called $q_0(t)$ and $\lim_{n \to \infty} q_n(t)$ respectively.   Then

$$q_n(t) = \int_{-\infty}^{\infty} q_0(t - \lambda) \phi(\lambda) d\lambda. \qquad (31)$$

where

$$\phi(\lambda) = \frac{1}{2\pi i} \int_{s+} w^{n+1} e^{z\lambda} dz. \qquad (32)$$

By the methods used under the discussion of convergence above it can then be shown that this expression exists and approaches zero as n increases indefinitely provided $q_0(t)$ exists and is equal to zero for $t < 0$. Equation (29) may therefore be written, subject to these conditions

$$s(t) = \frac{1}{2\pi i} \int_{s+} [F/(1-w)] e^{zt} dz. \qquad (33)$$

In the first place the integral is zero for negative values of t because the integrand approaches zero faster than the path of integration increases.   Moreover,

$$\int_I [F/(1-w)] e^{zt} dz \qquad (34)$$

exists for all values of t and approaches zero for large values of t if $1 - w$ does not equal zero on the imaginary axis.   Moreover, the integral

$$\int_C [F/(1-w)] e^{zt} dz \qquad (35)$$

exists because

1. Since F and w are both analytic within the curve the integrand does not have any essential singularity there,
2. The poles, if any, lie within a finite distance of the origin because· $w \to 0$ as $|z|$ increases, and
3. These two statements insure that the total number of poles is finite.

We shall next evaluate the integral for a very large value of t. It will suffice to take the C integral since the I integral approaches zero. Assume originally that $1 - w$ does not have a root on the imaginary axis and that $F(z)$ has the special value $w'(z)$. The integral may be written

$$\frac{1}{2\pi i} \int_C [w'/(1 - w)]e^{zt}dz. \tag{36}$$

Changing variables it becomes

$$\frac{1}{2\pi i} \int_D [1/(1 - w)]e^{zt}dw, \tag{37}$$

where z is a function of w and D is the curve in the w plane which corresponds to the curve C in the z plane. More specifically the imaginary axis becomes the locus $x = 0$ and the semicircle becomes a small curve which spirals around the origin. See Fig. 2. The function

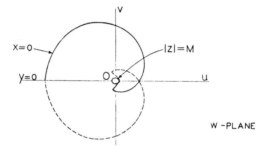

Fig. 2—Representative paths of integration in the w-plane corresponding to paths in Fig. 1.

z and, therefore, the integrand is, in general, multivalued and the curve of integration must be considered as carried out over the appropriate Riemann surface.[7]

Now let the path of integration shrink, taking care that it does not shrink across the pole at $w = 1$ and initially that it does not shrink across such branch points as interfere with its passage, if any. This shrinking does not alter the integral [8] because the integrand is analytic at all other points. At branch points which interfere with the passage of the path the branches stopped may be severed, transposed and connected in such a way that the shrinking may be continued past the branch point. This can be done without altering the value of the integral. Thus the curve can be shrunk until it becomes one or more very small circles surrounding the pole. The value of the total integral

[7] Osgood, loc. cit., Kap. 8.
[8] Osgood, loc. cit., Kap. 7, § 3, Satz 1.

(for very large values of t) is by the method of residues [9]

$$\sum_{j=1}^{n} r_j e^{z_j t}, \tag{38}$$

where z_j $(j = 1, 2 \cdots n)$ is a root of $1 - w = 0$ and r_j is its order. The real part of z_j is positive because the curve in Fig. 1 encloses points with $x > 0$ only. The system is therefore stable or unstable according to whether

$$\sum_{j=1}^{n} r_j$$

is equal to zero or not. But the latter expression is seen from the procedure just gone through to equal the number of times that the locus $x = 0$ encircles the point $w = 1$.

If F does not equal w' the calculation is somewhat longer but not essentially different. The integral then equals

$$\sum_{j=1}^{n} \frac{F(z_j)}{w(z_j)} e^{z_j t} \tag{39}$$

if all the roots of $1 - w = 0$ are distinct. If the roots are not distinct the expression becomes

$$\sum_{j=1}^{n} \sum_{k=1}^{r_j} A_{jk} t^{k-1} e^{z_j t}, \tag{40}$$

where A_{jr_j}, at least, is finite and different from zero for general values of F. It appears then that unless F is specially chosen the result is essentially the same as for $F = w'$. The circuit is stable if the point lies wholly outside the locus $x = 0$. It is unstable if the point is within the curve. It can also be shown that if the point is on the curve conditions are unstable. We may now enunciate the following

Rule: Plot plus and minus the imaginary part of $AJ(i\omega)$ against the real part for all frequencies from 0 to ∞. If the point $1 + i0$ lies completely outside this curve the system is stable; if not it is unstable.

In case of doubt as to whether a point is inside or outside the curve the following criterion may be used: Draw a line from the point $(u = 1, v = 0)$ to the point $z = -i\infty$. Keep one end of the line fixed at $(u = 1, v = 0)$ and let the other end describe the curve from $z = -i\infty$ to $z = i\infty$, these two points being the same in the w plane. If the net angle through which the line turns is zero the point $(u = 1, v = 0)$ is on the outside, otherwise it is on the inside.

If AJ be written $|AJ|(\cos \theta + i \sin \theta)$ and if the angle always

[9] Osgood, loc. cit., Kap. 7, § 11, Satz 1.

changes in the same direction with increasing ω, where ω is real, the rule can be stated as follows: The system is stable or unstable according to whether or not a real frequency exists for which the feed-back ratio is real and equal to or greater than unity.

In case $d\theta/d\omega$ changes sign we may have the case illustrated in Figs. 3 and 4. In these cases there are frequencies for which w is real and

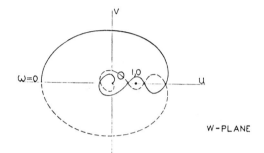

Fig. 3—Illustrating case where amplifying ratio is real and greater than unity for two frequencies, but where nevertheless the path of integration does not include the point 1, 0.

greater than 1. On the other hand, the point (1, 0) is outside of the locus $x = 0$ and, therefore, according to the rule there is a stable condition.

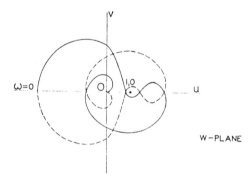

Fig. 4—Illustrating case where amplifying ratio is real and greater than unity for two frequencies, but where nevertheless the path of integration does not include the point 1, 0.

If networks of this type were used we should have the following interesting sequence of events: For low values of A the system is in a stable condition. Then as the gain is increased gradually, the system becomes unstable. Then as the gain is increased gradually still further, the system again becomes stable. As the gain is still further increased the system may again become unstable.

EXAMPLES

The following examples are intended to give a more detailed picture of certain rather simple special cases. They serve to illustrate the previous discussion. In all the cases F is taken equal to AJ so that f_0 is equal to AG. This simplifies the discussion but does not detract from the illustrative value.

1. Let the network be pure resistance except for the distortionless amplifier and a single bridged condenser, and let the amplifier be such that there is no reversal. We have

$$A J(i\omega) = \frac{B}{\alpha + i\omega}, \tag{41}$$

where A and α are real positive constants. In (18) [10]

$$f_n = \frac{1}{2\pi i} \int_I A^{n+1} J^{n+1}(i\omega) e^{i\omega t} d i\omega \tag{42}$$
$$= B e^{-\alpha t}(B^n t^n/n!).$$

$$s(t) = B e^{-\alpha t}(1 + Bt + B^2 t^2/2! + \cdots). \tag{43}$$

The successive terms f_0, f_1, etc., represent the impressed wave and the successive round trips. The whole series is the total current.

It is suggested that the reader should sketch the first few terms graphically for $B = \alpha$, and sketch the admittance diagrams for $B < \alpha$, and $B > \alpha$.

The expression in parentheses equals e^{Bt} and

$$s(t) = B e^{(B-\alpha)t}. \tag{44}$$

This expression will be seen to converge to 0 as t increases or fail to do so according to whether $B < \alpha$ or $B \geq \alpha$. This will be found to check the rule as applied to the admittance diagram.

2. Let the network be as in 1 except that the amplifier is so arranged that there is a reversal. Then

$$A J(i\omega) = \frac{-B}{\alpha + i\omega}. \tag{45}$$

$$f_n = (-1)^{n+1} B e^{-\alpha t}(B^n t^n/n!). \tag{46}$$

The solution is the same as in 1 except that every other term in the series has its sign reversed:

$$s(t) = -B e^{-\alpha t}(1 - Bt + B^2 t^2/2! + \cdots)$$
$$= -B e^{(-\alpha - B)t}. \tag{47}$$

[10] Campbell, loc. cit. Pair 105.

This converges to 0 as t increases regardless of how great B may be taken. If the admittance diagram is drawn this is again found to check the rule.

3. Let the network be as in 1 except that there are two separated condensers bridged across resistance circuits. Then

$$A J(i\omega) = \frac{B^2}{(\alpha + i\omega)^2}. \tag{48}$$

The solution for $s(t)$ is obtained most simply by taking every other term in the series obtained in 1.

$$s(t) = Be^{-\alpha t}(Bt + B^3 t^3/3! + \cdots)$$
$$= Be^{-\alpha t} \sinh Bt. \tag{49}$$

4. Let the network be as in 3 except that there is a reversal. Then

$$A J(i\omega) = \frac{- B^2}{(\alpha + i\omega)^2}. \tag{50}$$

The solution is obtained most directly by reversing the sign of every other term in the series obtained in 3.

$$s(t) = - Be^{-\alpha t}(Bt - B^3 t^3/3! + \cdots)$$
$$= - Be^{-\alpha t} \sin Bt. \tag{51}$$

This is a most instructive example. An approximate diagram has been made in Fig. 5, which shows that as the gain is increased the

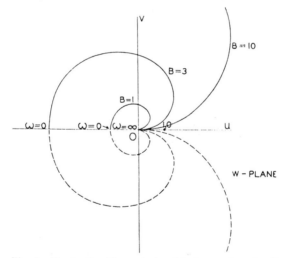

Fig. 5—Illustrating Example 4, with three values for B.

feed-back ratio may be made arbitrarily great and the angle arbitrarily small without the condition being unstable. This agrees with the expression just obtained, which shows that the only effect of increasing the gain is to increase the frequency of the resulting transient.

5. Let the conditions be as in 1 and 3 except for the fact that four separated condensers are used. Then

$$A J(i\omega) = \frac{B^4}{(\alpha + i\omega)^4} . \tag{52}$$

The solution is most readily obtained by selecting every fourth term in the series obtained in 1.

$$s(t) = Be^{-\alpha t}(B^3 t^3/3! + B^7 t^7/7! + \cdots)$$
$$= \tfrac{1}{2} Be^{-\alpha t} (\sinh Bt - \sin Bt). \tag{53}$$

This indicates a condition of instability when $B \geq \alpha$, agreeing with the result deducible from the admittance diagram.

6. Let the conditions be as in 5 except that there is a reversal. Then

$$Y = \frac{- B^4}{(\alpha + i\omega)^4} . \tag{54}$$

The solution is most readily obtained by changing the sign of every other term in the series obtained in 5.

$$s(t) = Be^{-\alpha t}(- B^3 t^3/3! + B^7 t^7/7! - \cdots). \tag{55}$$

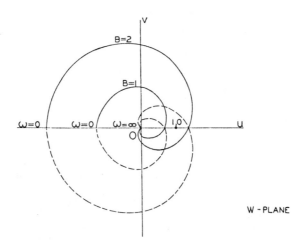

Fig. 6—Illustrating Example 6, with two values for B.

For large values of t this approaches

$$s(t) = - \tfrac{1}{2}Be^{(B/\sqrt{2}-\alpha)t} \sin (Bt/\sqrt{2} - \pi/4). \tag{56}$$

This example is interesting because it shows a case of instability although there is a reversal. Fig. 6 shows the admittance diagram for $B\sqrt{2} - \alpha < 0$ and for $B\sqrt{2} - \alpha > 0$.

7. Let

$$AG(t) = f_0(t) = A(1 - t), \qquad 0 \leq t \leq 1. \tag{57}$$

$$AG(t) = f_0(t) = 0, \qquad - \infty < t < 0, \qquad 1 < t < \infty. \tag{57'}$$

We have

$$AJ(i\omega) = A \int_0^1 (1 - t)e^{-i\omega t}dt$$

$$= A \left(\frac{1 - e^{-i\omega}}{\omega^2} + \frac{1}{i\omega} \right). \tag{58}$$

Fig. 7 is a plot of this case for $A = 1$.

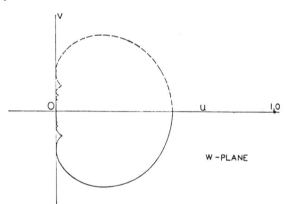

Fig. 7—Illustrating Example 7.

8. Let

$$AJ(i\omega) = \frac{A(1 + i\omega)}{(1 + i2\omega)}. \tag{59}$$

This is plotted on Fig. 8 for $A = 3$. It will be seen that the point 1 lies outside of the locus and for that reason we should expect that the system would be stable. We should expect from inspecting the diagram that the system would be stable for $A < 1$ and $A > 2$ and that it would be unstable for $1 \leq A \leq 2$. We have overlooked one fact, however; the expression for $AJ(i\omega)$ does not approach zero as ω

increases indefinitely.   Therefore, it does not come within restriction
(BI) and consequently the reasoning leading up to the rule does not
apply.

The admittance in question can be made up by bridging a capacity
in series with a resistance across a resistance line.   This admittance

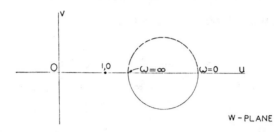

Fig. 8—Illustrating Example 8, without distributed constants.

obviously does not approach zero as the frequency increases.   In any
actual network there would, however, be a small amount of distributed
capacity which, as the frequency is increased indefinitely, would cause
the transmission through the network to approach zero.   This is
shown graphically in Fig. 9.   The effect of the distributed capacity is

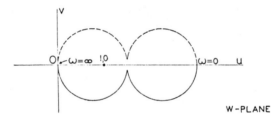

Fig. 9—Illustrating Example 8, with distributed constants.

essentially to cut a corridor from the circle in Fig. 8 to the origin, which
insures that the point lies inside the locus.

Appendix I

Alternative Procedure

In some cases $AJ(i\omega)$ may be given as an analytic expression in
$(i\omega)$.   In that case the analytic expression may be used to define w for
all values of z for which it exists.   If the value for $AJ(i\omega)$ satisfies all
the restrictions the value thus defined equals the w defined above for
$0 \leq x < \infty$ only.   For $-\infty < x < 0$ it equals the analytic continu-
ation of the function w defined above.   If there are no essential

singularities anywhere including at ∞, the integral in (33) may be evaluated by the theory of residues by completing the path of integration so that all the poles of the integrand are included. We then have

$$s(t) = \sum_{j=1}^{j=n} \sum_{k=1}^{r_j} A_{jk} t^{k-1} e^{z_j t}. \tag{60}$$

If the network is made up of a finite number of lumped constants there is no essential singularity and the preceding expression converges because it has only a finite number of terms. In other cases there is an infinite number of terms, but the expression may still be expected to converge, at least, in the usual case. Then the system is stable if all the roots of $1 - w = 0$ have $x < 0$. If some of the roots have $x \geq 0$ the system is unstable.

The calculation then divides into three parts:

1. The recognition that the impedance function is $1 - w$.[11]
2. The determination of whether the impedance function has zeros for which $x \geq 0$.[12]

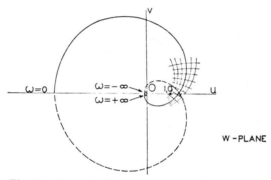

Fig. 10—Network of loci $x =$ const., and $y =$ const.

3. A deduction of a rule for determining whether there are roots for which $x \geq 0$. The actual solution of the equation is usually too laborious.

To proceed with the third step, plot the locus $x = 0$ in the w plane, i.e., plot the imaginary part of w against the real part for all the values of y, $-\infty < y < \infty$. See Fig. 10. Other loci representing

$$x = \text{const.} \tag{61}$$

and

$$y = \text{const.} \tag{62}$$

[11] Cf. H. W. Nichols, *Phys. Rev.*, vol. 10, pp. 171–193, 1917.
[12] Cf. Thompson and Tait, "Natural Philosophy," vol. I, § 344.

may be considered and are indicated by the network shown in the figure in fine lines. On one side of the curve x is positive and on the other it is negative. Consider the equation

$$w(z) - 1 = 0$$

and what happens to it as A increases from a very small to a very large value. At first the locus $x = 0$ lies wholly to the left of the point. For this case the roots must have $x < 0$. As A increases there may come a time when the curve or successive convolutions of it will sweep over the point $w = 1$. For every such crossing at least one of the roots changes the sign of its x. We conclude that if the point $w = 1$ lies inside the curve the system is unstable. It is now possible to enunciate the rule as given in the main part of the paper but there deduced with what appears to be a more general method.

APPENDIX II

Discussion of Restrictions

The purpose of this appendix is to discuss more fully the restrictions which are placed on the functions defining the network. A full discussion in the main text would have interrupted the main argument too much.

Define an additional function

$$n(z) = \frac{1}{2\pi i} \int_I \frac{AJ(i\lambda)}{i\lambda - z} d(i\lambda), \qquad -\infty < x < 0. \tag{63}$$

$$n(iy) = \lim_{x \to 0} n(z).$$

This definition is similar to that for $w(z)$ given previously. It is shown in the theorem [13] referred to that these functions are analytic for $x \neq 0$ if $AJ(i\omega)$ is continuous. We have not proved, as yet, that the restrictions placed on $G(t)$ necessarily imply that $J(i\omega)$ is continuous. For the time being we shall assume that $J(i\omega)$ may have finite discontinuities. The theorem need not be restricted to the case where $J(i\omega)$ is continuous. From an examination of the second proof it will be seen to be sufficient that $\int_I J(i\omega)d(i\omega)$ exist. Moreover, that proof can be slightly modified to include all cases where conditions (AI)–(AIII) are satisfied.

[13] Osgood, loc. cit.

For, from the equation at top of page 298 [13]

$$\left| \frac{w(z_0 - \Delta z) - w(z_0)}{\Delta z} - \frac{1}{2\pi i} \int_I \frac{AJ(i\lambda)}{(i\lambda - z_0)^2} d(i\lambda) \right|$$

$$\leq |\Delta z| \left| \frac{1}{2\pi i} \int_I \frac{AJ(i\lambda)d(i\lambda)}{(i\lambda - z_0 - \Delta z)(i\lambda - z_0)^2} \right|, \qquad x_0 > 0. \quad (64)$$

It is required to show that the integral exists. Now

$$\int_I \frac{AJ(i\lambda)d(i\lambda)}{(i\lambda - z_0 - \Delta z)(i\lambda - z_0)^2}$$

$$= \int_I \frac{AJ(i\lambda)d(i\lambda)}{(i\lambda - z_0)^3} \left(1 + \frac{\Delta z}{i\lambda - z_0} + \frac{\Delta z^2}{i\lambda - z_0} + \text{etc.} \right) \quad (65)$$

if Δz is taken small enough so the series converges. It will be sufficient to confine attention to the first term. Divide the path of integration into three parts,

$$-\infty < \lambda < -|z_0| - 1, \qquad -|z_0| - 1 < \lambda < |z_0| + 1, \qquad |z_0| + 1 < \lambda < \infty.$$

In the middle part the integral exists because both the integrand and the range of integration are finite. In the other ranges the integral exists if the integrand falls off sufficiently rapidly with increasing λ. It is sufficient for this purpose that condition (BI) be satisfied. The same proof applies to $n(z)$.

Next, consider $\lim_{x \to 0} w(z) = w(iy)$. If iy is a point where $J(iy)$ is continuous, a straightforward calculation yields

$$w(iy) = AJ(iy)/2 + P(iy). \quad (66a)$$

Likewise,

$$n(iy) = -AJ(iy)/2 + P(iy) \quad (66b)$$

where $P(iy)$ is the principal value [14] of the integral

$$\frac{1}{2\pi i} \int_I \frac{AJ(i\lambda)}{i\lambda - iy} d(i\lambda).$$

Subtracting

$$w(iy) - n(iy) = AJ(iy) \quad (67)$$

If (iy) is a point of discontinuity of $J(iy)$

$$|w| \text{ and } |n| \text{ increase indefinitely as } x \to 0. \quad (68)$$

Next, evaluate the integral

$$\frac{1}{2\pi i} \int_{I+I} w(z)e^{zt}dz,$$

[14] E. W. Hobson, "Functions of a Real Variable," vol. I, 3d edition, § 352.

where the path of integration is from $x - i\infty$ to $x + i\infty$ along the line $x = $ const. On account of the analytic nature of the integrand this integral is independent of x (for $x > 0$). It may be written then

$$\lim_{x \to 0} \frac{1}{2\pi i} \int_{x+I} w(z)e^{zt}dz = \lim_{x \to 0} \frac{1}{2\pi i} \int_{x+I} \frac{1}{2\pi i} \int_I \frac{A J(i\lambda)}{i\lambda - z} e^{zt}d(i\lambda)dz$$

$$= \lim_{x \to 0} \frac{1}{2\pi i} \int_{x+I} \frac{1}{2\pi i} \lim_{M \to \infty} \left[\int_{-iM}^{iy-i\delta} + \int_{iy-i\delta}^{iy+i\delta} + \int_{iy+i\delta}^{iM} \right] \frac{A J(i\lambda)}{i\lambda - z} e^{zt}d(i\lambda)dz$$

$$= \lim_{x \to 0} \left[\frac{1}{2\pi i} \int_{x+I} \frac{1}{2\pi i} \int_{iy-i\delta}^{iy+i\delta} \frac{A J(i\lambda)}{i\lambda - z} e^{zt}d(i\lambda)dz + Q(t, \delta) \right], \quad x > 0, \quad (69)$$

where δ is real and positive. The function Q defined by this equation exists for all values of t and for all values of δ. Similarly,

$$\lim_{x \to 0} \frac{1}{2\pi i} \int_{x+I} n(z)e^{zt}dz$$

$$= \left[\lim_{x \to 0} \frac{1}{2\pi i} \int_{x+I} \frac{1}{2\pi i} \int_{iy+i\delta}^{iy+i\delta} \frac{A J(i\lambda)}{i\lambda - z} e^{zt}d(i\lambda)dz + Q(t, \delta) \right], \quad x < 0, \quad (70)$$

Subtracting and dropping the limit designations

$$\frac{1}{2\pi i} \int_{x+I} w(z)e^{zt}dz - \frac{1}{2\pi i} \int_{x+I} n(z)e^{zt}dz = \frac{1}{2\pi i} \int_I A J(i\lambda)e^{i\lambda t}d(i\lambda). \quad (71)$$

The first integral is zero for $t < 0$ as can be seen by taking x sufficiently large. Likewise, the second is equal to zero for $t > 0$. Therefore,

$$\frac{1}{2\pi i} \int_{x+I} w(z)e^{zt}dz = \frac{1}{2\pi i} \int_I A J(i\omega)e^{i\omega t}d(i\omega) = AG(t), \quad 0 < t < \infty \quad (72)$$

$$-\frac{1}{2\pi i} \int_{x+I} n(z)e^{zt}dz$$

$$= \frac{1}{2\pi i} \int_I A J(i\omega)e^{i\omega t}d(i\omega) = AG(t) - \infty < t < 0. \quad (73)$$

We may now conclude that

$$\int_I n(iy)e^{iyt}d(iy) = 0, \qquad -\infty < t < \infty \qquad (74)$$

provided

$$G(t) = 0, \qquad -\infty < t < 0. \qquad (AII)$$

But (74) is equivalent to

$$n(z) = 0, \qquad (74')$$

which taken with (67) gives

$$w(iy) = A J(iy). \tag{BIII}$$

(BIII) is, therefore, a necessary consequence of (AII). (74') taken with (68) shows that

$$J(iy) \text{ is continuous.} \tag{BII}$$

It may be shown [15] that (BI) is a consequence of (AI). Consequently all the B conditions are deducible from the A conditions.

Conversely, it may be inquired whether the A conditions are deducible from the B conditions. This is of interest if $A J(i\omega)$ is given and is known to satisfy the B conditions, whereas nothing is known about G.

Condition AII is a consequence of BIII as may be seen from (67) and (74). On the other hand AI and AIII cannot be inferred from the B conditions. It can be shown by examining (5), however, that if the slightly more severe condition

$$\lim_{y \to \infty} y^\gamma J(iy) \text{ exists,} \qquad (\gamma > 1), \tag{BIa}$$

is satisfied then

$$G(t) \text{ exists,} \qquad -\infty < t < \infty, \tag{AIa}$$

which, together with AII, insures the validity of the reasoning.

It remains to show that the measured value of $J(i\omega)$ is equal to that defined by (6). The measurement consists essentially in applying a sinusoidal wave and determining the response after a long period. Let the impressed wave be

$$E = \text{real part of } e^{i\omega t}, \qquad t \geq 0. \tag{75}$$
$$E = 0, \qquad t < 0. \tag{75'}$$

The response is

$$\text{real part of } \int_0^t A G(\lambda) e^{i\omega(t-\lambda)} d\lambda$$

$$= \text{real part of } A e^{i\omega t} \int_0^t G(\lambda) e^{-i\omega\lambda} d\lambda. \tag{76}$$

For large values of t this approaches

$$\text{real part of } A e^{i\omega t} J(i\omega). \tag{77}$$

Consequently, the measurements yield the value $A J(i\omega)$.

[15] See Hobson, loc. cit., vol. II. 2d edition, § 335. It will be apparent that K depends on the total variation but is independent of the limits of integration.

Part V

CYBERNETICS

Editor's Comments
on Papers 22 Through 26

The word *cybernetics* first came into prominence in modern times with the publication of Norbert Wiener's well-known book with that title in 1948 (The Technology Press, and John Wiley & Sons, New York, second edition, corrected and revised, 1961). Wiener (1899–1964) adopted the name to denominate the science of communication and control in the animal and the machine. In his book he brought together for the first time reference to a host of phenomena from different branches of science in which the concept of control, particularly involving feedback, was central. In the three decades since Wiener's first publication the subject has proliferated enormously and enters into every aspect of human experience and activity. This has led to the production of a great body of literature devoted to cybernetics and its many applications.

It is of interest to note, however, that Wiener was not the first to employ the term cybernetics. This was done by the famous French physicist Andre Marie Ampere (1775–1836), who in his *Essai sur le philosophie des sciences* (1834) introduced the word *cybernetique* to denote the science of government in the social and political sense. He took the name from the Greek κυβερντεσ meaning the steersman of a ship. Wiener was evidently not aware of Ampere's priority in the use of the term when he prepared his book in 1946.

A volume of papers on the control of energy would be incomplete without a reference to cybernetics as the science that embraces control of all kinds. It is indeed difficult to see how there can be any control at all in the behavior of both animate and inanimate systems without the involvement of energy. All communication needed for control involves the transfer and/or transformation of energy. In many modern treatments of cybernetics this aspect is ignored for practical reasons, but from the fundamental standpoint it is always present.

In the five articles connected with cybernetics, which we reproduce here, we shall not be concerned with the multifarious applications of the subject. The literature is far too extensive to render such a course feasible. We shall limit ourselves to what might be called the historical background and the philosophical point of view.

We begin by reproducing in English translation as Paper 22 the very brief reference to cybernetics in the second volume of Ampere's celebrated essay (actually published in 1843 by Ampere's son, after his father's death). It should be emphasized that in his essay Ampere's aim was to organize all science in an elaborate series of hierarchies somewhat similar to the scheme developed by August Comte in his positivistic philosophy at about the same time (see Paper 23 for an illustration of Ampere's classification). There is no evidence that Ampere and Comte ever conferred about their separate schemes. It is worth noting that schemes for the classification of the various branches of science were rather common in France in the late 1700s and early 1800s. From the brief reference we are quoting it is apparent that Ampere intended cybernetics to refer to social and political control. The question of the extent to which Ampere may be said to have anticipated Wiener and 20th century cybernetics is an interesting one from the historical standpoint. This problem has recently been investigated by the late Belgian philosopher and scientific organization administrator Robert Caussin. We reproduce in English translation as Paper 23 the larger part of his article "Ampere et la Cybernetique" (1920). There remains some doubt indeed whether he has fully established his thesis claiming for Ampere an important role in the development of cybernetics. The paper is included, however, for historical and philosophical reasons and for the light it sheds on the ideas of both Wiener and Ampere. It exhibits Ampere, the famous physicist who established the fundamental principles of electrodynamics, in a light in which he is not commonly regarded by scientists. At the same time Caussin does full justice to Wiener's achievement. It seems that both deserve a distinguished position in the story of the development of the idea of control as a significant aspect of scientific and philosophical thought.

The reader who wishes to grasp the significance of Wienerian cybernetics in its manifold forms has an enormous literature to choose from. Even a bibliography of the material produced in this field in the past twenty-five years would fill a large book. The second edition of Wiener's volume referred to above is certainly worthy of careful study. An attractive popular little book on cybernetics is *What is Cybernetics* by the French cybernetician G. T. Guilbaud (Presses Universitaires de France, Paris, 1954 in French, and Heinemann, London, 1959 in English).

To pay due respect to Wiener we reproduce as Paper 24 his article in the initial issue of the new publication *Cybernetica*: "My connection with cybernetics. Its origins and its future" (1958). This provides an account of the historical development of Wiener's ideas, with emphasis on the mathematical, physical, and physiological background. It is interesting to compare the contents of this paper with what is said about Wiener's later views in Paper 23.

In view of the close association of the concept of communication or the transmission of information with that of control we have thought it useful to reproduce the remarkably well-written early pioneer paper by R. V. L. Hartley (1888–1970) "Transmission of Information" (1928). This had a very great influence on the subsequent development of cybernetics. Its inclusion in a volume of papers on the control of energy might at first thought seem questionable. However, as has already been emphasized earlier in this volume, the transmission of information does involve transfer of energy; on the other side, the control of energy, especially in feed back devices, is definitely dependent on the transmission of information. Hence Hartley's celebrated paper cannot be thought of as out of place in this volume. Hartley was from 1928–1950 a member of the Technical Staff of the Bell Telephone Laboratories.

To conclude this collection of papers we reproduce the philosophically oriented article by R. B. Lindsay: "The Larger Cybernetics" (1971). This emphasizes that the ultimate control of energy or everything else in human experience, for that matter, is exercised by Nature itself and described by the principles of thermodynamics. We here secure the advantage by completing our considerations with further emphasis on the concept of energy, the main theme of this volume.

22

INTRODUCTION OF THE WORD CYBERNETICS

André-Marie Ampere

*This excerpt was translated expressly for this Benchmark
volume by R. Bruce Lindsay, Brown University, from
pp. 140–141 of* Essai sur le philosophie des sciences, vol. 2,
Paris, 1834.

3. Cybernetics

The relations of people with other people, studied in the two preceding sciences, are only the least part of the objects to which a good government should devote its attention. The maintenance of public order, the execution of the laws, the just apportionment of taxes, the choice of men for employment, and everything that can contribute to the amelioration of the state of society demand its never ending concern. It has continually to choose, among different measures, those most appropriate for attaining its goal. It is only through the profound study and comparison of the various elements provided to it through the knowledge of everything relating to the nation being ruled, its character, its means of existence and prosperity, its organization and laws, that it is able to devise the general rules of conduct to guide it in each particular case. Hence we ought to place at the head of all the sciences involved in these different areas the one which is here in question and which I have named *cybernetics* from the Greek word *kubernetike*, which in its first restricted usage meant the art of steering a ship but later received the extended significance, even among the Greeks, of the art of governing in general.

23

AMPERE AND CYBERNETICS

Robert Caussin

*This article was translated expressly for this Benchmark
volume by R. Bruce Lindsay, Brown University, from
pp. 73–97 and 101–104 of "Ampere et la Cybernetique,"
in* Cybernetica **13**:73–104 (1970), *with the permission
of the publisher, the International Association for
Cybernetics.*

The present article has for its object a question of secondary interest, a historical point that has attracted the attention of some cyberneticians since it involves some controversy. The question is: Can Andre Marie Ampere justly be considered the founder of cybernetics?

THE OBJECT OF THE DISCUSSION

During 1969, under the patronage of the International Association of Cybernetics and at the instigation of some of its most active members, a limited number of qualified European personalities were invited, in preparation for a Round Table meeting, to present their ideas with respect to a systematic redefinition of cybernetics. The promoters of this project felt that the too extensive use of the term had led to confusion as to the exact meaning to be attributed to it.

The replies received, constituting a preliminary dossier of varied opinions about the meaning of the term cybernetics, showed that the confusion was real and that the envisaged meeting was timely.[1] A careful examination of one detailed point of the dossier relevant to the question raised can perhaps help to dissipate a part of the difficulty.

1. The first document of the dossier is an abstract of definitions established as a result of an inquiry of the French Society of Cybernetics in 1964. It reviews among others a definition produced in 1962 by the French Academy of Sciences, in which there occurs the following:

> To attain its goals cybernetics takes into account existing sciences and
> in particular those having to do with information; as its founder, Andre
> Marie Ampere stipulated, it takes its place along side the other sciences
> but without overlapping any of them.

The word "founder" in this extract invites reflection. In the view of most of the adherents of cybernetics the concepts involved in the current acceptation of the term are of relatively recent date; not going back before 1940. Are we then justified in considering A. M. Ampere (1775–1836) as the founder of cybernetics?

2. Others have also faced this question and have vigorously questioned the above assertions. An informed cybernetician, Robert J. Van Egten, promotor of the Round Table, writes:

> After Norbert Wiener had chosen to give a name of Greek origin to the
> new science of which he had laid the foundation, some have tried to
> steal his priority and with the use of some equivocal semantics have en-
> deavored to trace the concept back to Greek philosophers or some-
> what more modestly to the end of the 18th century. Has not the French
> Academy of Science designated Ampere as the founder of cybernetics?
> In going from absurdity to absurdity, have we not arrived at actual con-
> fusion?

This statement provides a clearcut discrepancy. It may be reasonable to
admit that the members of the Academy have allowed themselves to be
carried away by an admiration more passionate than rational for their illustrious former colleague Ampere. Should they not have provided an acceptable verification for their opinion?

3. Finally we quote a statement by Professor Jean Hoffmann of the Free
University of Brussels.

> According to Ampere, cybernetics is the art of control in the sense of
> regulating an engine, in particular a ship. In reality it is a question of the
> strategy involved in the optimum maneuverability of the device being
> controlled, taking into account the chance perturbations brought to
> bear on the device by the medium in which it operates and not being
> primarily concerned with the technical behavior of the mechanism.
> This conception places principal emphasis on the use of statistical pro-
> cedures in the treatment of the problem in question.[2]

Here we are led to the art of navigating in the proper sense, to a physical
problem in the larger sense. We then return to our third question: was this
precisely the thought of Ampere?

Without in the least wishing to oppose anyone of these different conceptions we are forced to admit that they present very divergent orienta-
tions. But they do possess the merit of inviting us to investigate further. Let
us therefore look at the sources and ask all these predecessors and especially Ampere himself what they really meant by cybernetics.

Cybernetics in Antiquity

It is often said that a word is the wrapping of an idea. But both the word
and the idea evolve in time. If from age to age a given concept can clothe
itself in different words, a word on its side can gradually acquire very different meanings. Cannot a vase hold in succession water and wine?

With documents in hand, then, let us see what has happened through
the years to the word cybernetics.

It comes to us from the Greek "kubernetes," meaning a pilot or helmsman. This is the original meaning. We note that the ancient Greeks called
"Cybernesies" the festivals instituted by Theseus in memory of the pilot
who had guided his vessels in his Cretan expedition.

But very soon the original meaning was elaborated. We are told in
"Grande Larousse Encyclopedique," Paris, 1966 that "the Greeks generalized the art of control and the word 'cybernetics' is even found in several
dialogues of Plato. This celebrated philosopher indeed amplified the
meaning of the verb "to govern" and provided many different examples,

all the way from the steering of a ship to the driving of a chariot and finally to the governing of people. On his side Xenophon sought to study in systematic fashion the art of government in the political sense."

Here are the more specific comments of Mario C. Losano of the University of Turin concerning the origin of the word cybernetics in antiquity.

> In Plato the term *kubernetike* or even *tokubernetikon* in the first restricted sense refers to the art of steering a ship (Gorgias 511, Politicus 299). This is the original and proper meaning of the word, which seems to have been derived from the Sanskrit *kubara* meaning tiller, whence comes also the Latin gubernum, destined to survive in the Romance languages as a political term (in French, however, in addition to *gouvernment*, there is also *gouvernail*, a tiller). In this sense the term was already in use at the time of Homer.
>
> But in Clitopho 408[b], Plato for the first time used the word to denote the art of governing in the political sense. The evidence shows that this innovation of Plato was not entirely original, since several earlier writers had already used words having the same root in a figurative sense. It suffices to cite, for example, Pindar, who used in the figurative sense *kubernao* (5.122), *kubernesis* [10.72], *kuberneter* [4.274] (in *le Pitiche*), and Euripides, who used *Kubernetes* [880] in *Suplices*.[3]

Here then we have reached a definite conclusion: among the Greeks, in the beginning the term cybernetics was used in its proper sense as the art of steering a vessel. Later it became used in a figurative sense: as the art of government in a political sense.

THE CYBERNETICS OF AMPERE

Let us now take up Ampere, a person much nearer to us. In 1834 Ampere used the term cybernetics in his "Essay on the Philosophy of the Sciences,"[4] which is really an essay on the classification and coordination of all the sciences known at his time.

In his nomenclature[5] Ampere at first divided the sciences into two great categories or kingdoms: the cosmological or natural sciences and the sciences of the mind (noological sciences). The cosmological sciences are further divided into two subkingdoms, i.e., the sciences dealing with inanimate objects and those that concern living things. The first subkingdom is composed of two branches: the mathematical sciences and the physical sciences. In pursuing this binary classification Ampere developed a table for the aggregate of the sciences and arts as follows:

2	kingdoms
4	subkingdoms
8	branches
16	subbranches
32	sciences of the first order
64	sciences of the second order
128	sciences of the third order

It must be confessed that Ampere arrived at his 128 sciences only by a process of splitting up sciences that had previously formed a single whole and transforming individual branches of a given science into sciences themselves. He also devised for these some rather singular names such as coenolbogy, cybernetics, terpnognosy, technesthetics, and so on.[6]

Of the sciences of the third order classified by Ampere in the second volume of his monumental work *(Sciences of the Mind)*, cybernetics is the next to the last. As is shown in the overall scheme in his appendix its place among the neighboring sciences is indicated as in the following table.[7]

Sciences of the first order	Sciences of the second order	Sciences of the third order
8. Politics	p. Syncimenics	81. Ethnodicy
	q. Politics proper	82. Diplomacy
		83. Cybernetics
		84. Theory of power

We see that Ampere, employing in his nomenclature the thought of the ancient Greeks in a way that leaves no room for doubt, ranks cybernetics under politics, a science of the first order, which he defines as follows: "Study of the causes which affect the prosperity of nations and prediction of the useful and determinental effects which can result from them, to provide guidance in the choice of measures relative to all aspects of the administration of the state."[8]

And here is the way in which he indicates his understanding of cybernetics:

> The relations of people with other people, studied in the two preceding sciences, are only the least part of the objects to which a good government should devote its attention. The maintenance of public order, the execution of the laws, the just apportionment of taxes, the choice of men for employment, and everything that can contribute to the amelioration of the state of society demand its never ending concern. It has continually to choose among different measures those most appropriate for attaining its goal. It is only through the profound study and comparison of the various elements provided to it through the knowledge of everything relating to the nation being ruled, its character, its customs, its opinions, its history, its religion, its means of existence and prosperity, its organization and laws, that it is able to devise the general rules of conduct to guide it in each particular case. Hence we ought to place at the head of all the sciences involved in these different areas the one which is here in question and which I have named *cybernetics* from the Greek word kubernetike, which in its first restricted usage meant the art of steering a ship but later received the extended significance, even among the Greeks, of the art of governing in general.[9]

A little further on Ampere, retracing in his binary nomenclature the sci-

327

ences of the third to the sciences of the second order and then to the first order, confirms his thought with a particularly explicit statement: "Cybernetics plays the same role with respect to the government of nations that strategy plays in the running of an army."[10]

Finally Ampere makes even more precise the general trend of his thought in the synthetic definition given in the Carmen mnemonicum,[11] a synoptic table placed at the end of his work. He went on to say: "Government ought to have recourse to cybernetics, *secura cives ut pace fruantur* (to secure citizens the fruits of peace)."

It is very clear that for Ampere, cybernetics is the art of control in general.

A Century of Oblivion

It would appear that the time has come to take a serious look at the whole matter.

1. As his biographies have shown, Ampere, a man with a very inquisitive encyclopedic mind, was strongly interested in mathematics (differential and integral calculus, mathematical probability), the physical sciences (electrodynamics), the natural sciences, and in philosophy and many special aspects of science in which his restless activity found a scope for itself.[12]

But it does not appear that Ampere was particularly attached to the detailed study of cybernetics as the art of control, even in its political sense.[13] Apparently what he really wanted to do was to give the field a special name, to describe its object and to indicate its relation to the other sciences.

2. In carrying out his aim on the semantic level he has followed classical suggestions in creating out of whole cloth a series of scientific terms whose etymology he has specified, with great care to make clear the object of the sciences thus named and to distinguish them as precisely as possible from the neighboring sciences.[14]

3. Following his plan of classification, in the last years of his life, Ampere made a remarkable effort at synthesis. Realizing the need for an orderly arrangement of new knowledge he established a complete terminology in which he proceeded to introduce a full range of categories, going progressively from the part to the whole, from the simple to the complex, in a spirit typical of his time, but which seems of rather debatable value today.[15]

Perhaps in his work of classification he was carried away by a love of system and a longing for symmetry, which appears somewhat astonishing to us today.

4. In his description of cybernetics Ampere did not at all show a desire to found a new theory, and in this domain his works have not involved anything that was not previously known. In this field of political cybernetics he does not seem to have had collaborators, disciples, or successors whose work was in any way connected with his ideas, as was the case in his researches in mathematics and physics.

5. After the death of Ampere on June 10, 1836, there was no retention of

the word cybernetics in common usage in the figurative meaning adopted by the Greeks and revived by Ampere. The word itself was abandoned or at any rate never became at all popular. In this domain the French language remained satisfied with the words which had been in use for a long time: *gouverner* and *gouvernail* (control) for the physical meaning (steering of vessels) and *gouvernement* and *gouverneur* (government and governor) for the figurative meaning in connection with the government of nations.

Littré, to whom we owe a biographical sketch of Ampere makes the following summary statement in his dictionary: *"Cybernetics.* Name given by Ampere to that part of politics concerned with the means of government." Pierre Larousse in the *Grand Dictionnaire Universel du XIXᵉ Siecle,* also limits himself to the very brief statement; *"Cybernetics.* Name of the part of political economy treating the art of government, in the classification of Ampere."

There ensued a long period of silence. The word fell into oblivion.

6. It could however happen that some unusually interested and patient investigators might one day discover in some forgotten 19th century author an allusion to the term created by Ampere, a new use of the word or some extension of its meaning. This hypothesis should not be disregarded.

Precisely along this line, Mario G. Lasano, in the work already cited[16] mentions a use of the word, which seems to have escaped the attention of most authors of works on cybernetics. This has been discovered recently by a German author, Alfred Müller.[17] The usage mentioned here is a rather specific one. We quote:

> In its political sense the term cybernetics is still used today to designate a branch of practical evangelical theology, relating to the organizational structure of churches. This ecclesiastical cybernetics embraces problems ranging all the way from ecumenism to the paying of tithes. Governing the relations between ecclesiastical rights and judicial institutions, it also expresses the demand that the churches should in complete autonomy have the right to designate their governors, that is to say that they should be self-governing and self-steering.

At first sight this might appear to the outsider as a new meaning of cybernetics, rather far removed from earlier versions. One might be tempted here to think of the "administration" of the church, since it seems to involve the relation between ecclesiastical rights and judicial institutions. One might also say that (almost in a literal sense) the "bark of St. Peter" needs to be governed, must have a pilot and a rudder. But the stated requirement as to an "electoral regime" for the church, a very real problem in fact, inclines one rather to think of an extension of the political meaning of cybernetics, which would thus be extended from the government of nations to the government of institutions.

It would be interesting at any rate to find out more about the origin of this particular use of the term cybernetics.

7. In order to come to a cybernetic theory somewhat closer to that of Wiener and his school, Losano goes on to say:[18]

It is necessary to go back to 1868, the year in which James Clerk Maxwell published the first scientific article on one of the most typical feedback mechanisms: the engine regulator of Watt. He called this mechanism a governor, thus emphasizing the Greek root of the word which Wiener chose as a name for his theory of information and control.[19] Looking back from where we stand today it is possible to find anticipations of cybernetics in writings earlier than those of Wiener. Thus we think, for example, of the physiologist Richard Wagner in 1925 and also of the article published by the mathematician Louis Couffignal in 1938. But it is only in the work of Wiener that these rather heterogeneous investigations became solidified in the new discipline whose name was suggested by the writings of Maxwell.

THE CYBERNETICS OF WIENER

The definition of cybernetics given by Norbert Wiener in the subtitle of his work *Cybernetics* (1948) was stated thus: *Control and Communication in the Animal and the Machine.*

Current dictionaries have the following entries:

"Comparative study of complex calculating machines and the human nervous system in order to understand better the functioning of the human brain." (Thorndike-Barnhart, New York, 1956).

"The comparative study of the automatic control system formed by the nervous system and brain, and by mechano-electrical communication systems and devices (as computing machines, thermostats, photoelectric sorters)." (Webster, New York, 1961).

"The comparative study of complex electronic calculating machines and the human nervous system." (Klein, Amsterdam, 1966).

"The science which studies the mechanisms of communication and control in machines and in living organisms." (Nouveau Petit Larousse, Paris, 1966).

Modern dictionaries, which are in a certain sense the recorders of usage, would appear on the whole to have rallied to the support of the terms employed by Wiener.

The meaning thus set forth, approximating our actual current conceptions, departs sensibly from the more general significance that Ampere gave to the word cybernetics in his terminology.

On the origin of the choice made by Wiener of the word cybernetics we note two texts of interest from the semantical point of view. Both are taken from American encyclopedias. In view of their interest we present them as in the original and not in translation.

The first is taken from the McGraw-Hill Encyclopedia of Science and Technology (1960):

The name Cybernetics was first applied to this science by Norbert WIENER in 1947. It is taken from a Greek word which is variously translated steersman or governor. For some years, a number of people, notably Norbert WIENER and Arturo ROSENBLUETH, had recognized an important degree of unity in the problems of communication and control wherever they arose, and had been hampered by the lack of unity in terminology applied to the same phenomena when discussed in the context of different fields. Their choice of a new name for the unified field was due largely to their desire to employ a term that would not indicate a bias toward any of the disciplines which they visualized as comprising the unified field. The beginnings of this field may be traced to a Macy Foundation Conference on cerebral inhibition in May 1942, at which some of the problems presented focused attention on the fact that the control engineer may have something to offer the neurologist. The early development of the field was greatly stimulated by a series of conferences sponsored by the Macy Foundation during the years 1946–1953. The transactions of the last five of these conferences are published.

The second extract is taken from the entry on cybernetics in the Encyclopedia Americana (1965):

When WIENER introduced the term, he was unaware that it had already had a considerable history, and that it had been used more than a century before by Andre AMPERE to cover the purely governmental side of such a theory, in the positivistic classification of scientific theories. The modern term was introduced because of the need to describe comprehensively a group of phenomena having a real community of ideas and appropriate methods of study, but belonging to conventionally different disciplines.

The above extract has its full value in view of the fact that it was signed by Norbert Wiener himself. That is why we reproduce here further extracts from the same entry.

Cybernetics includes the theory of information and its measurement; the concept of communication as a statistical problem in which messages not sent play an equal role with messages sent; the theory of the statistical prediction of sequences of events distributed in time; the theory of the relation between message and noise and their separation by wave filters; the theory of apparatus for control, and its design and application to servomechanisms; electrical computers; and the automatic factory.

It includes also the theory of apparatus that retains information in a sort of "memory" and that adapts its performance to improve its own efficiency by a sort of "learning" process; and the application of this idea to the lower animals and to man and his society to include the theory of Gestalt psychology. Closely related to it is the study of communication nets in which such nets settle down to an equilibrium of quasi-equilibrium of performance.

This congeries of sciences developed during World War II out of the need for putting mathematical and other scientific talents to work on practical problems of military design, which had not until then been considered to be of a purely scientific nature. This need was closely related to the further one of organizing such processes as the tracking down of airplanes, which by their very speed and complexity eluded the existing types of human intervention, through the use of automatic, mechanical, or electronical auxiliary devices.

Thus a field of investigation covering not only such mechanisms, but also their archetype, the brain and the nervous system, came into being and was treated by WIENER in his book on *Cybernetics, or Control and Communication in the Animal and the Machine* (1948). This book was the outgrowth of war work done by Julian BIGELOW and the author on automatic predictors for antiaircraft fire, of a long standing interest in computing machines, and of certain suggestions made by Arturo ROSENBLUETH concerning the human element in mixed human and mechanical fire control systems.

The cybernetic nexus of disciplines has interested neurophysiologists, psychologists and communication engineers, and there are writings emanating from all these groups which must be considered to be essentially of a cybernetic nature. In pure mathematics, it has the greatest repercussions on the students of probability.

This encyclopedia entry is particularly important for the history of cybernetics. It describes with great precision the current field of investigation covered by it without assigning too strict limits to its further development.

A third text, this time in French by P. de Latil, clarifies further Wiener's thinking on the subject of the word *cybernetics*.

In Wiener's group physiologists and mathematicians have been hampered by the absence of a vocabulary providing for satisfactory understanding. They had no term expressing the essential unity of the problems of communication and control in machines and living organisms, a unity of which they felt confident. All the words which had been proposed put too much emphasis on the side of the machine or on that of life, when on the contrary it was essential to express definitely the duality of the new science.

The word cybernetics has been admirably chosen through its evocation of the helmsman of a ship as well as the regulation of a machine and the *governor* of Watt. At some future date it may find again its Greek meaning of government; for the homeostat of Ashby suggests the future promise of machines that govern.

The definition of cybernetics follows then at once from the word itself: the science of government or perhaps one might even say 'self-government.'[20]

The thought of Wiener has been clarified even more recently by Mario G. Losano,[21] who adds the following to the consideration already cited above.

A definite bond unites the Wienerian cybernetics to the studies of Maxwell on machines. But a subtle and unexpressed bond also joins it to the conception of Ampere. We may say it was unexpressed since Wiener made special reference to Maxwell and not to Ampere. It was subtle because the bond rests not so much on the similarity of the words designating the field of research in question as on the function attributed to the research itself. Thus, for Ampere to govern a society is to maintain it in equilibrium, that is to say in peace.

A probably completely involuntary echo of this is found in a brief essay by Wiener devoted to *homeostasis*, that is to say the processes by which a living organism stabilizes in its interior the conditions for the maintenance for its health. After stating the theory of homeostasis (which its author, Claude Bernard, attributed to living organisms alone) Wiener concludes: "The physical body does not possess homeostasis adequate to its needs; the political body possesses even less!"

[*Editor's Note:* We should call attention here to the resurrection of the concept of homeostasis by Walter B. Cannon in the United States. We reprint a fundamental article by Cannon as Paper 19 in this volume.]

In passing from the physical to the social body Wiener gives it as his understanding that the technique for the conservation of social equilibrium is the law. But it must be admitted that its preoccupations are otherwise. As Wiener says: "The body politic is not without homeostasis, or at any rate not without an intention in this direction. That is what we mean, for example, when we call the Constitution of the United States a "constitution of control and equilibrium." And our intention that the administration should serve as a protection against any undue encroachment from the legislative side and that the legislative side should watch very attentively for any possible violation of its proper powers by the executive shows that the judiciary should control the actions of both legislative and executive branches and insure that they conform to national traditions. I have said *national traditions*. It is here that the difficulty resides![22]

In order that there should exist a social homeostasis it is necessary that the elements which are to be reciprocally in equilibrium should be known: past and present. Whence arises the Wienerian criticism that the social memory is too ephemeral to exert a real homeostatic effect in the interior of society.

Then again, cybernetic theory (homeostasis is really a form of feedback) amounts to the realization of that form of good which we call social equilibrium. This cybernetic equilibrium does not so much signify conservatism, in political language, as rather a knowledge of past and future. "To pay due respect to the future," says Wiener again, "we must be conscious of the past," [23] an assertion which in our day seems to be indissolubly connected with the *secura pax* evoked by Ampere."

The value of these different texts lies in the emphasis they lay not only on the complete objectivity of Wiener when he is said to have ignored the early use made by Ampere of the word cybernetics, but also the spirit of fair play in which he at the same time tries to show how the thought of Ampere and his own are to be distinguished so far as field of application is concerned and how in effect they are united in terms of fundamental concept: the equilibrium sought after in the action of a governor. This relation becomes particularly evident when it extends Wiener's conception of homeostasis to human society and assigns as the function of communication systems the effort to reach a state of equilibrium.

The Wienerian School

When the work of Wiener appeared immediately after the war the people in scientific-technological circles were all preoccupied with transferring wartime discoveries and inventions to the activities of peacetime and moneymaking efforts.

It is then not surprising that the center of gravity of ideas should at first have been fixed on applications. It was a question of transferring to civilian life the new possibilities which had been developed in the military sphere.

But this demanded the process of adaptation. The objectives of research were no longer the same. It was necessary to respond to other needs. The economic imperatives (the need to make money), which had taken a back seat during the war, now regained their true value.

Ideas naturally became centered on the comparison of man and the machine. Emphasis was placed in the first place on automatization or automation as it is now called and the transfer of human functions to the machine. Some impressive and indeed spectacular results have come to light in a mechanization and automation of human actions.

But after the necessities of the machine led to the incorporation of the essential notions of automation, of servomechanisms, of feedback, etc., one soon perceived that the practical results to be achieved along this line would be limited if one were not to integrate into the solutions newer and more profitable discoveries relative to materials, procedures involved in communication and control. New materials, a new chemistry, new sources of energy, electronics, and new manufacturing processes had to be put to work, thereby amounting essentially to a renovation of technology.

This evolution could not take place without closer attention to newly evolved scientific ideas and more advanced mathematics, inspired by the sciences and operations research, communication theory, probability, simulation techniques, etc.

The center of gravity has then moved in the direction of problems in information theory and its derivatives.

Simultaneously new applications have been made in fields of activity far removed from the point of departure: biology, medicine, education, law, psychology, economics, in short, in all branches of the humanistic sciences. We have found it desirable to measure, to employ formulas, and to quantify the most diverse things. This was a logical attitude and showed a laudable desire for progress, but while it opened up wider perspectives, it nevertheless had certain drawbacks associated with it.

Thus in this period there developed an extraordinary abundance of ideas claiming to be, for better or worse, associated with cybernetics and constituting what came to be called, in perhaps a pejorative sense, the Wienerian School.

People were led to attach to cybernetics specialized scientific elements which seemed to have little connection with it. Moreover, people have tried to use cybernetics to dress up other notions and to make unfortunate extensions, largely inspired by the thought of profitable applications, and this has led to the confused situation of which there is justified complaint today.

But the world of the cyberneticians, probably as a result of the proliferation of views produced within the International Association of Cybernetics, has witnessed the birth of a new current of thought showing particular promise.

It is now admitted that cybernetics is an interdisciplinary science, which only confirms the necessity of a redefinition.

Account has been taken of the fact that during the past twenty years the center of gravity of research activity in this general area has moved steadily in the direction of a science worthy of the name. Little by little we have drawn away from the "animal-machine" comparison, which in the beginning was the strong point for Wiener and his collaborators.

We have thus freed ourselves from applications, from mechanisms, from instrumentation that have, to be sure, provided useful education.

We have drawn away from specialized sciences that have often served as support for thought and have furnished points of comparison and landmarks for new stages of development.

At the same time we have drawn away from the political idea, from the concept of the government of nations and of institutions and various enterprises.

All these different notions that we have seen come up and then abandoned have, it is true, developed into specialized branches, sometimes very vigorously, such as automation, communication, information theory, legal cybernetics, etc., which pursue each on its own hook a fertile career, but all the same do not form "cybernetics."

We are thus led little by little to a more abstract conception of cybernetics properly speaking, based not only on objects, needs, and means, but also on "systems," a more philosophical or even metaphysical conception.

It does not behoove us either to prophesy the way this evolution will take place in the future or to find fault with its objective. We shall limit ourselves to the attempt to make a point along the general lines of our discussion.

What finally is it appropriate to retain from the intervention of Wiener and his school? It is indeed difficult to provide a synthesis in a few words.

Let us agree in any case that the word cybernetics was without question rediscovered (shall one say "resuscitated"?) by Wiener when he used it in 1948 in his first work on the technique of automatic systems. To quote Larousse again:

> In which he (Wiener) presented cybernetics as a kind of cross-roads science and freed it of general notions on the behavior of governing mechanisms. These views were intended to be the point of departure of a vast movement, signifying a veritable intellectual revolution in the logical analysis of the functions of superior beings and of processes permitting their artificial reproduction.
>
> Some cyberneticians think that social phenomena, from the fact that they result from exchange of information, should be studied by the methods of cybernetics, which enables them to foresee in the framework of a bold extrapolation the image of a future society commanded by machines that think and control.[24]

AMPERE AGAIN—NEW INVESTIGATIONS

We now find it worthwhile to return to a consideration of Ampere with minds illuminated by the new notions that we have encountered in more recent writers, in particular the adherents of the Wienerian School. We wish to see whether with the aid of these new views we cannot locate in the works of Ampere and his contemporaries some anticipations of cybernetics that escaped us in the first examination.

As we have seen, Ampere considered cybernetics as the art of governing in general. By this he understood the governing or control of peoples and of nations. His classification makes this clear. But did he also contemplate more specific applications of the art of control, such as control of groups and institutions, control of enterprise, of machines, of activities and movements of men and animals? These, of course, are aspects that have gained attention in the work of Wiener and his school.

In other words, did Ampere conceive of cybernetics in the general sense envisaged by Professor Hoffmann: "a strategy concerned with the optimal arrangements of the means of control, due account being taken of the chance perturbations brought about by the environment?" If he did, did he intend to extend his conception to systems of every kind?

Precisely what did he say?

1) At the outset we have investigated whether Ampere had extended the concept of "control of people and of nations" to other types of institutions or social organizations or indeed other sorts of groups.

We have not found, in those of his works to which we have had access, any particular mention of this subject.

Since however it is a question here of the part in place of the whole (the government of a province, a city, or a community) in the political sense, that is to say with a view of the maintenance of the peace or a state of equilibrium, it scarcely seems excessive to claim that the extension of the concept to this level was envisaged explicitly by Ampere.

2) Ampere was particularly explicit in his classification concerning the idea of control or direction of industries, as the following table shows.

Sciences of the first order	Sciences of the second order		Sciences of the third order
		1.	*Technography—* Knowledge of procedures, instruments, and machines
	Elementary Technology	2.	*Industrial finance—*To take account of profit and loss in an enterprise
Technology (Study of bodies to discover their possible utility for us)		3.	*Industrial economy—* Comparison of procedures, instruments and machines used at different times and in different places
	Comparative Technology	4.	*Industrial physics—* Application of theory to practice in the investigation of progress

But we must note that technology as understood by Ampere was put by him in the domain of the cosmological sciences, including the physical sciences, whereas for him cybernetics is a part of the mental sciences.

If on the eve of the first industrial revolution, an epoch in which the machine was still fundamentally something that moves and the machine tool was still embryonic, Ampere was already well aware of certain needs of industrial enterprise, he could scarcely predict the importance that the problems of social order and economics would take on, things that became well known only after his time.

3) Ampere does not seem to have noted particular features of the control of machines and of automatic control in general.

But at his time, as is clear from the works of his contemporaries, the machine tool was still in its infancy (the lathe was an instrument for blacksmiths or watchmakers, a source of entertainment to an earlier king). Moreover, scientists, the heirs of the encyclopedists and generally speaking the understudies of the philosophers, busied themselves with trying to reconcile the observation of nature with the speculation of the mind, introducing a multiplicity of comparisons of mind and matter, of man and animal, of animal and plant.

At that time the machine was still studied wholly as a motion-producing device. It is true indeed that in the technological works of that period one will find some curious, fragmentary anticipations of mechanisms that would ultimately assure the development of the industrial age.

4) What did Ampere think of the comparative study of control and communication systems in animals and machines?

Larousse's Dictionary mentions, as due to Ampere, a study published in 1824 by the Academy of Sciences (with no name attached), entitled "Philosophical Considerations on the Purpose of the Bodily and Nervous Systems of Vertebrate Animals."[25]

In view of the title, we had hoped to discover some relation between Ampere's thought and the investigations carried out by Wiener around 1940 with the help of physiologists and neurologists. But the reading of the work made this hope vain. It consists of reflections on the studies of comparative anatomy and physiology of the nervous system of vertebrate and invertebrate animals, published in the *Annales des Sciences Naturelles* by eminent physiologists like Cuvier, Geoffroy Saint-Hilaire, Serres, Flourens, Desmoulins, etc., without doubt very interesting to specialists, but far removed from the considerations of the present article.

5) Finally we have sought to find whether Ampere had in any explicit fashion anticipated the notion of homeostasis, of adaptation for the sake of the maintenance of equilibrium.

Here we have been somewhat more successful, for in addition to the relations stressed by Mario A. Losano between the social homeostasis mentioned by Wiener and the state of equilibrium (that is to say, of peace) evoked by Ampere, we have found in the latter a notion that on the level of individual man renews consideration of this for the maintenance of equilibrium. Here is what Ampere has to say in his classification of the sciences:

> *Phrenyegetique* (from the Greek φρεννγιεκ) is the study of the modification produced by moral causes in the organization of vital phenomena; such are the passions, the concentration of attention on certain ideas, sadness, gaiety, profound grief, the change in the ordinary relations of men with their surroundings, whether it be the result of the attainment of new social position or the result of the prescription of a physician.

It is then a question of "the effects of the soul or the mind on the vital organization and on health."[26] We note here that *phrenyegetics* is placed by Ampere in the kingdom of the cosmological sciences, in the subkingdom of the physical sciences, and in the branch of the medical sciences.

Somewhat further on Ampere noted "The influence of physics on morals, studied by the philosopher when he seeks to discover the causes which determine the characteristics, customs, and passions of men should, on the contrary, be placed in the second kingdom, where the truths relative to this activity will find a place in the science to which I have given the name Ethogeny."[27]

6) Finally we have tried to find out if Ampere, like Wiener, had handled the study of the motion of men and animals, as well as their general behavior.

If Ampere did not appear to be particularly preoccupied with the possibility of reproducing the behavior of men and animals with the aid of mechanical devices, as Vaucanson had already done, we do find in numerous places in his writings traces of a major concern of his, namely of the mechanisms of perception, of intelligence, and action. This would appear to come close to certain aspects of cybernetics and hence merits a special examination.

Analysis of the Act of Control

In addition to the principal works of Ampere that have already been examined, his writings include numerous studies and essays published in many places, as well as an abundant correspondence. In these one continually discovers a vocabulary of which cybernetics makes great use, such as observation, perception, impression, relations, comparisons and laws, etc. But it must be admitted that other sciences use this same terminology, which is besides found in the works of most contemporary scientists, since they clearly relate to the scientific method in general.

1. Scientific Method

In the scientific method as it was elaborated at the time of Ampere, one first observes facts (the natural sciences are of course the sciences of observation); one then seeks relations among the facts; one then evolves laws applicable to the given conditions. The next step is to trace the existence of causes (the philosophy or metaphysics of the situation). After answering the questions, *what, when,* and *how,* one goes on to the question *why.*

Ampere took a great interest in this line of research. He subdivided an effective scientific investigation into the following four steps:

—to gather and describe the facts
—to seek what is hidden under the facts
—to compare results and deduce general laws from them
—finally to seek the causes of the facts.

Ampere attached so much importance to this method that in his classification he subdivided each of the domains of knowledge into four parts corresponding to the four steps of the method; thus providing a total of 128 sciences of the third order.

Thus for politics he has

—the public law of nations
—knowledge of the origin of customs and characteristics
—the art of government
—the theory of power.

In similar fashion he subdivides technology into

—technography
—cerdoristics
—economy
—industrial physics.

In both cases the subdivision reflects the four steps: collection of facts, interpretation, comparison and search for causes.

2. Psychological Foundation

As we have already remarked the scanty fragments to be discovered in the work of Ampere are not in themselves sufficient to form a new science that might be called cybernetics. It is therefore necessary to look for something else. Can we perhaps find it in spiritual or psychological values, matters which are not so readily grasped?

Enlarging the domain of our investigations we have read biographies of Ampere and of his correspondents and commentators who knew him well, who have participated in his work, and who made up a part of what may be called his "familiar circle." Here we have been able to turn up some interesting pages.[28]

It becomes clear from the reading of this material that Ampere, in line with the spirit of the age in which he lived, was much preoccupied with the problem of the chief end or goal of man and carried out studies on (a) sense impressions and sensations, (b) the transmission of these to the brain and the reflection made on them there, and (c) the activity of the human organism, its acts and movements.

The goal of this work is the study of human faculties, of perception, intelligence, will, consciousness, in the light of the discovery of the consciousness of the individual.

Placing the chief end of the human being above practical needs the thought of Ampere seemed to be concentrated more on the soul than on the body, or to put it more exactly, Ampere seeks for the relation between the two. The thought of Ampere moves from the concrete to the abstract, resolutely toward philosophy and metaphysics.

Let us listen to what Sainte-Beuve had to say of him:

> In a summary of the psychological ideas of M. Ampere, edited by his friend M. Bredin of Lyon in 1811, I find 'All psychological phenomena can be related to three systems: the sensory, the cognitive and the intellectual.' The cognitive and intellectual systems, which would seem to be the same, are for him different in that it is to the cognitive system alone that he attributes the distinction between "I" and "not I," which comes from the proper activity of the human organism, according to M. de Biran. He reserves to the intellectual system, properly so called, the perception of all other relations.[29]

3. Feeling and Consciousness

At the outset of his observations Ampere distinguishes between voluntary and involuntary movements (the latter being those of internal organs). He wishes to investigate the connection between sense impressions (sight and touch, for example) and the impressions received by the brain. He distinguishes between the phenomena perceived by feeling and those perceived by consciousness.

According to Sainte-Beuve, Ampere goes on to say:

> From the intellectual point of view man has the faculty of acquisition and conservation. The faculty of acquisition is divided into three separate parts: man acquires information through his senses, by the deployment of motor activity to discover the causes of phenomena, and through reflection, which one can define as the faculty of perceiving relations which apply to the products of feeling and those of activity. We perceive relations among the first by comparison and among the second by observation of the effects produced by the causes. We may then divide all the phenomena our intelligence presents to us into four categories: (a) feeling, (b) activity, (c) comparison and (d) etiology (doctrine of causes).[30]

Ampere pursues his study with due attention to the operations which go on in the brain and to the comparison with preceding sensations, this confrontation of knowledge coming before the determination, itself preceding the act or voluntary movement. What he calls the "active" system or category is the mental representation of the received impressions.

[*Editor's Note:* The above is followed by a long discussion of Ampere's ideas, set forth in his correspondence with Maine de Biran, on what we normally call psychophysics and the philosophy of thinking, including emphasis on such topics as intelligence and will. Ampere was evidently much influenced in his psychological and philosophical views by his contemporaries de Tracy and de Biran. The discussion leads up to a consideration of the act of control in man, leading the author of this article to believe that Ampere and his contemporaries had grasped philosophically the idea which Norbert Wiener exploited in a more practical fashion. We omit the translation of this part, which is philosophically highly technical and resume the translation as follows.]

Comparison of the Views of Ampere and Wiener on the Thinking Process

Starting with the concept of "control in general," that is, "optimal maneuvering in the face of the chance variations in the environment," Ampere was led to the analysis of the act of control and thence to a philosophical point of view, the noumenal versus the phenomenal conception of the ego. With him the idea of the ultimate goal is the thing that prevails.

Wiener adopted an essentially different point of departure. He was faced with the necessity of solving very definite problems presented by the war. He was then concerned to find the means, apparatus, and instruments capable of more rapid perception, reaction, and organization than man could achieve by his unaided efforts. Thus, for example, he was definitely concerned with: (a) the location of the position of airplanes (radar), (b) the ability to plot their movements, (c) the calculation of their trajectories, (d) the pursuit of planes, (e) the means of defence against attack by planes and the means of attacking them, and (f) the launching and control of the flight of missiles.

From these practical problems the orientation of his work and that of his colleagues turned toward the study of the faculties of perception, communication, calculation and regulation, through a comparison with the reactions of the nervous systems of men and animals. Whether foreseen or not these works were bound to head up into industrialization.

On the basis of his studies and their practical realization, Wiener, after the war, was able to set up a body of laws and principles of scientific character and by developing his thinking little by little finally arrived at a philosophy that ultimately found its outlet in the government of nations. It then appears that Wiener finally attained to the philosophical position from which Ampere started out. He then rejoins

Ampere but with a gap in time of more than a century and with a gap in the level of the whole research undertaking.

To put it briefly we find in Ampere a predominance of pure thought, whereas Wiener is more concerned with the practical result. This is the whole difference between the man of science and the man of action.

Wiener not only benefited in his work from a whole century of science and technological progress (and what progress!), but he was able to undertake his researches under conditions of which Ampere could never dream: practically unlimited intellectual and technological means, abolition of all ideas of cost, the possibility of collaboration in many areas and access to all discoveries. Besides, in this modern age there is a great push for progress on all sides, needs are urgent and utility of results immediate with a quick payoff following success. Finally there exists the prospect of a profitable industrialization with new possibilities for the benefit of all mankind.

It is clear that the discoveries and achievements of Wiener and his school have been more spectacular from the standpoint of the general public than the abstract concepts elaborated by Ampere and the philosophers who shared his thought.

It seems indeed that at a certain moment people have desired to connect with the school of Wiener everything that even remotely touches on the rational sciences. As a matter of fact, in each of the latter one can apply the principle of the reorganization of the act of control, which remains the fundamental principle of the science of cybernetics, a genuine interdisciplinary science. In certain cases things have turned out fortunately. In certain others an impasse has been encountered. It has been necessary to recognize the limitations of the new knowledge and place its elements in their proper perspective.

Today things have settled down a bit, and it seems that under the influence of the new current of thought to which we have alluded we can accept the following conclusion as to what cybernetics is and what it ultimately can be:

> To begin with cybernetics is a logical science in the sense in which it provides a rational analysis of the act of control without facing the question of what it is that controls and how it does it. In this way it permits a vast theoretical classification of systems and machines, never before attempted. It is also the point of departure of many applications, since from its conclusions there follows the possible construction of control machines of all kinds, certain of which have already been extensively applied (e.g., automatic piloting of aircraft) and whose generalization will allow the systematic realization of automation in general.[31]

To sum up our considerations we have sensed both in the work of Wiener and that of Ampere that the mind is drawing away little by little from a concern for direct observation in specific fields and from material preoccupations in order to devote attention to interdisciplinary relations and ultimately to some sort of synthesis. It is seeking for a philosophy concerning the ultimate goal of man and all creation.

Here again, it seems clear that to whatever conclusions each of these two great minds has attained, they effectively come together on a great number of points. Both have served well the philosophy of cybernetics: Ampere, the old and Wiener, the new. It is the junction of their thought that is the important thing.

Is it not indeed the task of philosophy to reconcile matter and mind and to order facts and observations in the context of a higher goal?

The period of neglect in the development of the concepts of cybernetics between Ampere and Wiener, a whole century, has been devoted to a consideration of certain things which are not questions of cybernetics, or at any rate not thought of from this point of view. But the new current of thought, raising cybernetics to the level of a logical, interdisciplinary science, has revealed a real unity.

Ampere invented the name in French, whereas Wiener resurrected it in English. Ampere conceived the idea. Wiener rediscovered it a century later and, aided by propitious circumstances, gave it an importance that it had lost in the meantime.

May we not then legitimately pay joint homage to two great pioneers who at different epochs in their turn revived and enlarged an ancient Greek conception? Should we not pay tribute to the part that each played in the development of this science? After all, what is past is not the important thing; it is the future that really counts!

NOTES AND REFERENCES

1. The Round Table on cybernetics was held in Namur on December 13, 1969. Its proceedings constituted a report to the 7th International Congress on Cybernetics, 7–11 September, 1970.
2. Ibid., doc. XI, p. 30.
3. See Losano, Mario G. *Giuscibernetica*, Macchine e modelli cibernetici del diritto. Turin, 1969, p. 127.
4. Ampere, Andre Marie: "Essai sur la Philosophie du Science ou Exposition Analytique d'une Classification Naturelle de Toutes les Connaissances Humaines," Bachelier, Paris, 1838 and 1843. 2 volumes. A first edition appeared in 1834 and was in one volume restricted to the cosmological and natural sciences. The second volume was devoted to the noological and humane sciences.
5. In his essay Ampere seems to have been inspired to a certain extent by the works of some illustrious predecessors like Locke, with his "Essay on Human Understanding," Condillac, with his "Essay on the Origin of Human Knowledge," his "Treatise on Systems" and his "Treatise on Sensation"; Linnaeus with his "Natural Systems," which embraced not only botany, as is often thought, but also zoology and geology. The historian of science who wishes to trace the origin of the cybernetic idea in the sense in which it is used today will assuredly make some interesting discoveries in the works of the philosophers and scientists of the 18th century.
6. Larousse, Pierre. Grand Dictionnaire Universal. Paris, 1866–1876.
7. In the vocabulary that he uses and in large part was invented by him, Ampere assigns the following meaning to technical terms:

Syncimenics:	the relations among people
Ethnodicy:	the public law of nations

Diplomacy: knowledge of the circumstances which have given birth to customs and agreements and the spirit which presided at their formation.

8. Ampere, Andre Marie. Op. cit., Vol. II, p. 142.

9. Ibid., p. 104–141.

10. Ibid., p. 143.

11. This is a curious mnemonic table, dedicated to his son, in which Ampere summarizes in Latin verse the substance of his terminology of the sciences. In this matter he seems to have paid his respects to the taste of his time under the influence of certain eminent contemporaries, who are today largely forgotten; e.g., F. J. Gall, forerunner of brain histology and founder of phrenology; Destutt de Tracy, investigator of human faculties, particularly the memory; Aime Paris, who devised a method of education in music based on certain aspects of the perception of sound. This recourse to the technique of mnemonics might have appeared necessary at the time to assure memorization of the words invented to designate new sciences. Here is the verse treatment of sciences of the third order relevant to politics:

> Foedera tum noris[81], qua sint
>   servanda sagaci
> Arte[82], et secura cives ut pace
>   fruantur[83]
> Quae fluxa et quae sit mansura
>   potentia regum[84].

12. At the beginning of the second volume of the essay will be found excellent biographical notices of Ampere by Sainte-Beuve and Littré.

13. The great social problems which preoccupied the attention of Ampere formed a part of what he called coenolbogy, the science of public happiness.

14. It is worth recalling here that in his youth Ampere had laid the foundations of a universal language to replace in his thinking the existing idioms.

15. Ampere himself recognized the shortcomings of his system. In the second edition of his essay, which was not published until after his death, he introduced several changes.

16. Losano, Mario G. Op. cit. pp. 128–129.

17. Müller, Dedo (Alfred). *Aufgaben und Probleme der praktischen Theologie der Gegenwart*. Studium Generale, XIV, No. 2, 1961, p. 93.

18. Losano, Mario G. Op. cit. pp. 129–130.

19. Maxwell, James Clerk. *On Governors*. Proceedings of the Royal Society of London Volume XVI, 1868, pp. 170–83.

20. de Latil, Pierre. *La pensée artificielle*. Paris, Gallimard, Vol. 1, 1950, p. 23.

21. Losano, Mario G. Op. Cit., p. 130.

22. Wiener, Norbert, "Omeostasi individuale e sociale," in *Dio e Golem*. S.P.A., Boringhieri, Torino, 1967, p. 132.

23. Ibid., p. 134.

24. Grand Larousse Encyclopedique, Paris, 1966.

25. Academie des Sciences: *Annales des Sciences Naturelles*. Paris, Vol. II, 1824, pp. 295–310.

26. Ampere, Vol. I, pp. 146–147.

27. Ibid., pp. 150–151.

28. The principal sources are: a) Sainte-Beuve, Ch. *Notice sur M. Ampere*, second edition of Ampere's *Essai sur la Philosophie des Sciences*, Vol. II, 1843; b) Littré, E. *Notice sur les travaux scientifiques de M. Ampere*, ibid.; c) Sainte-Hilaire. *La philosophie des deux Ampere*, Paris, 1866; d) Maine de

Biran. *Journal Intime*; e) Correspondence d'Ampere, Gauthier-Villars, Paris, two volumes, 1936 (Societe des Amis d'Andre Marie Ampere).
29. Sainte-Beuve, Ch. Ibid., pp. 43 and 44.
30. Sainte-Beuve, Ch. Ibid., p. 43.
31. Grand Larousse Encyclopedique, Paris, 1966.

24

Reprinted from *Cybernetica* 1(1):1–14 (1958)

My connection with cybernetics.
Its origins and its future.

by Norbert WIENER,

*Professor of mathematics at the Massachusetts Institute of Technology
(Cambridge, U.S.A.)*

It has been suggested that now, some fourteen years after the writing of my book on cybernetics, I take up the history of the subject and a discussion of the lines of work which seem most interesting for the immediate and the remote future.

I wish to disclaim at the very start any attempt at complete comprehensiveness, for I have neither the ability nor inclination to give an encyclopedic account of all of the ramifications of the subject as a whole. Therefore this talk is going to be highly personal, and will constitute an account of my own relation to the beginnings of the subject and of the directions of work which interest me at present and which seem to me particularly tempting for the future.

My contact with cybernetic ideas goes back to 1919, just after the First World War when I had completed my military service and was looking around for significant problems to which to devote myself in my career as a mathematician. I had read rather broadly in the theory of the Lebesgue integral in Frechet's and Volterra's books on integral equations and the like, and had come to the conclusion that analysis was the branch of mathematics which was most tempting to me, and to which I should devote my career. At that time a colleague, Professor T. Barnett from the University of Cincinnati, happened to be in the American Cambridge, and I asked him to suggest to me what problem would seem to be coming into its critical stage of development. He called my attention to the problems of integration in function space and to the work of Gateaux and to P. J. Daniell. I started following up their work and obtained some minor results on an abstract basis, but these seemed to me thin and lacking in real significance. I therefore asked myself if there were any questions in physics and the other natural sciences in which the integration of functions of curves comes in naturally with a truly physical meaning.

It was in the summer of 1920 which I spent in Strasbourg working with Professor Frechet that a hint of the answer to this question came to me. The problem of the Brownian motion is one in
which random assemblages of curves naturally occur. Related to
this are a great many problems in statistical mechanics, and particularly in hydrodynamics. My office at the Massachusetts Institute
of Technology overlooked the basin of the Charles River, and I
had often reflected on the wave pattern of the surface under a wind
as another example of a functional entity belonging to a family
in which questions of distribution were important. I therefore
decided to study the problems of the distributions of functions with
this and other similar physical motivations as my basis.

This is not the place to go into the details of my subsequent work
which led to a successful theory of the Brownian motion. However
two comments are appropriate. One is that I came into contact
with the deeply significant work of Taylor, now Sir Geoffrey Taylor,
on turbulence in which the notion of autocorrelation played a
predominant role. The other is that in applying this notion to
problems arising out of the Brownian motion I was forced to
study a certain class of functions which had already been studied
by mathematicians, but which had been considered as more or
less pathological. These were the continuous non-differentiable
functions. I found that functions of this sort, far from being non-
physical, belonged to the very essence of the study of the Brownian
motion, and of distributions of curves in function space.

My physical interests led me to the physical interpretation of
the problem of integration in function space as one of probability.
Here I was much influenced by the fact that the ideas of Gibbs
in statistical mechanics, after a period in which they had been
foreign to the ways of thought of contemporary physicists, were
now coming back into vogue and were being really understood.
I therefore found myself definitely along the path of investigating
random physical phenomena from the probability point of view
with the aid of new technical methods which seemed to be precise
and promising.

At a very early stage of my work on this subject I became aware
that the harmonic analysis of random function was an essential
part of my program. I found that this harmonic analysis, although
it had been studied from a physical point of view by theoretical
workers in optics, and although Shuster had suggested certain
statistical aspects of it in his discussion of the periodogram, had
been neglected by the pure mathematicians, who, for the most
part, had confined their efforts to the study of phenomena which

were either strictly periodic, as in the case of the Fourier series, or definitely of limited duration in time, as in the case of the Fourier integral as it was then known. The harmonic analysis of continuing phenomena in time needed a new start, which I began in papers written in 1924 and which I brought to a successful conclusion in my paper on generalized harmonic analysis which appeared in Acta Matimatica in 1930. In these papers my work, which interacted most closely with that of Bochner, made constant appeal to the work I had done on the Brownian motion, and received its most significant applications in that field. Throughout this work I was compelled to consider non-differentiable continuous functions.

From the very beginning of this work, I was influenced by the contemporary investigation of Paul Lévy on random functions. It was Lévy who called my attention to the fact that my Brownian motion functions had already received a certain amount of study from Bachelier. Bachelier's work, however, while representing an excellent insight, had already been written at a time before the ideas of Lebesgue integration were available for the development of an adequate technique.

My work on random functions suggested to me that I might have a new approach to the problems of turbulence and of statistical mechanics in general. These concepts led me to a new series of papers on the theory of chaos. I found that my work, although unquestionably in the right direction, ran into many difficulties in this field which were intrinsic in the study of the dynamics of random processes. As this is a little off the main direction of cybernetics, I shall not go into them here, except to say that it became amply clear to me that without a really new body of ideas most of the mathematical developments in time occurring in the theory of turbulence and of random processes have the character that the series which one obtains will often tend to have a zero radius of convergence and to be asymptotic rather than convergent. Some recent work of Kolmogoroff has suggested to me the possibility of evading this difficulty, but the turning of these ideas into usable methods is still under way, and it is not yet the time to speak of the processes.

In the beginning of 1930 it had become amply clear that the work of Willard Gibbs on the ergodic hypothesis was intimately related to my field of interest, and that the Gibbsian form of the ergodic hypothesis that time averages in random systems were related to phase averages needed a new justification. This justification came from the work of Koopman and von Neumann, and particularly G. D. Birkhoff. Later on I was able to bring some of this work into closer relation with the theory of random processes than had been done by some of the other workers in the field. All this contributed to the growing technique of the field of work in which I had interested myself.

Here I must comment on the valuable discussions that I had at the time with Eberhardt Hopf who had come at that time from Germany to do some work at Harvard University. Hopf is an astronomer as well as a mathematician, and he turned my interest in the direction of certain integral equations which come up in the theory of the internal equilibrium of radiation. These integral equations, while they arose from an entirely different field of work, contributed significantly to the techniques which I was later to use in that field, and particularly in the theory of prediction. This theory of prediction, although it belongs to a later period of my work, is intimately associated with the prediction of weather, and therefore with the general statistical problem.

During the period of the late twenties and thirties, I received a very considerable amount of stimulus to and interest in engineering problems from my presence at the Massachusetts Institute of Technology, and was consulted more than once by my colleagues in connection with the task of reducing the unorthodox mathematical method of Heaviside to a rigorous basis. Here Vannevar Bush, in particular, pressed me to work in this direction and to collaborate with him in the writing of a book on electrical engineering operational methods. My interest in the operational calculus both received a great impetus from my studies of harmonic analysis and contributed to the broadening of my point of view in that work so that it forms an intrinsic part of the intellectual machinery which I was later to use in the study of random problems.

Moreover, Vannevar Bush was intensely interested at the time in an instrument known as the differential analyzer by means of which he was able to give a numerical solution to many problems arising in the study of differential equations, and in particular in operational calculus. This started a long-time interest on my part in the possibilities of mechanical computation by means of analogue devices. I, myself, was led by this contact with Bush to the design of an optical method of obtaining Fourier transformers.

Bush's work was of the utmost importance in the study of the solutions of ordinary differential equations in which a single variable such as on time is the independent variable. Bush was very much interested in extending this sort of work to the solution of partial differential equations, and asked me explicitly if I had any ideas as to how his instrument could be adapted to this purpose. It was clear to me at the beginning that the really difficult problem in the solution of partial differential equations was that of a manageable representation of functions to two or more variables so that they could become accessible to mechanical means of computation. I had already been much impressed with the incipient developments of television methods and with the idea of scanning which makes it possible to represent a function on two or more variables in the one variable of time. I suggested to Professor Bush that this technique of scanning should form an essential part of any mechanism for the study of partial differential equations. It was clear to me at the beginning that scanning methods involved an enormous compression of a great deal of information spread out in several variables into information in the one variable of time. I had already certain doubts as to the practicability of performing this compression by the use of analogue devices, and in my own mind I was already beginning to ask myself whether methods of a digital nature might not be more apropos. However, at this time the concept of digital computation was far in the future, and I did not go out of my way to implement these ideas by more practical speculations. Nevertheless, when just before the First World War the problems of the sort that Bush had raised had become of immediate practical importance, the idea of a digital computer and its relation to scanning was already prominent in my mind.

It was the Second World War, or rather the period of preparation just before the Second World War, which led me to an abrupt turn in the directions of my interest. Before America had entered the war it had become amply clear that the possibility of this entrance was very actual, and that the immediate problem that faced us was that of keeping England from succumbing in the Battle of Britain. Here two directions of work came to the fore. One was the development of radar which had already been initiated by British scientists, and which caused the need of a greatly refined technique of electrical engineering, particularly in matters concerning alternating current theory and the related theory of communication by fluctuating currents, which had already proved so fruitful in telephony.

I had already interested myself in telephone theory and in the design of wave filters, particularly through contact with Professor Y. W. Lee who had been a graduate student at the Massachusetts Institute of Technology in electrical engineering. Actually I came into contact with Lee because I had certain mathematical ideas on the use of Laguerre functions in the design of wave filters, and I needed a graduate student to work under me in this field. Lee contributed many essential ideas of the design of such filters. We proceeded to make the series of inventions which ultimately developed into patents and were sold to various firms, including the Bell Telephone Laboratories. When Lee had returned to China in the thirties, I received through him an invitation to lecture at Tsing Hua University at Peking. I accepted this invitation and we spent our time together there largely in the further elaboration of these inventions.

When the prospects of the United States entering the World War became imminent, I tried to see if any of my own work would prove relevant to the military effort. I actually thought of the work that I had already done on the mechanization of partial differential equations. This was about the time of the Mathematical Society meeting at Dartmouth in the summer of 1940. At that time we were shown a computing instrument made by the Bell Laboratories for the study of a complex algebra which is used in electrical circuit work, and this made use of the Binary System of notation. Putting all the ideas together that were in my mind, I came to the conclusion that my methods for the study of partial differential equations could be made practical, and that they would involve a digital computing machine rather than an analogy computing machine in order to obtain the speed that was necessary; that this digital computing machine should be based on the scale of 2 rather than of 10, so that the individual marks needed for the different digits should be of a « yes » or « no » character, which is suitable to electronic circuits, and that in some cases this machine might work on the basis of previous ideas of H. P. Phillips and myself where we found the situation of the potential problem to depend on an infinite sequence of averagings.

This sequence of averagings could readily be adapted to a scanning process, but the thing that struck me as most difficult about this process was the vast body of intermediate computations which it would require, and the enormous mass of subsequently useless data which would have to be written down. However, it became clear to me that a really practical machine would have to perform all its functions on the run. It would have to write quickly, read quickly what it had written down, and erase material already employed as quickly as possible so as to have its entire functioning immediately ready for new data. I gave a series of prescriptions of what would be required for such a machine, and I passed them on to professor Bush who was then in charge of the national scientific effort. Bush thought that my work was too far in the future to be immediately applicable, so that for the time being I abandoned this direction of effort. However, the series of five requirements which I laid down for my computing machine have all been found to be valid, and are still the basis of such work as that which has been done on the International Business Machines.

When the radiation laboratory was set up at the Massachusetts Institute of Technology in view of the emergency, I participated in their work and sought to introduce my colleagues to the concepts of filtering which we had already developed. In addition, I saw several opportunities for the introduction of the study of random functions into engineering techniques. I went a certain distance in this direction, but with relatively small immediate success. In fact my time was already heavily taken up with the other problem which was critical at the moment — that of the mechanical control of anti-aircraft fire.

I had in view a method of prediction based on some purely abstract mathematical considerations. It turned out that we could simulate the experiments necessary to verify the validity of this mathematical prediction on Bush's differential analyzer. We did this, and we found out that the method would function, provided that the curve which we wished to predict was sufficiently smooth. However, when the curve lacked this smoothness it became clear that our prediction apparatus would be highly unstable, and that after any abrupt turn in the curve there would be a considerable period before the machine would settle into an equilibrium which would make exact prediction possible. Thus we found ourselves confronted by two difficulties which worked in the opposite direction. The very improvement of a predictor which would make it valuable for curves of a given smoothness was associated with an instability which would make the prediction very dependent on the smoothness of the curves to which we should apply it. It became apparent to me that this dual difficulty of prediction had a certain analogy to the principle of indeterminacy in quantity, and that it probably belonged to the very nature of prediction itself.

Since this difficulty could not be eliminated by any method, it became necessary to consider how the errors of prediction could best be reduced in practical problems. Here it became clear to me that the reduction of the errors of prediction in an optimal manner was dependent on the particular statistics of the curves which we wished to predict. I was thus thrown back to my old ideas of the statistical distribution of curves.

One of the prediction problems was conceived as a minimization problem which could be set up mathematically, and there was some hope of solving it. I found that the minimization problem in question led to an integral equation of a type closely akin to that which Eberhardt Hopf and I had previously discussed. I was able to solve this problem which led to a program of the design of anti-aircraft predictors in which Dr. Julian Bigelow and I participated on behalf of the government. While this program did not lead to any single predictor design which was operated in practice, the ideas came to be applied to many other projects, and have carried over into the modern work on controlled missiles. The outcome of our work was a paper published for government use during the war, and reprinted after the war without restrictions, known as Extrapolation and Interpolation and Smoothing of Stationary Time Series with Engineering Applications.

This practical interest in computing machines led me to consider the general philosophy of the problem. On the one hand it became clear that the mechanism of a computation which depended on two value marks for the different digits could be easily adapted for the use of a machine to perform calculations of the algebra of logic, rather than numerical algebra. Here the two digital possibilities would correspond to the two possibilities of truth and falsity. Next we began to see that there was a certain analogy between digital computing machines and the human brain, particularly because of the fact that impulses in the nervous system seemed to be of an all or none nature, or in other words to involve two digital possibilities.

It must be borne in mind that our main work was the design of a computing machine to be used in the control of anti-aircraft fire, and that we were not only concerned with the ways in which decisions were made but in the ways in which these could be realized in action. This led inevitably to speculations concerning the way in which the human being or the animal performs purposive action. This was a problem which arose in the consideration (a) of how the observer following the airplane is able to keep his sights on the plane, and (b) in the study of how we could simulate the action of such an observer under laboratory conditions. Here I received suggestions from two quarters. On the one hand Mr. Bigelow took an active interest in the problem. On the other, through my friend and colleague, Professor Manuel Sandoval Vallarta, I had come some years before in contact with Dr. Arturo Rosenblueth who was then working on neurophysiology with Cannon at the Harvard Medical School, and who is now at the Instituto Nacional de Cardilogia in Mexico. Dr. Rosenblueth ran for some years an evening dinner seminar on neurological problems at the Harvard Medical School and I had participated in this. It was therefore Dr. Rosenblueth to whom I turned with the physiological implications of my problem.

My idea was this : in control apparatus one of the ways to stabilize the action consists of feeding back a quantity, depending on the success of the action, into the control apparatus as a new governing piece of information. Since any over-shoot in this feedback is compensated by a corrective action in the opposite direction, such a feedback is known as negative. It had occurred to Bigelow and myself that such simple human actions as driving a car were governed by negative feedbacks. We do not move the steering wheel on a car according to a set pattern, but rather in such a way that if we find ourselves too far to the left we make a correction to the right, and vice versa. Therefore we were convinced that negative feedback plays a part in the human control mechanism, and in particular in the mechanism by which we follow an airplane with our sights.

This idea struck me as capable of verification or contradiction. It is well known that the feedback in a control apparatus must be limited if it is to have a stabilizing effect. Otherwise with an excessive feedback the apparatus goes into a spontaneous oscillation which becomes more and more intense, and ultimately either destroys the apparatus or at any rate brings it widely out of control. The question which I asked Dr. Rosenblueth was the following : are there any human pathological conditions on which the attempt to complete a voluntary action leads instead of its efficient performance to a wild oscillatory error ?

Dr. Rosenblueth answered me that such conditions were indeed well known, and that they constituted what is called purpose tremor or cerebellum tremor because they seem to be associated with malfunction of the cerebellum. A patient with cerebellum tremor when he reaches out his hand to pick up a glass of water will go into wild oscillations, and either will spill the glass or be incapable of grasping it. This fact confirmed my conjecture that purposive action may take place by feedback, and that cerebellum tremor is merely a case of the general process of breakdown of an overloaded feedback.

The period of the war was a very busy one for me. After the war I found that an abrupt change in my mode of work was necessary. I found that the pressure of military or quasi-military work was not for me, and it was borne in upon me that the moral hazard of working in a field primarily devoted to destruction, and in which I would be subjected to the vicissitudes of secrecy and of the lack of any share in determining the use to be made of my work, rendered further pursuit in this direction impossible. I decided to work further with Dr. Rosenblueth who had gone back to Mexico, and received support from the Rockefeller Foundation for several years' joint effort with him. I was particularly interested in the study of clonus, and in general of the harmonic analysis of rhythmic physiological processes.

It was at about this time that I put together my ideas and those of several persons with whom I was in contact in the form of my book on Cybernetics. The book was bespoken by the late Mr. Freymann of Hermann et Co., and also received the support of the Technology Press and of John Wiley and Sons Inc. It represented a definite statement of my thesis that communication and control theory belonged together, both in the machine and in the living organism, and that the basis for this theory was probabilistic. I had already seen my probabilistic ideas taken up with great definiteness by my colleague, Claude Shannon, then of the Bell Telephone Laboratories.

The thesis which I made in this book had implications for the sociology of the age of automatization. It had become clear to me that the human brain gave some sort of an index of what automatic machinery could do and was subjected to the same principles. I saw that the digital computing machine was primarily a logical rather than a numerical machine, and could be adapted to the control of factory processes. It was necessary for me to take a definite point of view with regard to the moral problems posed by this new industrial revolution which was clearly under way. It was in this connection that I wrote my book on « The Human Use of Human Beings ».

Let me note that the whole nexus of ideas which I then introduced had since passed from the stage of merely speculative possibilities into that of actuality. Furthermore, in the United States, in Russia, and elsewhere, the point of view which I then expressed, that the problem of automatization was essentially a statistical problem involving the use of random functions of the Brownian motion type, has been taken up on both sides of the Iron Curtain.

Of late years my interest has become more and more devoted to the study of rhythmic processes in living organisms as generated by the response which is generally non-linear of such organisms to random inputs. Early in the thirties improved electrical techniques had made possible the precise study of electric potentials of nerve and muscle, and in particular certain fluctuating potentials generated by the brain and observable by means of electrodes placed on the scalp. This was the work on electroencephalography.

The early work in this field involved a maximum of experience and judgment in the reading of these brain waves and a minimum of mathematical techniques. I saw from the beginning that it was an appropriate field for the application of my ideas on generalized harmonic analysis, and in particular the use of the correlogram which I later saw to be the essential equivalent of the employment of the Michelson interferometer in optics. My colleagues at the Massachusetts Institute of Technology greatly furthered this work by the introduction of an appropriate instrumentation, and I pursued the field in collaboration with Dr. Walter Rosenblueth of the Massachusetts Institute of Technology and Dr. Mary Brazier of the Massachusetts General Hospital. In the course of this work it became clear that the brain waves had a fine structure which must escape the coarse meshes of the existing methods of analysis. I found that at the center of the alpha rhythm of about 10 cycles a second which seemed to be associated with visual processes there was a very narrow and sharp band of activity with a definite frequency pattern. In this pattern a sharp central line was associated with the depression of activity in its neighbourhood and represented a physiological index which promised to be of great theoretical and medical importance. When I had found this pattern, it occurred to me that I had obtained a general result which belonged to the study of non-linear processes stimulated by a random Brownian input.

I pursued this work both on an experimental and theoretical phase, and in the summer of 1957 the late Professor Aurel Wintner of Johns Hopkins University wrote a paper on the subject. It became clear to me that in this response of non-linear systems of random inputs I had a clue as to how physiological processes could organize themselves into a definite synergic activity. I am at present engaged in pursuing my ideas along these lines.

The problem of organization which I had also developed in a paper which I gave at the University of Southampton in 1955 was a subject with profound sociological as well as biological significance and was connected with the theory of information by the most intimate bonds. I also continued my work in this field during my stay in India in 1955-56 at the Statistical Institute in Calcutta, and I saw how it could give a definite rationale to the concepts of social and economic planning which were much in vogue there.

My concept of economic planning is the following. In any economic situation there are certain factors beyond our control which are given statistically. These include the weather, the fertility of the crops, and other factors of the sort. In addition there are certain factors which we can control. For example, the amount of seed grain to be planted, the rate of interest on agricultural loans, etc. The problem of planning is to optimize, or in other words minimize, some quantity depending on the controllable and the uncontrollable statistical factors in such a way that this minimization is maintained on the average. Such a problem is of a statistical character, and therefore of an informational character.

From the economic point of view, this is what we may call the statistically stable planning problem. If we have any planning situation which is meant to continue, it must of necessity lead to a statistically stable planning situation. On the other hand, for a planning situation to be statistically stable, it is not necessarily true that we can arrive at it from existing conditions in a statistically stable manner. In other words, this concept of statistically stable planning is only part of the social planning problem and must be supplemented by other conditions which enable us to make an effective planning of transient states and which arrive at a statistically stable situation.

It will be seen that cybernetics is leading me to a whole group of problems concerning organization, and it is in these fields that I think that an important part of the future of cybernetics lies. I have already mentioned the problem of self-organizing systems. The concept of self-organization is well known in biology, where there is a great deal of talk of material organizers of substances which in the embryo will cause different organs to come into being as for example in the case where a piece of an optic cup of a newt embryo inserted under the skin of the regenerating tail will cause an eyelid to form itself, and even possibly the rudiments of an organ of hearing. The aspect of self-organizing systems which has interested me most is that of systems which organize themselves into a rhythm. For example, in the formation of the vascular system of a vertebral embryo certain contractile cells form which very soon constitute a heart with a regular beat. How do these cells pull themselves into a concerted action ?

The situation has occurred to me that these cells lend a double status as organs of information. On the one hand they give out electric impulses which can affect other similar cells. On the other hand, they receive such impulses and their action is modified by their reception. If the relations between these organs as senders and as receivers were linear, then they could not modify the frequency of oscillation to one another. If, however, there is a tendency for the frequencies of two vibrating members to interact either to pull one another together or possibly to push one another apart, there is a possibility of organization. Such a system as it gathers greater and greater synchronism will emit an impulse which has a greater and greater tendency to synchronize oscillators which have not already been pulled into place until by a mass action they constitute a definite pulsating organ. We have. an example of this in electrical engineering systems where many alternators are connected to the same bus-bars. In this case the generators which tend to run fast or ahead of phase will carry a greater load than normal, and those running slow or behind phase will carry a smaller load. The result is to speed up the slow members and to slow down the fast ones. Even if for the individual members the speeding up and the slowing down is controlled by special governors for the individual generators, the whole system will contain a virtual governor more active than any of its component governors It is interesting to notice that this virtual governor is distributed over the whole system and cannot be located in any particular part of the system. This suggests that in many problems of organization, as in the case of the brain, we may have given way to an excessive tendency to suppose a sharp localization to function.

In the case of brain waves we have ample evidence of the existence of something of the nature of local oscillators. We also have ample evidence that these oscillations can act on one another in frequency. We know that the brain can be driven by flicker which will tend to pull the rhythm of the brain into phase and frequency with itself. Under these conditions the sort of hypothesis which we have here made concerning self-organization systems is quite reasonable. It can also be pointed out that in the case of such self-organizing systems it will be quite common to find a sharp emerging frequency surrounded by regions of less activity than we find in the immediate neighbourhood. This phenomenon, as I have said, is actually verified in the case of brain waves.

It might be interesting to speculate on other rhythmic phenomenon like brain waves which may have a similar explanation of oscillating organs pulling one another into the same frequency and the same phase. It has been observed, and also contradicted, that fireflies in a tree tend to flash in unison. Here, too, we have to deal with periodic organisms which act both as senders and receivers of messages. The firefly tends to flash in a more or less periodic manner, and at the same time it is entirely reasonable to suppose that the visual reception to flashes from other fireflies will affect its rate of flashing. Under these conditions it is not much to hope that we have an example of a self organizing activity which readily lends itself to experimental and theoretical study as by suggesting that further work be done in this direction.

So much for self-organizing systems. There is another group of cybernetic problems which interests me very greatly and which concerns the measurement of causality. If we have two series of events in time, when we study each one separately there is a certain amount of information given concerning its future, but its past is determined. When the past of the two series is simultaneously determined, we shall receive more information concerning the future of each than if we were studying them separately. This additional information may be regarded as a measure of the effectiveness of one time series in causing another. Without going into details, here is the source of a really metrical theory of causality.

These are some of the directions of further work in cybernetics which interest me at present. To carry them out in practice I have been forced to develop my theory of random functions considerably further than I had in the past along lines related to the work of Professor Friedrichs. There are certainly other directions of work in cybernetics which have great interest, and I do not wish to pretend that in giving these directions which have interested me personally I have any intention to dictate unnecessarily the future of developments in this field.

25

Transmission of Information[1]

By R. V. L. HARTLEY

SYNOPSIS: A quantitative measure of "information" is developed which is based on physical as contrasted with psychological considerations. How the rate of transmission of this information over a system is limited by the distortion resulting from storage of energy is discussed from the transient viewpoint. The relation between the transient and steady state viewpoints is reviewed. It is shown that when the storage of energy is used to restrict the steady state transmission to a limited range of frequencies the amount of information that can be transmitted is proportional to the product of the width of the frequency-range by the time it is available. Several illustrations of the application of this principle to practical systems are included. In the case of picture transmission and television the spacial variation of intensity is analyzed by a steady state method analogous to that commonly used for variations with time.

WHILE the frequency relations involved in electrical communication are interesting in themselves, I should hardly be justified in discussing them on this occasion unless we could deduce from them something of fairly general practical application to the engineering of communication systems. What I hope to accomplish in this direction is to set up a quantitative measure whereby the capacities of various systems to transmit information may be compared. In doing this I shall discuss its application to systems of telegraphy, telephony, picture transmission and television over both wire and radio paths. It will, of course, be found that in very many cases it is not economically practical to make use of the full physical possibilities of a system. Such a criterion is, however, often useful for estimating the possible increase in performance which may be expected to result from improvements in apparatus or circuits, and also for detecting fallacies in the theory of operation of a proposed system.

Inasmuch as the results to be obtained are to represent the limits of what may be expected under rather idealized conditions, it will be permissible to simplify the discussion by neglecting certain factors which, while often important in practice, have the effect only of causing the performance to fall somewhat further short of the ideal. For example, external interference, which can never be entirely eliminated in practice, always reduces the effectiveness of the system. We may, however, arbitrarily assume it to be absent, and consider the limitations which still remain due to the transmission system itself.

In order to lay the groundwork for the more practical applications of these frequency relationships, it will first be necessary to discuss a few somewhat abstract considerations.

[1] Presented at the International Congress of Telegraphy and Telephony, Lake Como, Italy, September 1927.

The Measurement of Information

When we speak of the capacity of a system to transmit information we imply some sort of quantitative measure of information. As commonly used, information is a very elastic term, and it will first be necessary to set up for it a more specific meaning as applied to the present discussion. As a starting place for this let us consider what factors are involved in communication; whether conducted by wire, direct speech, writing, or any other method. In the first place, there must be a group of physical symbols, such as words, dots and dashes or the like, which by general agreement convey certain meanings to the parties communicating. In any given communication the sender mentally selects a particular symbol and by some bodily motion, as of his vocal mechanism, causes the attention of the receiver to be directed to that particular symbol. By successive selections a sequence of symbols is brought to the listener's attention. At each selection there are eliminated all of the other symbols which might have been chosen. As the selections proceed more and more possible symbol sequences are eliminated, and we say that the information becomes more precise. For example, in the sentence, "Apples are red," the first word eliminates other kinds of fruit and all other objects in general. The second directs attention to some property or condition of apples, and the third eliminates other possible colors. It does not, however, eliminate possibilities regarding the size of apples, and this further information may be conveyed by subsequent selections.

Inasmuch as the precision of the information depends upon what other symbol sequences might have been chosen it would seem reasonable to hope to find in the number of these sequences the desired quantitative measure of information. The number of symbols available at any one selection obviously varies widely with the type of symbols used, with the particular communicators and with the degree of previous understanding existing between them. For two persons who speak different languages the number of symbols available is negligible as compared with that for persons who speak the same language. It is desirable therefore to eliminate the psychological factors involved and to establish a measure of information in terms of purely physical quantities.

Elimination of Psychological Factors

To illustrate how this may be done consider a hand-operated submarine telegraph cable system in which an oscillographic recorder traces the received message on a photosensitive tape. Suppose the sending operator has at his disposal three positions of a sending key which correspond to applied voltages of the two polarities and to no applied voltage. In making a selection he decides to direct attention to one of the three voltage conditions or symbols by throwing the key to the position corresponding to that symbol. The disturbance transmitted over the cable is then the result of a series of conscious selections. However, a similar sequence of arbitrarily chosen symbols might have been sent by an automatic mechanism which controlled the position of the key in accordance with the results of a series of chance operations such as a ball rolling into one of three pockets.

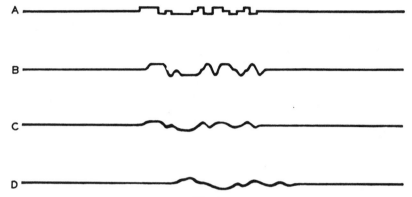

Fig. 1

Owing to the distortion of the cable the results of the various selections as exhibited to the receiver by the recorder trace are not as clearly distinguishable as they were in the positions of the sending key. Fig. 1 shows at A the sequence of key positions, and at B, C and D the traces made by the recorder when receiving over an artificial cable of progressively increasing length. For the shortest cable B the reconstruction of the original sequence is a simple matter. For the intermediate length C, however, more care is needed to distinguish just which key position a particular part of the record represents. In D the symbols have become hopelessly indistinguishable. The capacity of a system to transmit a particular sequence of symbols depends upon the possibility of distinguishing at the receiving end between the results of the various selections made at the sending end. The operation of recognizing from the received record the sequence of symbols selected at the sending end may be carried out by those of us who are not familiar with the Morse code. We would do this equally well for a sequence representing a consciously chosen message and for one sent out by the automatic selecting device already referred to. A trained operator, however, would say that the sequence sent out by the automatic device was not intelligible. The reason for this is that only a limited number of the possible sequences have been assigned meanings common to him and the sending operator. Thus the number of symbols available to the sending operator at certain of his selections is here limited by psychological rather than physical considerations. Other operators using other codes might make other selections. Hence in estimating the capacity of the physical system to transmit information we should ignore the question of interpretation, make each selection perfectly arbitrary, and base our result on the possibility of the receiver's distinguishing the result of selecting any one symbol from that of selecting any other. By this means the psychological factors and their variations are eliminated and it becomes possible to set up a definite quantitative measure of information based on physical considerations alone.

Quantitative Expression for Information

At each selection there are available three possible symbols. Two successive selections make possible 3^2, or 9, different permutations or symbol sequences. Similarly n selections make possible 3^n different sequences. Suppose that instead of this system, in which three current values are used, one is provided in which any arbitrary number s of different current values can be applied to the line and distinguished from each other at the receiving end. Then the number of symbols available at each selection is s and the number of distinguishable sequences is s^n.

Consider the case of a printing telegraph system of the Baudot type, in which the operator selects letters or other characters each of which when transmitted consists of a sequence of symbols (usually five in number). We may think of the various current values as primary symbols and the various sequences of these which represent characters as secondary symbols. The selection may then be made at the sending end among either primary or secondary symbols. Let the operator select a sequence of n_2 characters each made up of a sequence of n_1 primary selections. At each selection he will have available as many different secondary symbols as there are different sequences that can result from making n_1 selections from among the s primary symbols. If we call this number of secondary symbols s_2, then

$$s_2 = s^{n_1}. \tag{1}$$

For the Baudot System

$$s_2 = 2^5 = 32 \text{ characters.} \tag{2}$$

The number of possible sequences of secondary symbols that can result from n_2 secondary selections is

$$s_2{}^{n_2} = s^{n_1 n_2}.$$ (3)

Now $n_1 n_2$ is the number n of selections of primary symbols that would have been necessary to produce the same sequence had there been no mechanism for grouping the primary symbols into secondary symbols. Thus we see that the total number of possible sequences is s^n regardless of whether or not the primary symbols are grouped for purposes of interpretation.

This number s^n is then the number of possible sequences which we set out to find in the hope that it could be used as a measure of the information involved. Let us see how well it meets the requirements of such a measure.

For a particular system and mode of operation s may be assumed to be fixed and the number of selections n increases as the communication proceeds. Hence with this measure the amount of information transmitted would increase exponentially with the number of selections and the contribution of a single selection to the total information transmitted would progressively increase. Doubtless some such increase does often occur in communication as viewed from the psychological standpoint. For example, the single word "yes" or "no," when coming at the end of a protracted discussion, may have an extraordinarily great significance. However, such cases are the exception rather than the rule. The constant changing of the subject of discussion, and even of the individuals involved, has the effect in practice of confining the cumulative action of this exponential relation to comparatively short periods.

Moreover we are setting up a measure which is to be independent of psychological factors. When we consider a physical transmission system we find no such exponential increase in the facilities necessary for transmitting the results of successive selections. The various primary symbols involved are just as distinguishable at the receiving end for one primary selection as for another. A telegraph system finds one ten-word message no more difficult to transmit than the one which preceded it. A telephone system which transmits speech successfully now will continue to do so as long as the system remains unchanged. In order then for a measure of information to be of practical engineering value it should be of such a nature that the information is proportional to the number of selections. The number of possible sequences is therefore not suitable for use directly as a measure of information.

We may, however, use it as the basis for a derived measure which does meet the practical requirements. To do this we arbitrarily put the amount of information proportional to the number of selections and so choose the factor of proportionality as to make equal amounts of information correspond to equal numbers of possible sequences. For a particular system let the amount of information associated with n selections be

$$H = Kn, \tag{4}$$

where K is a constant which depends on the number s of symbols available at each selection. Take any two systems for which s has the values s_1 and s_2 and let the corresponding constants be K_1 and K_2. We then define these constants by the condition that whenever the numbers of selections n_1 and n_2 for the two systems are such that the number of possible sequences is the same for both systems, then the amount of information is also the same for both; that is to say, when

$$s_1^{n_1} = s_2^{n_2}, \tag{5}$$

$$H = K_1 n_1 = K_2 n_2, \tag{6}$$

from which

$$\frac{K_1}{\log s_1} = \frac{K_2}{\log s_2}. \tag{7}$$

This relation will hold for all values of s only if K is connected with s by the relation

$$K = K_0 \log s, \tag{8}$$

where K_0 is the same for all systems. Since K_0 is arbitrary, we may omit it if we make the logarithmic base arbitrary. The particular base selected fixes the size of the unit of information. Putting this value of K in (4),

$$H = n \log s \tag{9}$$

$$= \log s^n. \tag{10}$$

What we have done then is to take as our practical measure of information the logarithm of the number of possible symbol sequences.

The situation is similar to that involved in measuring the transmission loss due to the insertion of a piece of apparatus in a telephone system. The effect of the insertion is to alter in a certain ratio the power delivered to the receiver. This ratio might be taken as a measure of the loss. It is found more convenient, however, to take the logarithm of the power ratio as a measure of the transmission loss.

If we put n equal to unity, we see that the information associated with a single selection is the logarithm of the number of symbols available; for example, in the Baudot System referred to above, the number s of primary symbols or current values is 2 and the information content of one selection is log 2; that of a character which involves 5 selections is 5 log 2. The same result is obtained if we regard a character as a secondary symbol and take the logarithm of the number of these symbols, that is, log 2^5, or 5 log 2. The information associated with 100 characters will be 500 log 2. The numerical value of the information will depend upon the system of logarithms used. Increasing the number of current values from 2 to say 10, that is, in the ratio 5, would increase the information content of a given number of selections in the ratio $\dfrac{\log 10}{\log 2}$, or 3.3. Its effect on the rate of transmission will depend upon how the rate of making selections is affected. This will be discussed later.

When, as in the case just considered, the secondary symbols all involve the same number of primary selections, the relations are quite simple. When a telegraph system is used which employs a non-uniform code they are rather more complicated. A difficulty, more apparent than real, arises from the fact that a given number of secondary or character selections may necessitate widely different numbers of primary selections, depending on the particular characters chosen. This would seem to indicate that the values of information deduced from the primary and secondary symbols would be different. It may easily be shown, however, that this does not necessarily follow.

If the sender is at all times free to choose any secondary symbol, he may make all of his selections from among those containing the greatest number of primary symbols. The secondary symbols will then all be of equal length, and, just as for the uniform code, the number of primary symbols will be the product of the number of characters by the maximum number of primary selections per character. If the number of primary selections for a given number of characters is to be kept to some smaller value than this, some restriction must be placed on the freedom of selection of the secondary symbols. Such a restriction is imposed when, in computing the average number of dots per character for a non-uniform code, we take account of the average frequency of occurrence of the various characters in telegraph messages. If this allotted number of dots per character is not to be exceeded in sending a message, the operator must, on the average, refrain from selecting the longer characters more often than their average rate of occurrence. In the language

366

of the present discussion we would say that for certain of the n_2 secondary selections the value of s_2, the number of secondary symbols, is so reduced that a summation of the information content over all the characters gives a value equal to that derived from the total number of primary selections involved.   This may be written

$$\sum_{1}^{n_2} \log s_2 = n \log s, \qquad (11)$$

where n is the total number of primary symbols or dot lengths assigned to n_2 characters.   This suggests that the primary symbols furnish the most convenient basis for evaluating information.

The discussion so far has dealt largely with telegraphy.   When we attempt to extend this idea to other forms of communication certain generalizations need to be made.   In speech, for example, we might assume the primary selections to represent the choice of successive words.   On that basis s would represent the number of available words.   For the first word of a conversation this would correspond to the number of words in the language.   For subsequent selections the number would ordinarily be reduced because subsequent words would have to combine in intelligible fashion with those preceding. Such limitations, however, are limitations of interpretation only and the system would be just as capable of transmitting a communication in which all possible permutations of the words of the language were intelligible.   Moreover, a telephone system may be just as capable of transmitting speech in one language as in another.   Each word may be spoken in a variety of ways and sung in a still greater variety. This very large amount of information associated with the selection of a single spoken word suggests that the word may better be regarded as a secondary symbol, or sequence of primary symbols.   Let us see where this point of view leads us.

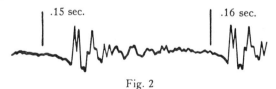

.15 sec.   .16 sec.

Fig. 2

The actual physical embodiment of the word consists of an acoustic or electrical disturbance which may be expressed as a magnitude-time function as in Fig. 2, which shows an oscillographic record of a speech sound.   Such functions are also typical of other modes of communication, as will be discussed in more detail later.   We have then to examine the ability of such a continuous function to convey information. Obviously over any given time interval the magnitude may vary in accordance with an infinite number of such functions. This would mean an infinite number of possible secondary symbols, and hence an infinite amount of information. In practice, however, the information contained is finite for the reason that the sender is unable to control the form of the function with complete accuracy, and any distortion of its form tends to cause it to be confused with some other function.

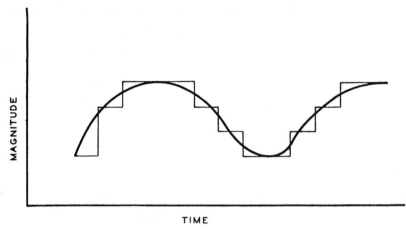

Fig. 3

A continuous curve may be thought of as the limit approached by a curve made up of successive steps, as shown in Fig. 3, when the interval between the steps is made infinitesimal. An imperfectly defined curve may then be thought of as one in which the interval between the steps is finite. The steps then represent primary selections. The number of selections in a finite time is finite. Also the change made at each step is to be thought of as limited to one of a finite number of values. This means that the number of available symbols is kept finite. If this were not the case, the curve would be defined with complete exactness at each of the steps, which would mean that an observation made at any one step would offer the possibility of distinguishing among an infinite number of possible values. The following illustration may serve to bring out the relation between the discrete selections and the corresponding continuous curve. We may think of a bicycle equipped with a peculiar type of steering device which permits the rider to set the front wheel in only a limited number of fixed positions. On such a machine he attempts to ride in such a manner that the front wheel shall follow an irregularly curved line.   The accuracy with which he is able to accomplish this will depend upon how far he goes between adjustments of the steering mechanism and upon the number of positions in which he is able to set it.

By this more or less artificial device the continuous magnitude-time function as used in telephony is made subject to the same type of treatment as the succession of discrete selections involved in telegraphy.

RATE OF COMMUNICATION

So far then we have derived an expression for the information content of the symbols at the sending end and have shown that we may evaluate a transmission system in terms of how well the wave as received over it permits distinguishing between the various possible symbols which are available for each selection.   Let us consider next how the distortion of the system limits the rate of selection for which these distinctions between symbols may be made with certainty.

Limitation by Intersymbol Interference

We shall assume the system to be free from external interference and to be such that its current-voltage relations are linear.   In such a system the form of the transmitted wave may be altered due to the storage of energy in reactive elements such as inductances and capacities, and its subsequent release.   To evaluate the effect of such distortion in making it impossible to determine correctly which one of the available symbols had been selected, we may think of this distortion in terms of "intersymbol interference."   In order to determine the result of any one selection an observation is made at such time that the disturbance resulting from that selection has its maximum effect at the receiving end.   Superposed on this effect there will be a disturbance which is the resultant of the effects of all the other symbols as prolonged by the storage of energy in the system. This resultant superposed disturbance is what is meant by intersymbol interference.   Obviously if this disturbance is greater than half the difference between the effects produced by two of the values available for selection at the sending end, the wave resulting from one of those values will be taken as representing the other.   Thus a criterion for successful transmission is that in no case shall the intersymbol interference exceed half the difference between the values of the wave at the receiving end which correspond to the selection of different values at the sending end.

Obviously the magnitude of the intersymbol interference which affects any one symbol depends on the particular sequence of symbols which precedes it. However, it is always possible for the sending operator so to make his selections that any one selection is preceded by that sequence which causes the maximum possible interference. Hence every selection must be separated from those preceding it by at least a certain interval which is determined by the worst condition of interference. If longer intervals than this are used, the transmission is unnecessarily retarded. Hence to secure the maximum rate of transmission the selection should be made at a constant rate. It might appear at first sight that the selections could be made at shorter intervals near the beginning of the message where there are fewer preceding symbols to cause interference. This assumes, however, that the system has previously been idle. Actually the previous user may have finished his message with that sequence which causes maximum intersymbol interference.

Relation to Damping Constant

How the intersymbol interference limits the rate of communication over the system depends upon the properties of the particular system. The relations involved are very complex, and no attempt will be made to obtain a complete or rigorous solution of the problem. We may, however, by treating a very simple case, arrive at an interesting relation. Consider a resistance in series with a capacity. Let one terminal be connected to one terminal of a battery made up of a very large number of cells of negligible internal resistance. Let the other terminal be connected to the battery through a switch. This switch is so arranged that by pressing any one of s keys the circuit terminal may be moved up along the battery by any number of cells from zero to $s - 1$. Let the sending operator make selections among the s keys at regular intervals, and let the receiving operator observe the current through the resistance. The most advantageous time for this observation is at the instant [1] at which the key is pressed, since the current has then its maximum value. The finest distinction to be made by the receiving operator is that between two currents which result from battery changes that differ from each other by one cell. The difference between two such currents is equal to the initial current which flows when one cell is introduced into the circuit. This is

$$i_s = \frac{E}{R},\tag{12}$$

where E is the electromotive force of one cell and R the resistance of the circuit.

---

[1] Results identical with those which follow may be obtained if he observes the average current over a period beginning when the key is pressed and lasting not longer than the interval between selections.

The intersymbol interference will consist of currents resulting from all of the preceding symbols. The contribution of any one symbol will depend on its size, that is, on the number of added cells it represents, and on how long it preceded the symbol in question. For a given rate of selection the resultant of these contributions will be a maximum for that particular sequence of symbols for which at every selection preceding the one in question the operator had selected the largest

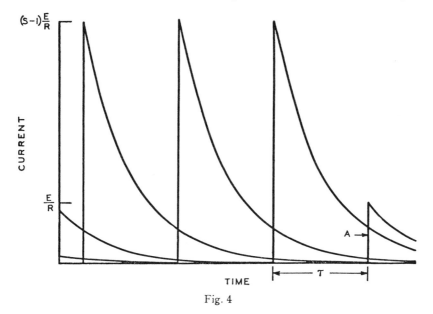

Fig. 4

possible symbol, that is, a voltage change of $(s - 1)E$. The form of the received current is then as shown on Fig. 4, where A represents the disturbed symbol. The curves are drawn for s equal to five. The current resulting from one such change occurring at time zero is

$$i = (s - 1)\frac{E}{R}e^{-\alpha t}, \tag{13}$$

where the damping constant,

$$\alpha = \frac{1}{RC}. \tag{14}$$

If the interval between selections is τ, then the time during which any one interfering current is damped out before it makes its contribution to the interference with the disturbed symbol is $q\tau$ where q is the number of selections by which it precedes the disturbed symbol. The magnitude of its contribution is therefore from (13)

$$i_q = (s - 1)\frac{E}{R}e^{-q\alpha\tau}. \tag{15}$$

If we sum this expression for all values of q from one to infinity, we get the combined effect of all the preceding symbols, that is, the intersymbol interference. Calling this i_i,

$$i_i = (s - 1)\frac{E}{R}\sum_{q=1}^{q=\infty}e^{-q\alpha\tau} \tag{16}$$

$$= (s - 1)\frac{E}{R}\frac{1}{e^{\alpha\tau} - 1}. \tag{17}$$

This obviously increases as the interval τ between selections is decreased. If this interval is made small enough, the intersymbol interference may cause confusion between symbols. Since the interference is here always of one sign it can cause confusion only when it becomes as large as the minimum difference, i_s, between symbols. Placing these two quantities equal, we get from (12) and (17) as the minimum permissible value of τ,

$$\tau_1 = \frac{\log s}{\alpha}. \tag{18}$$

The maximum number, n, of selections that may be made in t seconds is given by

$$n = \frac{t}{\tau_1}. \tag{19}$$

From (18) and (19)

$$\frac{n\log s}{t} = \alpha. \tag{20}$$

Here the numerator is, in accordance with our measure of information, the amount of information contained in n selections, so the left-hand member is the information per unit time or the rate of communication. This is equal to the damping constant of the circuit. We therefore conclude that for this particular case the possible rate of communication is fixed solely by the damping constant of the circuit and is independent of the number of symbols available at each selection. It is, of course, true that the larger this number the more susceptible will the system be to the effects of external interference.

Probably the practical system which most nearly approaches this idealized one is the non-loaded submarine telegraph cable when operated at such low speeds that its inductance may be neglected. It is of considerable historical interest to note that Lord Kelvin's study of such cables led him to the conclusion that the extent to which the cable limited the dotting speed was given by KR, that is, the product of the total capacity and total resistance. Had he stated his results in terms of permissible speed he would have had the reciprocal of this quantity, which corresponds very closely to the damping constant which we arrived at as a measure of the rate of communication. It should be noted, however, that his consideration was limited to a fixed number of symbols, and did not involve the relation here developed between this number and the dotting speed.

The more complicated systems are similar to the simple case just treated in that the contribution of any one symbol, *a*, to the interference with any other symbol, *b*, is determined by the free vibration of the system which results from the change applied to it in the production of symbol *a*. This free vibration, instead of being expressible by a single exponential function as in the case just considered, may be the resultant of a large number of more or less damped oscillatory components corresponding to the various natural modes of the system. The total interference with any one symbol is the resultant of a series of these complex vibrations, one for each interfering symbol. The instantaneous values of the various components of the interference are so dependent upon their phases at the particular instant of observation that it is difficult to draw any general conclusions as to the magnitude of the total interference. It is equally difficult therefore to draw any general conclusions as to the relation between the rate of transmission over a particular circuit and the number of available symbols.

Relation to Storage of Energy

Even though for any one system there exists a number of available symbols for which the rate of communication is greater than for any other number, it is still possible to make a generalization with respect to the storage of energy in the system and its effect on the rate of transmission which is of considerable practical importance.

Each of the natural modes of vibration of a linear system has the general form

$$i = A e^{-\alpha t} \cos (\omega t - \theta), \qquad (21)$$

where the natural frequency ω and damping constant α are characteristic of the system and the amplitude A and phase θ depend on the conditions of excitation. Wherever the time appears in this expression it is multiplied by either the damping constant α or the frequency ω. Consequently if both α and ω be changed, say in the ratio k, the instantaneous value of this mode of vibration in the new system will be the same at a time t/k that it was at time t in the original system. If the same change is made in the damping constant and frequency of each one of the modes of free vibration, their resultant, or the wave set up by any one symbol, will also be so changed that any particular value occurs at time t/k instead of t. Suppose also that the interval τ between selections be changed to τ/k. Then any two symbols originally separated by a time t_1 will be separated by t_1/k. The value of the interfering wave at the time t_1/k when the disturbed symbol occurs will be the same as it was at the corresponding time t_1 when it occurred in the original system. Hence the contribution of this wave to the intersymbol interference is unchanged. Since this relation holds for all of the interfering symbols, the total intersymbol interference remains unchanged, and so the number of possible symbols that may be distinguished is unaltered. The rate of making selections is changed in the ratio k, and hence the maximum rate of communication is changed in the same ratio as the damping constants and natural frequencies.

Now let us consider what physical changes must be made in the system to bring about the assumed changes in the damping constants and natural frequencies of the various modes. Take the simple case of an inductance, capacity and resistance connected in series. Here we have the well-known relations

$$\alpha = \frac{R}{2L}, \tag{22}$$

$$\omega = \sqrt{\frac{1}{LC} - \left(\frac{R}{2L}\right)^2}. \tag{23}$$

If R remains fixed and L is changed to L/k, α becomes $k\alpha$. If, in addition, C is changed to C/k, ω becomes $k\omega$. What we have done is to leave the energy-dissipating element, R, unchanged and change both the energy-storing elements, L and C, in the inverse ratio in which the rate of communication is changed. For more complicated systems the expressions for α and ω are correspondingly complicated. In every case, however, it will be found that if all of the dissipating and storing elements are treated as in the simple case just considered all of the damping constants and natural frequencies will be similarly altered. Where mechanical systems are concerned we are to substitute for electrical resistances their mechanical equivalents, and for inductances and capacities, inertias and compliances. This generalization that a proportionate change in all of the energy-storing elements of the system with no accompanying change in the dissipating elements produces an inverse change in the possible rate of transmission will be used later.

Steady State and Transient Viewpoints

So far very little has been said about frequencies, and in fact nothing in the sense of the term in which our results are to be stated. By this I mean the use of the word "frequency" as applied to an alternating current or other sinusoidal disturbance in the so-called "steady state." The steady state viewpoint has proven very useful in certain branches of communication, notably telephony. During the past few years much progress has been made in establishing relations between steady state phenomena and what might be called transient phenomena of the sort which we have just been discussing. Before proceeding with the main argument I shall attempt to review in non-mathematical language the relationship of these two points of view to each other.

As its name implies, steady state analysis deals with continuing conditions. If a sustained sinusoidal electromotive force be applied at the sending end of a system, a sinusoidal current of the same frequency flows at the receiving end. The vector ratio of the received current to the sending electromotive force is known as the transfer admittance of the system at that frequency. It is assumed ideally that the driving electromotive force has been acting from the beginning of time, and practically that it has been acting so long that the results are indistinguishable from what would be obtained in the ideal case. The absolute magnitude of the transfer admittance gives the amplitude of the received current which results from a driving electromotive force of unit amplitude, and its phase angle gives the phase of the current relative to that of the driving electromotive force. The curves which represent this amplitude and phase as functions of the frequency constitute a steady state description of the transmission properties of the system. For a system which is free from energy storage such as a circuit containing resistances only, the transfer admittance is the same for all frequencies. The amplitude-frequency curve is a horizontal line whose position depends on the magnitude and arrangement of the resistances while the phase-frequency curve coincides with the frequency axis. The storage of energy in the system and its subsequent release cause the admittance-frequency curves to take other forms. If the only storage is that which occurs in a dissipationless medium, a condition which is approximated when a sound wave traverses the open air, the only effect is to make the phase-frequency curve a straight line passing through the origin and having a slope proportional to the time of transmission through the medium. Other forms of energy storage give to the admittance-frequency curves shapes which are characteristic of the particular system. This alteration of these curves is commonly spoken of as frequency distortion. Their form may in most cases be deduced fairly readily from the values of the energy-storing and energy-dissipating elements of the system. This fact makes such a description of the system particularly useful for design purposes.

This physical description is, of course, useful as a criterion of performance only in so far as it can be related to the satisfactoriness with which the system performs its primary function of transmitting information. In the case of telephony it has been found practical to establish such a correlation by purely empirical means. Until fairly recently adequate results have been obtained by considering the amplitude-frequency function only. With the use of lines of increasing length and with increasingly severe standards of performance it is coming to be necessary to take account of the phase-frequency function as well.

In attempting to extend this method of treatment to telegraphy it was not found desirable to establish the correlation between steady state properties and overall performance by purely empirical methods. One reason for this was that a considerable fund of information has been accumulated with reference to the correlation between the overall performance and the transient properties of the system. A correlation therefore between steady state and transient properties would offer a means of bringing this empirical information to bear on the design of apparatus and systems on a steady state basis. For bridging this gap between steady state and transient phenomena there was already available one arch in the form of the Fourier Integral. This integral may be thought of as a mathematical fiction for expressing a transient phenomenon in terms of steady state phenomena. It permits the determination, for any magnitude-time function, of the relative amplitudes and the phases of an infinite succession of sustained sinusoids whose resultant is at any instant equal to the magnitude of the function at that instant. The amplitudes of the sinusoids are infinitesimal and the frequencies of successive components differ from each other by infinitesimal increments. The relative amplitudes and the phases of these components expressed as functions of the frequency constitute a steady state description of the magnitude-time function.

Suppose then we have given the magnitude-time function representing an impressed transient driving force and wish to obtain the magnitude-time function of the received current. We deduce the steady state description of the driving force, modify the amplitudes and phases of its various components in accordance with the known admittance-frequency functions of the system, and obtain the amplitude- and phase-frequency curves which represent the steady state description of the received current. From these we deduce the magnitude-time function representing the received wave.

A slightly different point of view, however, leads to results which fit in better with our method of measuring information. We may apply the method just outlined to deduce the magnitude-time function which results when the applied wave consists merely of an instantaneous change in a steadily applied electromotive force from one value which may be zero to another value differing from it by one unit. The resulting wave form is characteristic of the system and has been called by J. R. Carson its indicial admittance. We may think of this as a transient description of the system. If we regard a continuously varying applied wave as being formed by a succession of steps, we may think of the received wave at any instant as being the resultant of a series of waves each corresponding to a single step. The wave form of each is that of the indicial admittance, its magnitude is proportional to the size of the particular step and its location on the time axis is determined by the time at which the particular step or selection was made. When the steps are made infinitely close together, a summation of these components becomes a process of integration whereby the resulting magnitude-time function may be accurately determined from the applied function. For the incompletely determined waves involved in communication where the separation of the steps is finite a corresponding summation of the indicial admittance curves resulting from all selections other than the one being observed gives a measure of the intersymbol interference.

Still another viewpoint, while it has, perhaps, less direct application to the present problem, is of interest in that it brings out the significance from the transient standpoint of the steady state characteristics of the system. If we take as the applied wave a mathematical impulse, that is to say, a disturbance which lasts for an infinitesimal time, we find that the amplitudes of its steady state components are the same for all frequencies. If the impulse occurs at zero time the phase-frequency curve coincides with the axis of frequency, and if not it is a straight line through the origin whose slope is proportional to the time of occurrence. In order to find the current resulting from such an impulse applied at zero time we multiply the constant amplitude-frequency curve of its steady state components by the amplitude-frequency curve of the system and obtain as the amplitude-frequency curve of the received wave a function of the same form as the amplitude-frequency curve of the system. Similarly we add to the phase-frequency curve of the impressed wave, which is zero at all frequencies, the phase-frequency curve of the system and obtain for the received wave a phase-frequency curve identical with that of the system. The corresponding magnitude-time function gives the instantaneous value of the received current resulting from the impressed impulse. Thus we see that the steady state transfer admittance of a system is identical with the steady state description of the wave which is received over the system when it is subjected to an impulsive driving force. Once the form of this received wave is known the received wave resulting from any applied wave may be deduced by assuming the applied wave to consist of an infinite succession of impulses infinitesimally close together whose magnitudes vary with time in accordance with the given magnitude-time function. Methods for integrating the effect of this infinite succession of responses to impulses so as to obtain the transmitted wave have been developed.

From this review it is evident that the so-called frequency distortion and transient distortion are merely two methods of describing the same changes in wave form which result from the storage of energy in parts of the transmission system.

Significance of Product of Frequency-Range by Time

Distortion of this sort with its accompanying intersymbol interference may be unavoidable in the design of the system, or it may be deliberately introduced. The use of electrical filters to obtain multi-

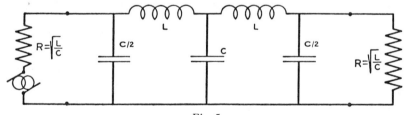

Fig. 5

plex operation, as in carrier systems, is an example of its deliberate use. Consider the effect of introducing a low pass filter, as shown in Fig. 5, into an otherwise distortionless transmission system. If the impedances of the circuits to which the filter is connected are approximately pure resistances of the values indicated in the figure, steady state frequencies above a critical value known as the cut-off frequency are so reduced as to be made practically negligible, while frequencies below this value are transmitted with very little distortion. The transient distortion corresponding to this steady state distortion must result in intersymbol interference; hence it places a limit on the rate at which distinguishable symbols may be selected, that is, on the rate of transmitting information.

It does not necessarily follow, however, that the rate of transmission with such a system is the maximum attainable for systems whose transmission is limited to the frequency-range determined by the cut-off of the filter. It is conceivable that by the introduction of additional energy-storing elements the transfer admittance curves for frequencies within the transmitted range may be altered in such a way as to reduce the total intersymbol interference and so permit an increased rate of selection. The maximum rate of transmission of information which can be secured by such methods represents the maximum rate corresponding to that range of frequencies.

Let us consider next the way in which this possible rate of transmission varies with the cut-off frequency of the filter. The theory of filter design teaches us that the cut-off frequency may be changed without altering the required terminating resistances if we change all inductances and capacities in the inverse ratio of the desired change in cut-off frequency. Suppose this change in energy-storing elements, with no change in dissipative elements, is made not only for the filter but for the entire system. We have already seen that such a modification changes the rate of transmission in the inverse ratio of the change in energy-storing elements; that is, in the direct ratio of the change in cut-off frequency in the present case. That the new rate is the maximum for the new frequency-range is evident when we consider that the transfer admittance curves of the new system bear the same relation to its cut-off frequency as held in the original system.

This brings us to the important conclusion that the maximum rate at which information may be transmitted over a system whose transmission is limited to frequencies lying in a restricted range is proportional to the extent of this frequency-range. From this it follows that *the total amount of information which may be transmitted over such a system is proportional to the product of the frequency-range which it transmits by the time during which it is available for the transmission.* This product of transmitted frequency-range by time available is the quantitative criterion for comparing transmission systems to which I referred at the beginning of this discussion. The significance of this criterion can perhaps best be brought out by applying it to some typical situations.

Fitting the Messages to the Lines

To facilitate this discussion it seems desirable to introduce and explain a few terms. For transmitting a sequence of symbols various sorts of media may be available, such as a wire line, an air path, as in direct speech, or the ether, as in radio communication. For convenience we shall group all of these under the general name of "line." Each such medium is generally characterized by a range of frequencies over which transmission may be carried on with reasonable freedom from distortion and external interference. This will be called the "line-frequency-range." Similarly the symbol sequences corresponding to the various modes of communication such as telegraph and telephone, will be designated as "messages." Each of these will, in general, be characterized by a "message frequency-range." This may be thought of as being determined by the frequency-range of that line which will just transmit the type of message satisfactorily, or we may think of it as that part of the frequency scale within which it is necessary to preserve the steady state components of the message wave in order to permit distinguishing the various symbols as they appear in the transmitted wave.

When we set up practical communication systems it is often found that the message-frequency-range and the line-frequency-range do not coincide either in magnitude or in position on the frequency scale. If then we are to make use of the full transmission capacity of the line, or lines, we must introduce means for altering the frequency-ranges required by the messages. Two such means are available, which together offer the theoretical possibility of accomplishing the desired end of making the message-frequency-ranges fit the available line-frequency-ranges.

The process of modulation so widely used in radio systems and in carrier transmission over wires makes it possible to shift the frequency-range of any message to a new location on the frequency scale without altering the width of the range. This follows at once from the well-known fact that the steady state description of the wave which results from the modulation of a carrier wave by a symbol wave includes a pair of side-bands in each of which there is a component corresponding to each steady state component of the original wave. The frequency of each component of the side-band differs from the carrier frequency by the frequency of the corresponding component of the symbol wave. The elimination of one of these side-bands results in a wave which retains the information embodied in the original symbol wave and occupies a frequency-range of the same width as the original but displaced to a new position on the frequency scale determined by the carrier frequency. The interval which must be allowed between these displaced messages in carrier operation is determined by the selectivity of the filters which are available for their separation. The imperfection of practical filters tends to make the message-frequency-range which may be transmitted less than the line-frequency-range which the messages occupy. The time for which the line is used to transmit a given amount of information is the same as the duration of the message conveying it. Thus the sum of the products of frequency-range by time for the messages is always equal to or less than the corresponding sum of the products of line-frequency-range by time.

In case the line-range available is less than the message-range, as would be the case in attempting to transmit speech over a submarine telegraph cable, it is still possible, if enough lines are available, to accomplish the transmission. The message wave may, by suitable filters, be separated into a plurality of waves each made up of those components of the original which lie in a portion of the message-range which is no wider than the line-range. Each of these portions of the message may then, by modulation, be transferred down to the frequency-range of the line and each transmitted over a separate line. A reversal of the process at the receiving end restores the original message.

While it is theoretically possible, if enough messages and lines are available, to fit the message-ranges to the line-ranges by modulation and subdivision of message-frequency-ranges, it is not always practical. It is sometimes more desirable to utilize the second method of transformation already referred to. This consists in making a record of the symbol sequence and reproducing it at a different speed in order to secure the wave used in transmission. The tape used in sending telegraph messages may be used in this manner. Here the symbol sequence represents a series of selections of secondary symbols. These selections are made at a rate at which it is convenient for the operator to manipulate the keys of the tape-punching machine. The electric wave impressed on the line by the holes in the tape represents a corresponding sequence of primary symbols. The rate at which these are applied to the line is determined by the velocity of the tape in reproduction. Since for a given number of different primary symbols the frequency-range required is proportional to the rate of making selections, it is obvious that the frequency-range of the message as reproduced from the tape may be made to fit whatever line-frequency-range is available, at least so far as width of the range is concerned.

Modulation may, of course, be necessary to bring the message-range to the proper part of the frequency scale. The time required for the reproduction of a message involving a given number of selections varies inversely as the velocity of the tape in reproduction, and therefore also inversely as the frequency-range required by the reproduced sequence. Thus the product of frequency-range by time for the reproduced message, which is also the required product for the line, is independent of the rate of reproduction, and depends only on the information content of the message in its original form.

In case the available line range calls for reproduction at a considerably increased speed a single operator cannot conveniently keep the sending apparatus supplied with tape. Multiplex operation may then be employed in which the line is used by the various operators in rotation. It is interesting to note that this distributor type of multiplex utilizes the frequency-range of the line as efficiently as would a single printing telegraph channel using the same dotting speed, and more efficiently than does the carrier multiplex method. By the distributor method each operator utilizes the full frequency-range of the line during the time allotted to him and there is no time wasted in separating the channels from each other. In the carrier multiplex, on the other hand, while each operator uses the line for the full time it is available, a part of the frequency-range is wasted in separating the channels because of the departure of physical filters from the ideal. Also both side-bands are generally transmitted in telegraphy, in which case a still greater line-frequency-range is required for the carrier method.

If the message is produced originally as a continuous time function, as in speech, the same method may be used by substituting for the tape a phonographic record. That here also the required line-frequency-range varies directly as the speed of reproduction and inversely as the time of reproduction is obvious when we consider an imperfectly defined wave as equivalent to a succession of finite steps or a perfectly defined wave as a succession of infinitesimal steps. From the steady state viewpoint, all of the component frequencies are altered in the ratio of the reproducing and recording velocities, and hence the range which they occupy is altered in the same ratio.

Thus we see that for all forms of communication which are carried on by means of magnitude-time functions an upper limit to the amount of information which may be transmitted is set by the sum for the various available lines of the product of the line-frequency-range of each by the time during which it is available for use.

Application to Picture Transmission

However, if in order to utilize fully the line-frequency-range we introduce the process of recording, our message no longer exists throughout its transmission as a magnitude-*time* function, but becomes a magnitude-*space* function. Also in the case of picture transmission the information to be transmitted exists originally as a magnitude-space function. We may, of course, regard either a phonograph record or a picture as a secondary symbol, and say that the information transmitted consists of the sender's selection of a particular record

Fig. 6

or picture to which he desires to call the attention of the receiver. The information involved in such a selection is then measured by the logarithm of the number of different records or pictures which he might have selected. The problem then is to analyze the magnitude-space function which constitutes the secondary symbol into a sequence of primary symbols. This may be done in a manner similar to that already employed for magnitude-time functions.

The case of a phonograph record is directly analogous to those already considered in that the magnitude is a function of the distance along a single line. This distance is therefore analogous to time and the information content may be found exactly as it would be from the pressure-time curve of the air vibration.   In a picture, on the other hand, two dimensions are involved.   We may, however, reduce this to a single dimension by dividing the area into a succession of strips of uniform width, as is done by the scanning aperture which is used in the electrical transmission of pictures.   Figure 6 shows this scanning mechanism.   The picture is mounted on a revolving cylinder which at each revolution is advanced by a spiral screw by the width of the desired strip.   This scanning operation is equivalent to making an arbitrary number of selections in a direction at right angles to the strips.   The number of these determines the degree of resolution in that direction.   If the resolution is to be the same in both directions, we may consider the magnitude-distance function along the strip to be made up of the same number of selections per unit length.   The total number of primary selections will then be equal to the number of elementary squares into which the picture is thus divided.   These elementary areas differ from each other in their average intensity. The number of different intensities which may be correctly distinguished from each other in each elementary area of the reproduced picture represents the number of primary symbols available at each selection.   Hence the total information content of the picture is given by the number of elementary areas times the logarithm of the number of distinguishable intensities.

In an actual picture the intensity as a function of distance along what we may call the line of scanning is a definite continuous function of the distance, but if there is any blurring of the picture as reproduced this function loses some of its definiteness.   This blurring may be thought of as a form of intersymbol interference, since the intensity at one point in the distorted picture depends upon the original intensity at neighboring points.   The similarity of this type of distortion to the intersymbol interference occurring in magnitude-time functions as a result of energy storage suggests that the picture distortion may also be treated on a steady state basis.   We may think of the magnitude-distance function representing the picture as being analyzed into sustained components in each of which the intensity is a sinusoidal function of the distance.   We may visualize such a single component in terms of the mechanism employed for recording and reproducing speech by means of a motion picture film.   The intensity of the light transmitted by the developed film varies along its length in accordance with the magnitude of the electric wave resulting from the speech sound.   If the speech wave be replaced by a sustained alternating current, there will result on the film a sinusoidal variation

384

in intensity with distance.   The distance between successive maxima, or the wave-length, will vary inversely with the frequency of the applied alternating current.   Figure 7 shows such a record of a speech wave and of sinusoidal waves of two different frequencies.   The variations are superposed on a uniform component so as to avoid the difficulty of negative light.

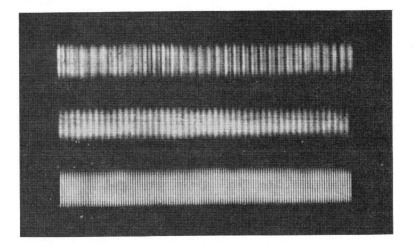

Fig. 7

The frequency of an alternating current is defined as the number of complete cycles which it executes in unit time.   The analog of frequency in the corresponding alternating space wave is therefore the number of complete cycles or waves executed in unit distance. This is the reciprocal of the wave-length just as the frequency is the reciprocal of the period.   Inasmuch as the term wave-number has been used by physicists to designate the reciprocal of wave-length, I shall use that term to designate the quantity corresponding to frequency in the steady state analysis of a magnitude-distance function. The distortion suffered by a picture in transmission may therefore be expressed in terms of the steady state amplitude and phase distortions as functions of wave number.   Just as the transmission of a given amount of information requires a given product of frequency-range by time, so the preservation of a given amount of information in a picture requires a corresponding product of wave-number-range by distance.   To illustrate, consider the effect of enlarging a picture without changing its detail or fineness of intensity discrimination. Suppose the enlargement to be made in two steps.   In the first the horizontal dimension is increased and the vertical dimension left unchanged. Let the scanning strips run in a horizontal direction. If we consider the magnitude-distance function representing the variation along any horizontal strip, the effect of the enlargement is to increase the wave-length of each steady state component in the ratio of the increase in linear dimension. The wave number of each component is therefore decreased in this ratio, and so the wave-number-range is also decreased in the same ratio. The product of the wave-number-range by the length of the strip remains constant, as does also the sum of the products for all of the strips, that is, for the entire picture. The second step consists in increasing the vertical dimensions with the horizontal dimensions fixed. By considering the scanning strips as running vertically in this case it follows at once that the product of wave-number-range by distance remains constant during this operation also.

Since the information transmitted is measured by the product of frequency-range by time when it is in electrical form and by the product of wave-number-range by distance when it is in graphic form, we should expect that when a record such as a picture or phonographic record is converted into an electric current, or vice versa, the corresponding products for the two should be equal regardless of the velocity of reproduction. That this is true may be easily shown. Let v be the velocity with which the recorder or reproducer is moved relative to the record. Let the wave-number-range of the record extend between the limits w_1 and w_2. If we consider any one component of the distance function which has a wave-length λ, the time required for the reproducer to traverse a complete cycle is λ/v, or $1/vw$. This is the period of the resulting component of the time wave, so the frequency f of the latter is the reciprocal of this, or vw. The frequency-range is therefore given by

$$f_2 - f_1 = v(w_2 - w_1). \tag{24}$$

If D is the length of the record, then the time required to reproduce it is

$$T = \frac{D}{v}, \tag{25}$$

from which

$$(f_2 - f_1)T = (w_2 - w_1)D. \tag{26}$$

This shows that the two products are numerically equal regardless of the velocity.

Application to Television

As our first illustration was drawn from one of the earliest forms of electrical communication, the submarine cable, it may be fitting to use as the last what is probably the newest form, namely, television. Here the information to be transmitted exists originally in the form of a magnitude which is a continuous function of both space and time. In order to determine what line facilities are needed to maintain a constant view of the distant scene we wish to determine the line-frequency-range required. This we know to be measured by the total information to be transmitted per unit time.

In the systems of television which have been most successful the method has been similar to that of the motion picture in that a succession of separate representations of the scene is placed before the observer and the persistence of vision is relied upon to convert the intermittent illumination into an apparently continuous variation with time. The first step in determining the required frequency-range is to determine the information content of a single one of the successive views of the scene. This may be determined exactly as for a still picture. The required degree of resolution into elementary areas and the required accuracy of reproduction of the intensity within each area determine an effective number of selections and a number of primary symbols available at each selection. These determine a minimum product of wave-number-range by distance. This in turn is equal to the product of line-frequency-range by time which must be available for the transmission of a single view of the scene. The time available is set by the fact that flicker becomes objectionable if the interval between successive pictures exceeds about one sixteenth of a second. Thus we have only to divide the product of wave-number-range by distance for a single picture by one sixteenth to obtain the line-frequency-range necessary to maintain a continuous view.

In the result just obtained an important factor is the interval necessary to prevent flicker. The tendency to flicker is, however, the result of the particular method of transmission. If it were practical to eliminate this factor, the required frequency-range might be somewhat different. We might, for example, imagine a system more like that of direct vision in which the magnitude-time function representing the intensity variation of each individual elementary area is transmitted over an independent line and used to produce a continuously varying illumination of the corresponding area of the reproduced scene. The frequency-range required on any one of these individual lines would then be determined by the extent to which the intensity at any one instant could be permitted to be distorted by the inter-symbol interference from the light intensities at neighboring times; that is to say, the frequency-range necessary would depend upon a blurring in time analogous to the blurring in space which is used to set the wave-number-range for a single picture. It seems probable that the total frequency-range required would be somewhat less for such a system than for one in which flicker is a factor.

Conclusion

At the opening of this discussion I proposed to set up a quantitative measure for comparing the capacities of various systems to transmit information. This measure has been shown to be the product of the width of the frequency-range over which steady state alternating currents are transmitted with sensibly uniform efficiency and the time during which the system is available. While the most convenient method of operation does not always make the fullest use of the frequency-range of the line, as is the case in double side-band transmission, a comparison of the frequency-range actually used with that which would be required on the basis of the actual information content of the material transmitted gives an idea of what may be gained in the cost of lines by making sacrifices in the convenience or cost of terminal equipment. Finally the point of view developed is useful in that it provides a ready means of checking whether or not claims made for the transmission possibilities of a complicated system lie within the range of physical possibility. To do this we determine, for each message which the system is said to handle, the necessary product of frequency-range by time and add together these products for whatever messages are involved. Similarly for each line we take the product of its transmission frequency-range by the time it is used and add together these products. If this sum is less than the corresponding sum for the messages, we may say at once that the system is inoperative.

26

Reprinted from *Zygon* **6**(2):126–134 (1971)

THE LARGER CYBERNETICS

by R. B. Lindsay

It has not been uncommon for men who have made great contributions to special fields of science to take an interest in broader issues as they grow older and, indeed, become in a measure what we may properly call philosophers. Examples abound: Poincaré in mathematics; Planck, Einstein, and Bohr in physics; Haeckel in biology; William James in psychology—all come to mind. Here I should like to call attention to a great French scientist who followed a similar route. Some may be surprised at the mention of the name of the physicist known to all from the practical unit of electrical current: André Marie Ampère (1775–1836)—one of the founders of electromagnetism. His profound mathematical memoirs of 1820–25, amply confirmed by careful experimental research, laid the foundations of what came to be called electrodynamics and, when supplemented by the later discoveries of Michael Faraday and Joseph Henry, provided the technological basis of our modern, electrically oriented civilization.

Later in his career (in 1834) Ampère wrote a brilliant document, "Essai sur le philosophie des sciences," in which he took a particularly broad view of the philosophy of science, including social and political studies as well as the better-established natural sciences, in his discussion. It was in this memoir that Ampère first introduced the term *cybernetique* to refer to the science of government. He evidently felt that this was appropriate terminology since κγβερντεσ is the Greek for helmsman or governor, the one who controls the direction of the ship. This may be considered the beginning of the formal recognition of the science of control, though it does not appear that Ampère's definition gained much attention in the nineteenth century, nor in our own century for that matter, until Norbert Wiener resurrected the term in his book called *Cybernetics,* published in 1948, and attempted to put the subject on a more formal basis.

From the humanistic point of view, I am naturally tempted to pause here to consider the possible relation of cybernetics as the science of

R. B. Lindsay is professor emeritus of physics, Brown University. This paper was presented at the symposium on science and human values during the annual meeting of the American Association for the Advancement of Science, Chicago, Illinois, December 29, 1970.

control to the motto of Phi Beta Kappa, the honorary scholastic society now approaching the bicentennial of its founding. For this motto, *Philosophia Biu Kubernates*, is usually translated: "The love of wisdom is the helmsman of life." There is little doubt that the founders of this fraternity, so celebrated in American collegiate annals, felt very strongly the appropriateness of the motto. And on the whole, down through the past two centuries it has seemed to have a meaning for those who, by devotion to academic studies, have convinced themselves that such studies can lead to that wisdom which serves as the best guide to the conduct of life. It is true that in some academic quarters today doubts seem to be arising about the validity of the motto.

TRANSFER AND TRANSFORMATION OF ENERGY

However, it is not my intention here to dwell further on Phi Beta Kappa, its significance and problems. What I wish to do is to emphasize that the fundamental idea of helmsmanship or control is one of the vital factors in the whole of human experience. From the standpoint of the scientist the reason for this is simple. It has to do with the prime position in the interpretation of human experience of the concept of energy, the most important idea in the whole of science. Representing in its simplest form the notion of constancy in the midst of change, it has come to pervade all aspects of life. There is nothing in our experience which cannot ultimately be described in terms of the transfer of energy from one place to another and for the transformation of energy from one form to another.

An illustration or two will remind you of this important scientific generalization. When you start the engine in your car, you are initiating a transformation of chemical energy (the rapid combustion of a gas mixture) into thermal or heat energy and thus in turn into the mechanical energy of the pistons in the cylinders. This energy of motion, however, does not remain localized in the engine but is transferred by the crankshaft to the rear wheels, thus enabling them to move and make the car go. The details may be complicated but the fundamental idea is simple.

A second illustration is provided by the processes going on inside our bodies as we digest our meals. We avoid details but merely emphasize that in substance what happens in metabolism is the transformation of the energy of chemical reaction, in which certain substances are changed into other substances which are then conveyed mechanically by the blood to the various parts of the body, where they provide for the so-called life of the cells. A large part of this chemical energy is converted into heat to keep our bodies warm; some into the mechanical energy of bodily movement, which, as it ceases, is transformed ultimately into heat. There is of course the additional complication of the conversion of some of the energy into electrical energy in the muscles and nerves. But the fundamental idea behind the whole business is that of the transformation into that random form of energy that we call heat. It is exemplified by the popular admonition that you must watch your *calories,* for the calorie is the standard physical unit of heat.

Appropriately enough, the first evaluation of the important numerical constant governing the transformation of heat into work and work into heat (the so-called mechanical equivalent of heat) was made not by a physical scientist but by a physician, Julius Robert Mayer, one of the scientific immortals of the nineteenth century and one of the founders of thermodynamics, the greatest physical theory ever concocted by the mind of man.

There seems to be no reason to doubt that not only is the physical behavior of human beings describable in terms of energy, but that the same is true of mental and emotional behavior commonly ascribed to the nervous system. It is not necessary to give further examples from the inorganic realm: they are all around us in terms of the various engines and other devices which, as we say, do the work of the world. In sum, all that goes on in the civilization that man has invented, including all the activities of living things, can be described most simply in terms of the transfer and transformation of energy.

Control of Energy Changes

It takes little reflection to convince oneself that life itself could never have developed nor could civilization have been invented without the *control* of all the energy changes involved. The study of such control is the province of cybernetics. Let us consider a couple of simple illustrations. One of the most important devices for the transformation of heat into mechanical energy or work was and still is, for that matter, the steam engine. An early problem of some technical difficulty involved in such an engine was to keep it running smoothly in spite of variations in the load on it. Without attention to this matter, if the device being driven by the engine demanded energy at too great a rate, the engine would stall; on the other hand, if the load dropped to nothing, the engine would race. Obviously some kind of control was needed. At first this was taken care of manually. A human being was deputed to control the flow of steam into the cylinder by somehow keeping track of the variations in the load. This was a clumsy kind of control. It was obvious that someone would think of something more clever than that. This happened when James Watt, the canny Scot, invented what he quite appropriately called a governor, a device attached to the engine which employs a small amount of the energy being transformed by the engine for controlling the rate of this transformation. The details are not important. What is significant is that the governor used a little of the engine's energy to ascertain when the load demanded more or less power and communicated the relevant message to the throttle, telling it in effect to let in more or less steam as needed. This is a process known in cybernetics as feedback. Watt was one of the first applied cyberneticians and, though he did not use the modern terminology, evidently had a very firm grasp of the basic principle of feedback control.

To return to your car which you have now put into motion, thanks to the transformation of energy you have initiated, your prime necessity is to steer it properly and control its speed. These are cybernetic activities in which you use a very small amount of energy, namely, that in your eye motions and the motions of your hands and feet to control the relatively large amount of energy involved in the motion of the car. You are indeed the steersman and, in modern parlance, a cybernetician.

Another familiar example is the operation of the simple thermostat which regulates the temperature of our living spaces. This ingenious device utilizes a very small portion of the heat energy of the room to inform the burner in the heater in the basement to transform more chemical energy into heat in order to raise the temperature of the room or to stop the transformation in order to bring the temperature down. Here is again a process of feedback control or the control of relatively large amounts of energy transformation by the use of a relatively minute amount.

Even more important is the vital feedback control in our bodies, without the presence of which human life would become impossible. Everyone must have heard of homeostasis, a word made famous by the late Dr. Walter B. Cannon of the Harvard Medical School in his book *The Wisdom of the Body.* It means among other things the regulation of bodily temperature by an elaborate control system in which a small amount of energy (i.e., heat loss or gain) in the skin sends a message to the thalamus, which in turn induces mechanical energy increase in skeletal muscle or in the blood vessel endings in the skin. The details of the process are complicated, but the fundamental basis for the automatic body regulation of temperature is again feedback control or an example of cybernetics: the wise governor that the process of evolution has developed for enabling so-called warm-blooded animals to face the thermal vicissitudes of their environment.

Still another example of cybernetic control is involved in speech communication in human beings. Think of the enormous amount of energy transformation that can be brought about by the utterance of a single word! Examples are unnecessary! Without such communication there could have been no development of human civilization as we know it. What is worth emphasizing is the relatively small amount of speech energy needed to control the transformation of enormous amounts. If I were presenting the contents of this essay orally to an audience, my voice would need to provide speech energy only at the rate of about 1/10,000 of a watt. This would be the power being dissipated as the mechanical power in the sound radiated from my mouth as a speaker. Let us suppose you can buy energy from the electric power company at about five cents per kilowatt hour. It is simple arithmetic to figure out that I should have to keep on talking at that pace for over one thousand years in order to provide a nickel's worth of energy or enough to keep a three hundred–watt lamp going for a bit over three hours. You have always suspected that talk is cheap. This simple calculation should convince anyone!

Nature—the Great Cybernetician

The point here is that Nature, the name we give to the content of our experience which we interpret as being independent of ourselves, also seems to exert control over the transformation of energy. Nature is indeed in the last analysis the great cybernetician. Its control is exercised in terms of the two great principles of thermodynamics, already mentioned as the science which deals with energy. The first principle, established by Mayer and confirmed by the Englishmen Joule and others, says that the total energy in the universe of our experience is constant. All we can do is to transfer or transform it; we can never change its total amount.

Application of this fundamental principle is indeed part of the accumulated wisdom of our race. In fact the authors of the sacred writings of the Judeo-Christian tradition included it in the scriptures. For in the third chapter of the Book of Genesis, we learn of God admonishing the sinners in the Garden and saying: "In the sweat of thy face shalt thou eat bread." This is the first law of thermodynamics. The clever ones have indeed through the ages tried to arrange matters so that someone else's sweat was required, but that does not invalidate the principle that someone has to sweat to provide subsistence for the race and that, in our universe, on net balance you do not get something for nothing, or as some modern wit has put it, you cannot win! Other clever ones have tried to extract work out of so-called perpetual motion machines but have never succeeded. Here is one of Nature's means of control over the transformation of energy; there seems to be little we can do except put up with it. As Mayer liked to phrase it: *ex nihilo nil fit.*

As a matter of fact, one of the best expressions of the first law of thermodynamics is the famous essay of Ralph Waldo Emerson, *Compensation.* The shrewd Yankee moralist was keenly aware of that inscrutable duality of human experience involved in the fact that there is never a gain without a compensatory loss, never a plus without a minus, never a good without a connected evil. It may be objected that this is an unnecessarily pessimistic view of experience, but I think it is just as valid for the optimist. All he can truthfully say is that for every minus there is a plus, while a pessimist merely says that for every plus there is a minus. As Emerson himself might have said, it comes to the same thing in the end.

But now I must emphasize that the first law of thermodynamics, the principle of the conservation of energy, is not the only natural control in our environment; not the only grip that Nature has on all our energy transactions, whether in the marketplace, the home, or even the racetrack. Nature appears to have another trick up its sleeve, in many ways a cruel trick. It reinforces its control over energy transformation by relentlessly arranging that every time any energy transformation takes place under natural conditions the chance of the repetition of the transformation is reduced by a measurable amount. Every time an engine goes through a cycle there is an inevitable degradation of the availability of energy to repeat this process. Every time you light up a cigarette you decrease by a certain amount the chance of lighting another one. This is the principle of the increase of entropy, the measure (introduced by the German physicist Rudolf Clausius) of the steady decrease in the *availability* of the constant amount of energy in our universe. This is the second law of thermodynamics: Nature's ultimate control, the cybernetics of the universe. It leads to the conclusion that the ultimate end of our universe is the heat death, in which, though the total amount of energy remains the same, no part of it is in a condition for further transformation. Nothing further can happen so far as energy transformation is concerned.

It must be emphasized that the heat death as a conclusion from the second law of thermodynamics is based on the extension of the principles of thermodynamics to the whole universe. Clausius was content with nothing less. In his famous 1885 memoir he expressed the principles in the simple form:

> Die Energie der Welt ist constant.
> Die Entropie der Welt strebt einem Maximum zu.

In this extension of thermodynamics to the universe we part company with the orthodox thermodynamics of the chemists and engineers, who insist on restricting the application of the principles to closed or isolated systems (or to those which, if open, can effectively be closed by considering finite portions of the environment). The attitude is understandable, since by this restriction they are able to calculate energy and entropy changes in specific systems of practical interest. The extrapolation to the whole universe seems to be a dangerous one from the standpoint of strict logical clarity. Nevertheless it is a tempting procedure, and many physicists have found illumination in it. This point of view rests heavily on Ludwig Boltzmann's interpretation of entropy as proportional to the logarithm of the statistical probability, which in turn is the number of ways in which a given system can be distributed over the possible energy states available to it. On this view the second law merely means that the total state of the universe tends to move toward a state of greater probability, a kind of tautology. Some have sought to rephrase this by introducing a concept of order and saying that the natural tendency of the universe is to move from order to disorder. This has been criticized on the ground that the concept of order is difficult to define precisely so as to make it fit logically all cases. Fortunately for the cybernetic implications of thermodynamics as applied to the universe, we do not really need to use the order-disorder notion. All we need to do is to say that the second law implies the steady increase in entropy in the universe, meaning thereby the corresponding degradation of the availability of energy for further transformation.

This means that it will do no good affixing stickers to our cars with the slogan: "Save America from Entropy." Here is a war we cannot win. All experience teaches us that the steady decrease in the availability of energy for new transformations is an irresistible and irrevocable one. All experience shows that this transition, which for the individual living organism we call the downhill march to death, is the fate of each one of us. The second law assures us that it is the fate of our universe as a whole.

THE THERMODYNAMIC IMPERATIVE

To prevent unhealthy concern, a few words of encouragement are in order. The first relates to the time scale. Relative to the time range of human history, the degenerative process described by the second law is a slow one; it is just as well to relax. Lots can happen before the final heat death. In fact some astronomers consider that the sun on which all life on earth depends may in due course become a supernova and

blow up. In this case our paltry little planet may wind up its independent existence with a bang, even though the universe rolls on
serenely with its steady entropy increase toward the eventual end as
T. S. Eliot's whimper, an appropriate one-word metaphor for the heat
death.

C. P. Snow has complained that humanists have been content to remain ignorant of the second law of thermodynamics. He seems to have
overlooked Eliot and, somewhat more surprisingly, the Victorian poet
A. C. Swinburne, the final stanzas of whose "Garden of Proserpine" constitute one of the best versions of the heat death implied by the second
law we could ask for:

> Then sun nor star shall waken
> Nor any change of light
> Nor sound of waters shaken
> Nor any sound or sight
>
> Nor wintry leaves nor vernal
> Nor days nor things diurnal
> Only the sleep eternal
> In an eternal night.

How much more appealing this is than Clausius's statement quoted
above! We can be fairly sure that many modern poets have tried to
express the same thing that Swinburne did in his somewhat jingly verse.
At any rate, Swinburne's verse has the merit of being intelligible,
whereas in most modern poetry you need a computer to extract the
signal from the noise.

Another silver lining in the apocalyptic gloom of the second law is
that life itself constitutes what may legitimately be called a fight against
the increase in entropy. For life is a local consumption of entropy in
the midst of the sea of its production. This is true in the sense that the
production and maintenance of a living cell represent the local increase
in the availability of energy for transformation. It does not mean
that the living cell contravenes the second law, for in its environment
the entropy still tends to increase irreversibly. But it does mean
that life constitutes a local consumption of entropy in the same sense
as the operation of the domestic mechanical refrigerator, which makes
heat flow from a cold to a warmer region in the face of the natural
tendency for heat to flow the other way. Locally, the refrigeration
process consumes entropy, though there is still an overall production
of entropy represented practically in the monthly electric power bill.

More important still, life in the form of human beings has through
thousands of years waged a relentless struggle to build a civilization
which represents an attempt to increase the availability of energy for

transformation or, as some would prefer to put it, to fashion order out of disorder. Even though the second law guarantees that the struggle will not avail and that we must all go down ultimately to defeat, the challenge to fight on is still there. To me it conveys the distinct suggestion that we as individuals should endeavor to consume as much entropy as possible to increase the order in our environment. This is the thermodynamic imperative, possibly not unworthy to rank alongside the categorical imperative of Kant or even the Golden Rule.

The relevance of the thermodynamic imperative is particularly great at this late twentieth-century period of social disorder and threatened educational disruption. We think here not merely of the urge to destroy the visible material signs of the civilization which has been painfully built up over the ages, but the denial by many people that rational thinking has any part to play in our confrontation with the hard facts of human experience. Now rational thought as represented not only by science but also by philosophy, religion, and indeed all the humanities, constitutes the crux of man's endeavor to associate meaning with his experience and establish order in it. The denigration of rational thinking in our education of the young would constitute an abdication of our fight against the second law. Man's lot has been and is a hard one. We shall not make it easier by abandoning the only way in which we can cope even in limited fashion with Nature's ultimate cybernetic control.

AUTHOR CITATION INDEX

SUBJECT INDEX